Manufacturing, Characterization and Modeling of Advanced Materials

Manufacturing, Characterization and Modeling of Advanced Materials

Guest Editors

Charles Lu

Madhav Baral

MDPI

Basel • Beijing • Wuhan • Barcelona • Belgrade • Novi Sad • Cluj • Manchester

Guest Editors

Charles Lu
Department of Mechanical and Aerospace Engineering
University of Kentucky
Lexington
USA

Madhav Baral
Department of Mechanical and Aerospace Engineering
University of Kentucky
Lexington
USA

Editorial Office
MDPI AG
Grosspeteranlage 5
4052 Basel, Switzerland

This is a reprint of the Special Issue, published open access by the journal *Materials* (ISSN 1996-1944), freely accessible at: https://www.mdpi.com/journal/materials/special_issues/WN0MQ395Q4.

For citation purposes, cite each article independently as indicated on the article page online and as indicated below:

Lastname, A.A.; Lastname, B.B. Article Title. *Journal Name* **Year**, *Volume Number*, Page Range.

ISBN 978-3-7258-7735-5 (Hbk)
ISBN 978-3-7258-7736-2 (PDF)
https://doi.org/10.3390/books978-3-7258-7736-2

Contents

About the Editors

Charles Lu

Charles Lu is a faculty member in the Department of Mechanical and Aerospace Engineering at the University of Kentucky Paducah campus. His research interests span composites, advanced materials, mechanics, and finite element modeling. With expertise in both experimental and computational approaches, his work contributes to advancing the understanding and application of next-generation materials.

Madhav Baral

Madhav Baral is a faculty member in the Department of Mechanical and Aerospace Engineering at the University of Kentucky Paducah campus. His research focuses on plasticity, constitutive modeling, ductile fracture, and the integration of experimental and numerical methods to study material behavior. He is particularly interested in sheet metal and tube forming, material characterization, and advanced manufacturing processes.

Preface

This Special Issue highlights recent advances across the manufacturing, characterization, and constitutive modeling of advanced materials, integrating experimental and theoretical perspectives to link process, structure, property, and performance. The collection of works includes topics in nanomaterials and biomaterials, their mechanical response under diverse loading, and fatigue and fracture behavior. By uniting pioneering methods and understandings, this collection aims to accelerate the development of next-generation materials and advance the broader field of materials science and engineering.

Charles Lu and Madhav Baral
Guest Editors

materials

MDPI

Article

Epoxy Adhesive Materials as Protective Coatings: Strength Property Analysis Using Machine Learning Algorithms

Izabela Miturska-Barańska [1] and Katarzyna Antosz [2,*]

[1] Department of Production Computerisation and Robotisation, Faculty of Mechanical Engineering, Lublin University of Technology, Nadbystrzycka 36, 20-618 Lublin, Poland; i.miturska@pollub.pl
[2] Faculty of Mechanical Engineering and Aeronautics, Rzeszow University of Technology, 35-959 Rzeszow, Poland
* Correspondence: katarzyna.antosz@prz.edu.pl

Abstract: This study analyzed the mechanical properties of epoxy adhesive materials used as functional coatings, focusing on how physical modifications impact their microstructure and strength. Compositions based on Epidian 5, 53 and 57 resins were cured using TFF, Z-1, or PAC curing agents and modified with various fillers: mineral ($CaCO_3$ calcium carbonate), active (activated carbon filler, CWZ-22), and nanostructured (montmorillonite, ZR-2) fillers. The best results were achieved with calcium carbonate (10–20 wt%) in Epidian 5 or 53 resins cured with TFF or Z-1, yielding tensile strength up to 64 MPa, compressive strength up to 145 MPa, and bending strength up to 123 MPa. Activated carbon and nanofillers showed moderate improvements, particularly in more flexible matrices. To support property prediction, machine learning algorithms were applied and successfully modeled the mechanical behavior based on composition data. The most accurate models reached R^2 values of 0.93–0.95 for compression and bending strength. While the models for compression and bending strength demonstrated high accuracy, the tensile strength model yielded lower predictive performance, indicating that further refinement and expanded input features are necessary. Shapley analysis further identified curing agents and fillers as key predictive features. This integrated experimental and data-driven approach offers an effective framework for optimizing epoxy-based coatings in industrial applications.

Keywords: material characterization; SEM; machine learning algorithms; SVM; NN algorithm

1. Introduction

Modern engineering structures increasingly demand materials that can simultaneously fulfill structural and protective roles. In mechanically and chemically aggressive environments, coatings serve as a vital barrier to moisture, corrosion, UV radiation, and mechanical damage. Depending on their chemical composition and technical function, coatings can perform barrier, adhesive, electro-insulating, or self-healing functions [1,2].

Among these, epoxy adhesive systems are particularly versatile. Traditionally used as structural adhesives, they are also employed as protective coatings for substrates such as steel, concrete, and composites. These two-component polymeric systems form cross-linked networks with excellent adhesion, chemical resistance, and thermal stability after curing. Their properties can be tailored through appropriate selection of curing agents and fillers [3,4].

In various engineering applications—including construction, marine, automotive, and energy—epoxy coatings are valued for their adaptability and durability. They also serve as the foundation for smart coatings and hybrid materials [5].

One established method for enhancing epoxy performance is physical modification using mineral, active, or nanostructured fillers. These improve characteristics such as modulus, toughness, and barrier performance. Fillers like calcium carbonate, activated carbon, and montmorillonite have demonstrated measurable influence on microstructure and mechanical behavior [4–6]. However, the effectiveness of such modification depends on filler dispersion and interaction with both the resin and curing agent.

Thus, the use of various types of fillers—both macro-, micro-, and nano-structural, which in specific proportions can significantly improve the mechanical, thermal, and tribological properties of composite systems—is a key strategy for modifying epoxy resins. Studies by Karthikeyan and Naveen [4] have shown that mineral additives such as $CaCO_3$ can improve elastic modulus and compressive strength, while active carbon fillers (e.g., CWZ-22) can have beneficial effects on vibration damping and thermal conductivity. On the other hand, the use of nanofillers—such as montmorillonite of the ZR-2 NanoBent type—can achieve better barrier properties and significant reinforcement at a low mass proportion (1–5%) [7].

Scanning electron microscopy (SEM) is often used to assess the resin–filler interface, as poor dispersion can result in porosity or reduced mechanical performance [8]. At the same time, traditional mechanical testing of new formulations is often time-consuming and expensive.

Another challenge remains the rapid and reliable evaluation of the mechanical properties of new compositions, which traditionally requires time-consuming and expensive experimental testing. To address this, machine learning (ML) algorithms are increasingly employed to predict and optimize material properties. These include linear models, random forests, and neural networks, which can identify key formulation parameters and forecast strength, durability, or failure [9–11]. For example, Yan et al. [9] used the ML model to predict the tribological properties of epoxy coatings modified with nanoparticles, achieving high prediction accuracy. Other researchers [10] developed regression models to evaluate the strength of adhesive bonds depending on resin type, filler type, and curing method, indicating the potential of AI in the design of new materials.

Despite the numerous studies on epoxy resins and the widespread interest in the topic of modification of polymeric materials, there is still a lack of research that comprehensively combines experimental strength analysis, microstructural characterization, and predictive modeling of mechanical properties for epoxy protective coatings. This is particularly true for resins in the Epidian group, which are widely used in industry, but insufficiently described in the context of ML-supported physicochemical modifications.

Due to the challenges in directly testing thin coatings, model samples of standardized geometry were used, allowing consistent evaluation of compositional effects. This type of approach allows unambiguous evaluation of the influence of epoxy resin, curing agent, and filler type on mechanical properties, eliminating uncontrolled factors related to the substrate, uneven film thickness, or curing conditions in thin films. Although the samples used do not fully replicate the in-situ operating conditions of the coatings, they provide a reliable model to evaluate the material properties as such. A similar approach is commonly used in the literature on adhesion and coating testing of epoxy materials [12–14].

In response to the indicated research gap, the purpose of this study was to conduct a multivariate analysis of epoxy adhesive materials used as protective coatings, taking into account the following:

- the effect of resin type, curing agent type, and filler concentration and type on mechanical properties;
- evaluation of microstructure using scanning electron microscopy (SEM);
- construction and validation of regression models that enable prediction of strength parameters based on material composition.

The overarching goal of the paper is therefore to integrate experimental approaches with machine learning algorithms in the context of functional adhesive coatings based on Epidian epoxy resins.

2. Numerical Methods and Data Algorithms in Epoxy Materials Engineering

Advances in epoxy materials today are no longer solely the result of experimental research. Increasingly, the development of adhesion and coating technologies is based on the integration of classical laboratory approaches with computational methods and data analysis. In particular, numerical methods and machine learning algorithms are finding application in adhesion materials engineering as tools for predicting, optimizing, and deeply understanding the effects of chemical composition and microstructure on the mechanical and performance properties of epoxy composites.

The development of this field is made possible by the increasing availability of experimental data, the digitization of research, and the growth of computing power. Of key importance here are regression algorithms, classification methods, and data mining techniques that allow identification of non-obvious relationships between variables such as resin type, curing agent type, filler concentration, or composite preparation method. The most commonly used algorithms are linear and multivariate regression, Ridge and LASSO regression, decision trees, and ensemble models such as random forests (random forest). More advanced applications also use deep neural networks (DNNs) and convolutional neural networks (CNNs), especially for microscopic image analysis (e.g., SEM) [15–17].

In the engineering of epoxy adhesion materials, data-driven predictive models make it possible, among other things, to predict tensile strength, flexural and compressive strength, and to assess degradability under service conditions. In a study by Yan et al. [9], the use of a machine learning model made it possible to predict the tribological properties of epoxy coatings with high accuracy based on material composition and curing conditions. Another work (e.g., Jang et al. [10]) has demonstrated the effectiveness of regression in evaluating the effect of fillers on the mechanical parameters of structural adhesives.

Particularly important in the case of epoxy resins is to take into account the microstructural variability resulting from the type of fillers (mineral, active, nano) and their dispersion in the polymer matrix. The use of numerical methods makes it possible to quantitatively analyze this variability and account for it in model predictions. Data-based models can also be coupled with microscopic observations, allowing not only the classification of structural defects but also their automatic assignment to technological parameters [6,18]. Figure 1 shows a modern approach to the analysis and design of epoxy materials using computational tools and artificial intelligence methods. The diagram illustrates how extensively numerical methods and data algorithms support modern epoxy engineering—from property analysis and prediction to intelligent design of materials with preplanned parameters. It also points to the growing integration of experimental approaches with digital analytical tools as part of the so-called digital transformation of materials research.

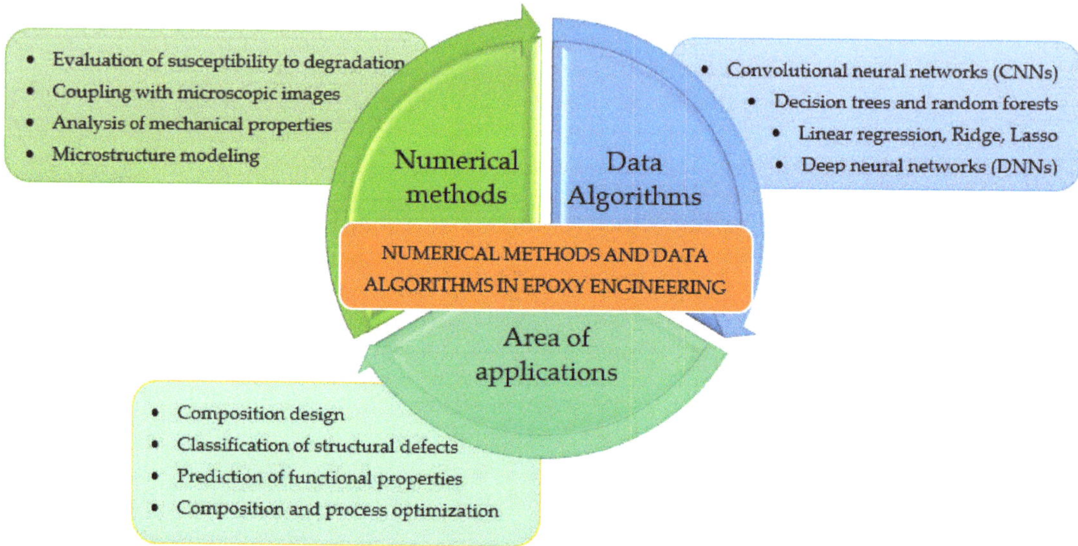

Figure 1. Numerical methods and data algorithms in epoxy engineering.

In this study, machine learning models were used to support the analysis of the influence of input variables on the mechanical properties of epoxy composites. Such a choice, despite its simplicity, is fully justified—especially for studies where sample size is limited and the goal is to understand the influence of specific formulation parameters. These models, properly calibrated and validated, make it possible to predict strength properties based on material composition, supporting both experimental analysis and practical application of materials in surface engineering.

The use of numerical methods and data algorithms in the engineering of epoxy adhesive materials not only speeds up the testing process but also enables the development of functional coatings with engineered properties tailored to the requirements of specific working environments. This represents a clear step toward so-called smart materials design and the digital transformation of materials research.

3. Materials and Methods

3.1. Materials and Their Properties

The study used epoxy composite compositions based on three different Epidian resins: Epidian 5, Epidian 53, and Epidian 57 (CIECH S.A., Sarzyna, Poland), which differ in the degree of chemical modification, viscosity, and elasticity after curing. Epidian 5 epoxy resin is a classic, low-viscosity bisphenol resin with a wide range of engineering applications. Epidian 53 epoxy resin, which is an ester-modified resin, is characterized by higher hardness and heat rejection. Epidian 57 epoxy resin, on the other hand, contains additives that improve the plasticity of the composition, making it more flexible after crosslinking.

Three different curing agents were used: TFF Mannich's principal, Z-1 aminoamide and the PAC traditional polyamide curing agent (CIECH S.A., Sarzyna, Poland). Each of them is characterized by a different crosslinking mechanism and reaction time, which translates into different final properties of the material. Crosslinking reactions were carried out under room conditions, according to the manufacturers' recommended ratios, keeping the appropriate gelation and curing times.

The properties of the used materials are summarized in Tables 1 and 2.

Table 1. Physical and chemical properties of the epoxy resins used in the study [19–23].

Properties	Epidian 5 Epoxy Resin	Epidian 53 Epoxy Resin	Epidian 57 Epoxy Resin
Epoxy number	0.48–0.52 mol/100 g	≥0.41 mol/100 g	≥0.40 mol/100 g
pH value	ok. 7	ok. 7	ok. 7
Viscosity at 25 °C	20,000–30,000 mPa·s	900–1500 mPa·s	13,000–19,000 mPa·s
Density in 20 °C	1.16 g/cm^3	1.11–1.15 g/cm^3	1.14–1.17 g/cm^3
Flash temperature	266 °C	58 °C	134 °C
Temperature auto-ignition	490 °C	460 °C	455 °C
Melting point	30–50 °C	Not applicable	Not applicable
Initial boiling point	not marked—distribution	141 °C	>215 °C does not boil

Table 2. Properties of the curing agents used in the tests [23,24].

Properties	Mannich's Principle (TFF Curing Agent)	Amine Curing Agent (Z-1 Curing Agent)	Polyamide Curing Agent (PAC Curing Agent)
Viscosity at 25 °C	max. 10,000 mPa·s	20–30 mPa·s	10,000–25,000 mPa·s
Density in 20 °C	1.15–1.20 g/cm^3	0.978–0.983 g/cm^3	1.10–1.20 g/cm^3
Amine number	500–700 mg KOH/g	min. 1100 mg KOH/g	290–360 mg KOH/g
Gelation time (example for composition with Epidian 5 at 20 °C, for 100 g sample)	17 min	33 min	180 min

The content in the epoxy compositions of the curing agents used was selected on the basis of the recommended stoichiometric ratios (Table 3), according to the type of epoxy resin.

Table 3. Used amount of curing agents in epoxy compositions.

Epoxy Resin	Curing Agent (Parts by Weight per 100 Parts by Weight of Epoxy Resin)		
	TFF Epoxy Resin	Z-1 Epoxy Resin	PAC Epoxy Resin
Epidian 5 (E5)	26 part-weight.	12 part-weight.	80 part-weight.
Epidian 53 (E53)	22 part-weight.	10 part-weight.	80 part-weight.
Epidian 57 (E57)	22 part-weight.	10 part-weight.	80 part-weight.

In order to evaluate the effect of physical modification of the epoxy compositions, the compositions were enriched with three types of fillers: nanofiller (montmorillonite type NanoBent ZR2) (Zakłady Górniczo-Metalowe "Zębiec" S.A., Zębiec, Poland), mineral (calcium carbonate—$CaCO_3$) (Zakłady Przemysłu Wapienniczego Trzuskawica S.A., Siatkówka, Poland) and activated (activated carbon CWZ-22) (PPH STANLAB SP. Z O.O., Lublin, Poland). Fillers were introduced into the resins at different ratios: 5%, 10% and 20% by weight for $CaCO_3$ and CWZ-22, and 1%, 3% and 5% by weight for montmorillonite. Their selection was based on the previous literature reports indicating their cor- cussive effects on the mechanical and barrier properties of the epoxies.

The properties of the fillers used in the study are summarized in Tables 4–6. Tables 7–9 present the compositions of the compositions tested in the study.

Table 4. Properties of NanoBent ZR2 montmorillonite [25].

Properties	ZR2 NanoBent
Form	cream-colored lamellar powder
Water content	≤3.0% weights
Roasting loss at 650 °C	25–30% weights
Swelling in Xylene	>20% volume
Vapor sorption of white spirit 48 h	>20% weights
Bulk density	<0.5 g/cm^3
CEC ion exchange capacity of bentonite raw material	min. 80 mmol/100 g dry bentonite raw material

Table 5. Basic physical and chemical properties of $CaCO_3$ used in the study [26].

Properties	$CaCO_3$ Calcium Carbonate
Form	light gray solid of various sizes: lumps or fine powder
Fragrance	odorless
pH	9.2 (at 25 °C)
Melting point	>450 °C (decomposition temperature—825 °C)
Flammability	non-flammable
Explosive limits	non-explosive (free of any chemical structures associated with explosive properties)
Relative density	2.711 g/cm^3 (at 20 °C)
Solubility in water	14 mg/dm^3 (at 25 °C)
Viscosity	not applicable (solid with melting point > 450 °C)
Explosive properties	non-explosive (free of any chemical structures associated with explosive properties)
Oxidizing properties	non-oxidizing (based on chemical structure, the substance does not contain excess oxygen or any structural group tending to react exothermically with combustible material)
Decomposition temperature	825 °C
Bulk density	(0.7–1.4)·10^6 g/m^3 (at 20 °C)
Electrostatic properties	the substance does not generate electrostatic charges

Table 6. Basic properties of CWZ-22 activated carbon [27].

Properties	CWZ-22 Activated Carbon
Form	solid, dusty black color
Fragrance	odorless
pH	about 6 (50 g/L H_2O as suspension, 20 °C)
Melting point	no data, sublimation about 3700 °C
Explosive limits	no data
Relative density	about 2 g/cm^3
Solubility in water	in water: insoluble in organic solvents: no data
Bulk density	about 400·10^3 g/m^3

Table 7. Tested compositions based on epoxy resin Epidian 5.

No	Epoxy Resin			Curing Agent				Filler	
	Epidian 5	Epidian 53	Epidian 57	TFF	Z-1	PAC	ZR-2	$CaCO_3$	CWZ-22
1	100	0	0	26	0	0	0	0	0
2	100	0	0	26	0	0	1	0	0
3	100	0	0	26	0	0	3	0	0
4	100	0	0	26	0	0	5	0	0
5	100	0	0	26	0	0	0	5	0
6	100	0	0	26	0	0	0	10	0
7	100	0	0	26	0	0	0	20	0
8	100	0	0	26	0	0	0	0	5
9	100	0	0	26	0	0	0	0	10
10	100	0	0	26	0	0	0	0	20
11	100	0	0	0	12	0	0	0	0
12	100	0	0	0	12	0	1	0	0
13	100	0	0	0	12	0	3	0	0
14	100	0	0	0	12	0	5	0	0
15	100	0	0	0	12	0	0	5	0
16	100	0	0	0	12	0	0	10	0
17	100	0	0	0	12	0	0	20	0
18	100	0	0	0	12	0	0	0	5
19	100	0	0	0	12	0	0	0	10
20	100	0	0	0	12	0	0	0	20
21	100	0	0	0	0	80	0	0	0
22	100	0	0	0	0	80	1	0	0
23	100	0	0	0	0	80	3	0	0
24	100	0	0	0	0	80	5	0	0
25	100	0	0	0	0	80	0	5	0
26	100	0	0	0	0	80	0	10	0
27	100	0	0	0	0	80	0	20	0
28	100	0	0	0	0	80	0	0	5
29	100	0	0	0	0	80	0	0	10
30	100	0	0	0	0	80	0	0	20

Table 8. Tested compositions based on epoxy resin Epidian 53.

No	Epoxy Resin			Curing Agent				Filler	
	Epidian 5	Epidian 53	Epidian 57	TFF	Z-1	PAC	ZR-2	$CaCO_3$	CWZ-22
1	0	100	0	22	0	0	0	0	0
2	0	100	0	22	0	0	1	0	0
3	0	100	0	22	0	0	3	0	0
4	0	100	0	22	0	0	5	0	0
5	0	100	0	22	0	0	0	5	0
6	0	100	0	22	0	0	0	10	0
7	0	100	0	22	0	0	0	20	0
8	0	100	0	22	0	0	0	0	5
9	0	100	0	22	0	0	0	0	10
10	0	100	0	22	0	0	0	0	20
11	0	100	0	0	10	0	0	0	0
12	0	100	0	0	10	0	1	0	0
13	0	100	0	0	10	0	3	0	0

Table 8. *Cont.*

No	Epoxy Resin			Curing Agent				Filler	
	Epidian 5	Epidian 53	Epidian 57	TFF	Z-1	PAC	ZR-2	$CaCO_3$	CWZ-22
14	0	100	0	0	10	0	5	0	0
15	0	100	0	0	10	0	0	5	0
16	0	100	0	0	10	0	0	10	0
17	0	100	0	0	10	0	0	20	0
18	0	100	0	0	10	0	0	0	5
19	0	100	0	0	10	0	0	0	10
20	0	100	0	0	10	0	0	0	20
21	0	100	0	0	0	80	0	0	0
22	0	100	0	0	0	80	1	0	0
23	0	100	0	0	0	80	3	0	0
24	0	100	0	0	0	80	5	0	0
25	0	100	0	0	0	80	0	5	0
26	0	100	0	0	0	80	0	10	0
27	0	100	0	0	0	80	0	20	0
28	0	100	0	0	0	80	0	0	5
29	0	100	0	0	0	80	0	0	10
30	0	100	0	0	0	80	0	0	20

Table 9. Tested compositions based on epoxy resin Epidian 57.

No	Epoxy Resin			Curing Agent				Filler	
	Epidian 5	Epidian 53	Epidian 57	TFF	Z-1	PAC	ZR-2	$CaCO_3$	CWZ-22
1	0	0	100	22	0	0	0	0	0
2	0	0	100	22	0	0	1	0	0
3	0	0	100	22	0	0	3	0	0
4	0	0	100	22	0	0	5	0	0
5	0	0	100	22	0	0	0	5	0
6	0	0	100	22	0	0	0	10	0
7	0	0	100	22	0	0	0	20	0
8	0	0	100	22	0	0	0	0	5
9	0	0	100	22	0	0	0	0	10
10	0	0	100	22	0	0	0	0	20
11	0	0	100	0	10	0	0	0	0
12	0	0	100	0	10	0	1	0	0
13	0	0	100	0	10	0	3	0	0
14	0	0	100	0	10	0	5	0	0
15	0	0	100	0	10	0	0	5	0
16	0	0	100	0	10	0	0	10	0
17	0	0	100	0	10	0	0	20	0
18	0	0	100	0	10	0	0	0	5
19	0	0	100	0	10	0	0	0	10
20	0	0	100	0	10	0	0	0	20
21	0	0	100	0	0	80	0	0	0
22	0	0	100	0	0	80	1	0	0
23	0	0	100	0	0	80	3	0	0
24	0	0	100	0	0	80	5	0	0
25	0	0	100	0	0	80	0	5	0
26	0	0	100	0	0	80	0	10	0
27	0	0	100	0	0	80	0	20	0
28	0	0	100	0	0	80	0	0	5
29	0	0	100	0	0	80	0	0	10
30	0	0	100	0	0	80	0	0	20

So, in this study, compositions labeled as in Table 7 were prepared and tested.

3.2. Preparing Samples for Testing

Preparation of the compositions was carried out using a mechanical mixer (Güde, Wolpertshausen, Germany) with a dispersing disc mixer. The flow of the process of preparation of epoxy compositions was carried out according to the following methodology (Figure 2):

- Weighing the specified amount of resin;
- Adding a filler (for modified compositions) to the epoxy resin in the appropriate amount;
- Mechanical stirring of the composition components for a period of 3 min at a speed of 460 rpm;
- Adding curing agent to the resin mixed with the filler in appropriate amounts;
- Mechanical mixing of the components of the composition at a speed of 460 rpm for a period of 3 min;
- Venting the epoxy composition over a period of 3 min.

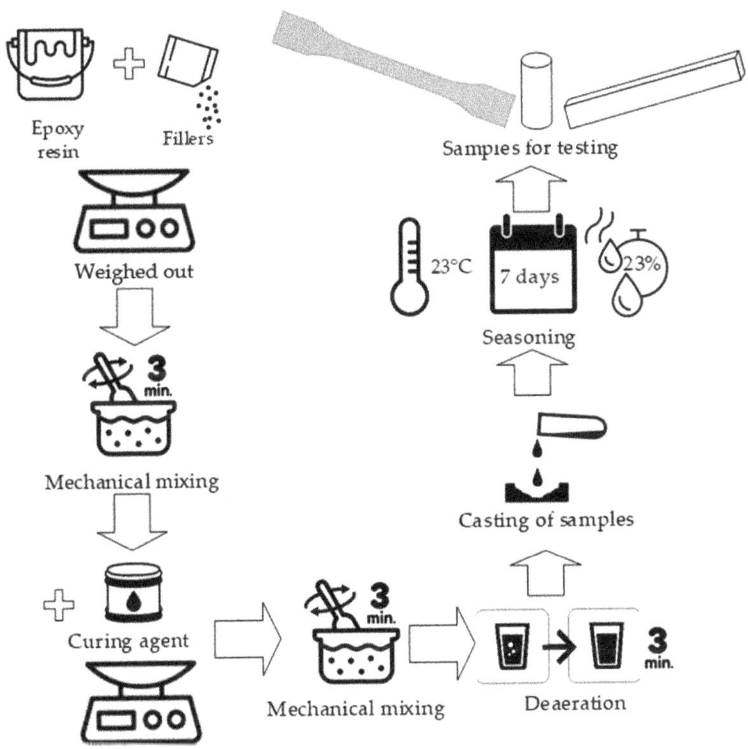

Figure 2. Flowchart of the test sample preparation process.

For unmodified compositions, the step of introducing the filler into the resin was omitted.

Weighing of the components of the epoxy compositions was performed using a KERN CKE 3600-2 laboratory balance (Kern, Albstadt, Germany) with a measurement accuracy of 0.01 g.

For practical and metrological reasons, strength tests were carried out on model specimens with controlled geometries (prisms and rollers), conforming to PN-EN ISO standards. This approach allows an unambiguous assessment of the influence of the type of epoxy resin, curing agent, and filler on mechanical properties, while eliminating variables related to the type of substrate, uneven application, or film thickness. The resulting data reflect the actual material parameters and are widely used in the literature as authoritative predictors of functional coating behavior under operating conditions [28–30].

The compositions were then vented in a vacuum chamber to eliminate air bubbles. Venting was carried out using a CPS model VP6D two-stage vacuum pump (CPS, Miami, FL, USA). The finished mixtures were poured into silicone molds, conforming to the dimensions of the samples specified in the applicable PN-EN ISO standards. The dimensions and geometry of the prepared samples are shown in Figure 3.

(a)

(b)

(c)

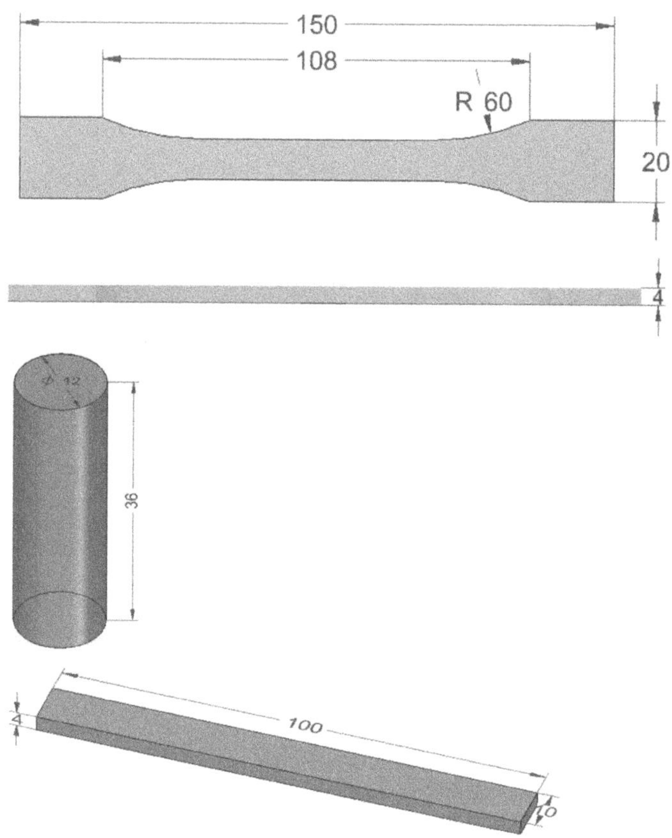

Figure 3. Shape and dimensions of the sample of epoxy compositions for testing (**a**) tensile strength (according to PN EN ISO 527-2 standard) [31], (**b**) compressive strength (according to ISO 604 standard) [32], and (**c**) bending strength (according to ISO 178:2003 standard) [33].

After one-step curing under laboratory conditions, the samples were seasoned for seven days invariably at 23 ± 2 °C at 23 ± 3% humidity to stabilize the network structure. After this time, strength tests were carried out and SEM photos were taken.

A series of 5 samples was prepared for each adhesive composition.

3.3. Material Testing

During the research, the strength properties of the epoxy compositions were tested. Tensile strength, compressive strength, and bending strength were tested and their structure was analyzed using scanning electron microscopy.

Tensile tests of epoxy compositions were carried out on a Zwick Roell Z150 testing machine (Zwick/Roell, Wroclaw, Poland) in accordance with PN EN ISO 527-1 (Plastics—Determination of mechanical properties in static tension) [31]. The crosshead traverse speed during the test was 5 mm/min, and the tensile modulus test speed was 1 mm/min. The initial test force was 30 N.

Compressive strength tests of the epoxy compositions were also performed on a Zwick Roell Z150 testing machine (Zwick/Roell, Wroclaw, Poland). These tests were carried out in accordance with ISO 604 (Plastics—Determination of Compressive Properties). This standard describes a method for determining, under certain conditions, the compression properties of plastics [32]. The assumed crosshead traverse speed during the test was 10 mm/min, and the compression modulus test speed was 1 mm/min. The initial test force was 20 N.

Bending strength tests were performed on a Zwick Roell Z2.5 testing machine (Zwick/Roell, Wroclaw, Poland) in accordance with DIN-EN ISO 178 (Plastics—determination of bending properties). This standard refers to a freely supported beam loaded at mid-span, i.e., three-point bending [33]. The test speed was 10 mm/min, and the initial test force was 5 N.

Microscopic examination of epoxy composition samples was studied using a TESCAN MIRA 3 GMX SE scanning electron microscope (SEM) (Tescan Orsay Holding, Brno-Kohoutovice, Czech Republic). The accelerating voltage was 10 kV.

3.4. Machine Learning Approach

This study employed a machine learning approach to support the prediction and analysis of the mechanical properties of epoxy adhesive compositions. Several supervised learning algorithms were tested to identify the most effective predictive models for tensile strength, compressive strength, and bending strength. The initial modeling phase involved developing and comparing multiple machine learning methods, including linear regression, support vector machines (SVMs), decision trees, random forests, and artificial neural networks (ANNs). Each algorithm was trained and validated using experimental data, including information on resin type, curing agent, filler type and concentration, and corresponding mechanical strength results.

For each model type, a range of hyperparameters were explored using grid search and cross-validation techniques to optimize performance. For neural networks, variations in the number of layers, the number of neurons per layer, and the activation functions (e.g., ReLU) were tested, as well as different regularization strengths. For SVM models, kernel types (linear, polynomial, and radial basis function (RBF)) and regularization parameters (C and gamma) were tuned.

While various algorithms were investigated, the study focuses on the models that performed the best prediction for each mechanical property. Specifically, neural network models achieved the highest accuracy in predicting tensile and bending strength, whereas the SVM model was most effective in predicting compression strength.

Support vector machine (SVM) is a supervised learning algorithm that constructs an optimal hyperplane to separate data points in a high-dimensional space. In regression tasks (SVR), the objective is to identify a function that differs from the actual target values by no more than a specified margin (ε), while maintaining the flattest possible model. SVMs are particularly effective at capturing non-linear relationships by using kernel functions (e.g., radial basis functions), which project input data into higher-dimensional feature spaces. In this study, the SVM model produced the most accurate predictions for compression strength, demonstrating its effectiveness in datasets involving complex and context-dependent interactions between variables [34,35].

Artificial neural networks (ANNs) are computational models inspired by the structure of the human brain. They consist of layers of interconnected nodes (neurons) that transform input data via weighted connections and activation functions. ANNs are well suited to modeling nonlinear relationships and capturing patterns in large, complex datasets. In this research, neural networks with different architectures (two and three hidden layers) and ReLU activation were used [34,35].

These models were selected based on the different indicators: R^2, the mean absolute error (MAE), the mean absolute percentage error (MAPE), the mean squared error (MSE), and the root mean squared error (RMSE) [34–36].

The coefficient of determination (R^2) shows what proportion of the variation of the dependent variable is explained by the model, and we estimate as follows:

$$R^2 = 1 - \frac{\sum_{i=1}^{k}(y_i - \hat{y}_i)^2}{\sum_{i=1}^{k}(y_i - \bar{y})^2} \tag{1}$$

where $\bar{y} = \frac{1}{k}\sum_{i=1}^{k} y_i$.

The mean absolute error (MAE) quantifies the average magnitude of errors between predicted and actual values, measured using the l_1 norm. It is computed using the following formula:

$$MAE = \frac{1}{k}\sum_{i=1}^{k} |y_i - \hat{y}_i| \tag{2}$$

The mean absolute percentage error (MAPE) quantifies the average magnitude of absolute errors as a percentage of the actual values and is computed using the following formula:

$$MAPE = \frac{1}{k}\sum_{i=1}^{k} \frac{|y_i - \hat{y}_i|}{\max(\epsilon, |y_i|)} \tag{3}$$

where $\epsilon > 0$ is a small positive constant introduced to prevent undefined behavior when y is zero.

The Mean Squared Error (MSE) quantifies the average of the squared differences between predicted and actual values, serving as a standard metric for evaluating model accuracy. It is computed using the following formula:

$$MSE = \frac{1}{k}\sum_{i=1}^{k}(y_i - \hat{y}_i)^2 \tag{4}$$

The root mean squared error (RMSE) quantifies the typical magnitude of error by measuring the average deviation between predicted and actual values.

$$RMSE = \sqrt{\frac{1}{k}\sum_{i=1}^{k}(y_i - \hat{y}_i)^2} \tag{5}$$

The final results and the interpretation of feature importance, including Shapley analysis, were based exclusively on these top-performing models [37]. This methodological approach ensured both model transparency and relevance to the physical behavior of the tested materials.

4. Results and Discussion

In this chapter, the obtained results of the strength tests will be presented, as well as the analysis of the taken SEM images of the breakthroughs of the samples.

4.1. Strength Test Results

Epoxy compositions used as protective layers—regardless of the substrate (metallic, concrete, or composite)—must meet several key conditions: provide adhesion, resistance to cracking and deformation, and impermeability to aggressive agents. Therefore, mechanical analysis, including tension, bending, and compression, provides information not only on structural strength but also on the material's behavior as a functional barrier under operating conditions. The purpose of the strength study was to determine the effect of epoxy resin type, curing agent type, and filler grade and concentration on the mechanical properties of epoxy compositions that can act as protective coatings.

The mechanical properties of the tested epoxy compositions are shown in Figures 4–6.

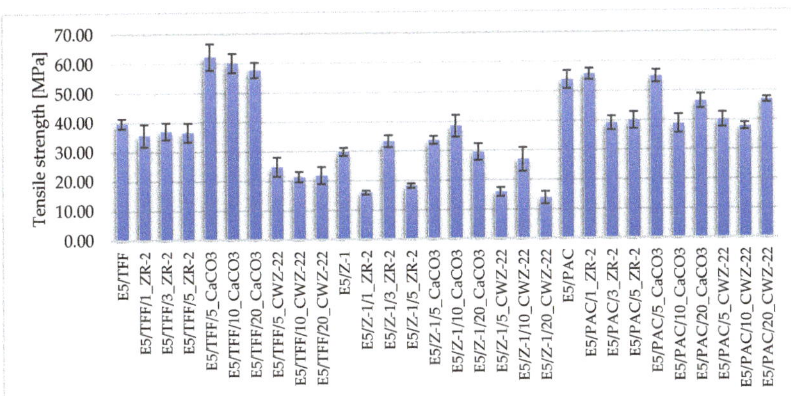

(a)

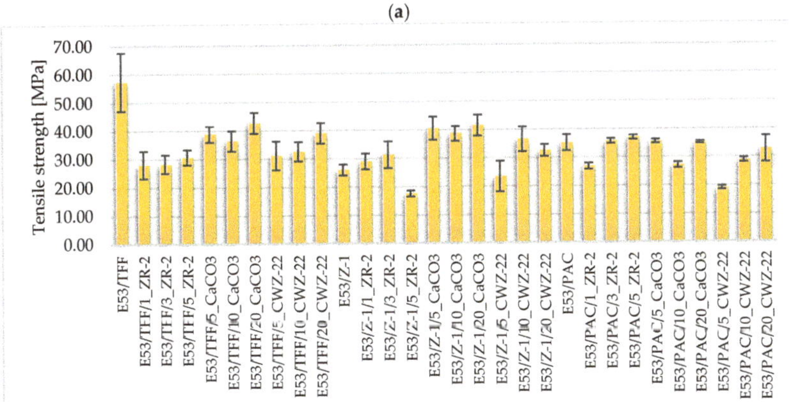

(b)

Figure 4. *Cont.*

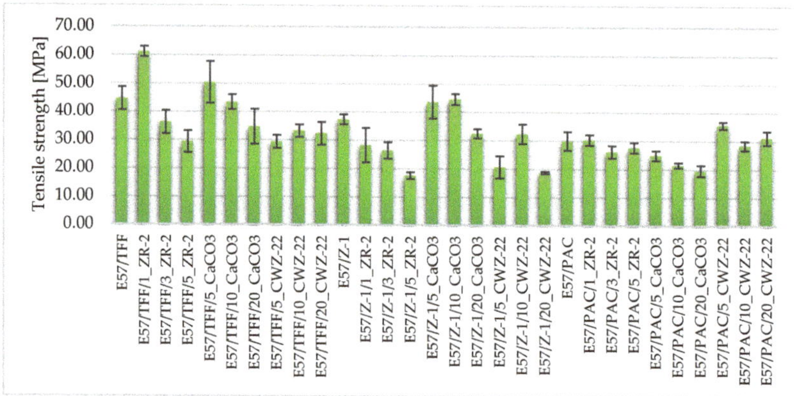

(**c**)

Figure 4. Tensile strength of epoxy compositions (**a**) based on Epidian 5 epoxy resin, (**b**) based on Epidian 53 epoxy resin, and (**c**) based on Epidian 57 epoxy resin.

Figure 4 shows the tensile strength of the epoxy compositions. The best parameters were achieved for compositions containing Epidian 5 resin and TFF curing agent, especially in the presence of 10–20 wt% calcium carbonate ($CaCO_3$). Tensile strengths exceeding 60 MPa in these compositions indicate effective stress transfer between the matrix and filler phases, which agrees with the results of Nwoye et al. [38], who confirmed an increase in mechanical resistance as a result of $CaCO_3$ in epoxy systems.

In the case of compositions containing active fillers (CWZ-22), despite the good dispersion observed in SEM analysis, lower strength values were obtained. This indicates that activated carbons, despite their developed specific surface area and favorable contact with the matrix [39–41], do not provide mechanical strengthening of the structure. Their effect is mainly limited to improving functional properties such as thermal conductivity or resistance to chemicals.

Compositions with montmorillonite nanofiller ZR-2 showed a different behavior. At low concentrations (1–3%), especially in flexible matrices (e.g., E57/PAC), a moderate increase in strength was observed, which can be explained by better dispersion of the particles in the resin structure. This phenomenon has also been described by Foroutani et al. [13], indicating a relationship between mechanical properties and dispersion quality of nano-aluminosilicates. For higher concentrations of ZR-2 (5%), however, the effects were less favorable—agglomeration dominated, causing local weakening.

Compression test results (Figure 5) confirmed the high efficiency of the mineral reinforcement. The highest values—reaching up to 145 MPa—were obtained for compositions E5/TFF/20$CaCO_3$ and E53/Z1/20$CaCO_3$. The presence of rigid $CaCO_3$ particles largely reduced local deformation, promoting the transfer of loads in a distributed manner, which corresponds to the classical model of passive reinforcement [42,43]. Importantly, these compositions, despite their higher stiffness, showed no signs of brittleness, thanks to the favorable configuration of interfaces and filler distribution in the matrix.

Fillers in the form of activated carbon (CWZ-22) did not improve compressive strength. Moreover, for compositions containing 20% CWZ-22 in harder resins (e.g., E53/Z1), a decrease in mechanical resistance was observed. This may be due to the presence of micropores around the carbon particles, visible in the SEM, and the fact that CWZ-22 activated carbon is not a structurally load-bearing filler but a functional modifier [13,44].

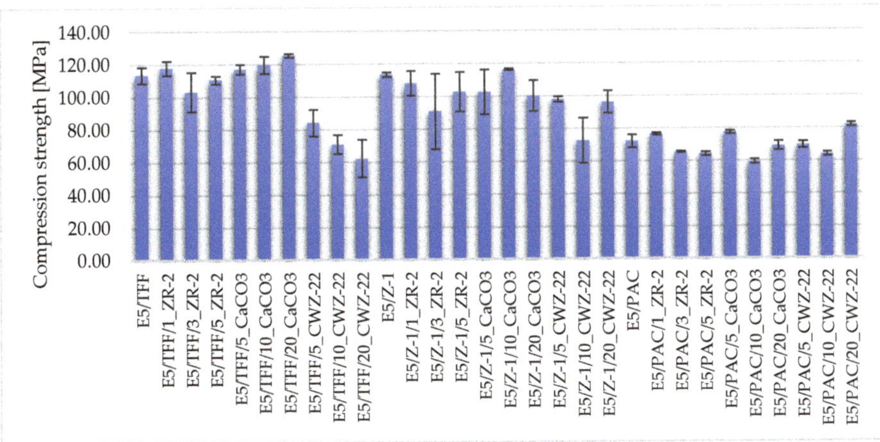

(**a**)

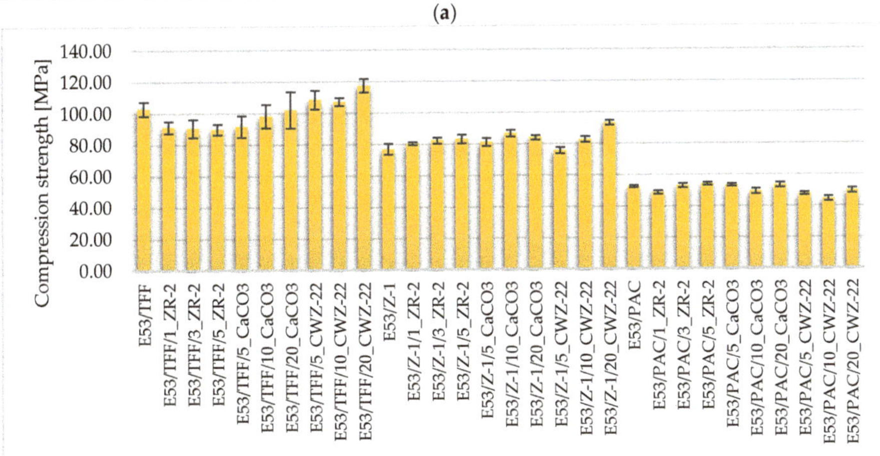

(**b**)

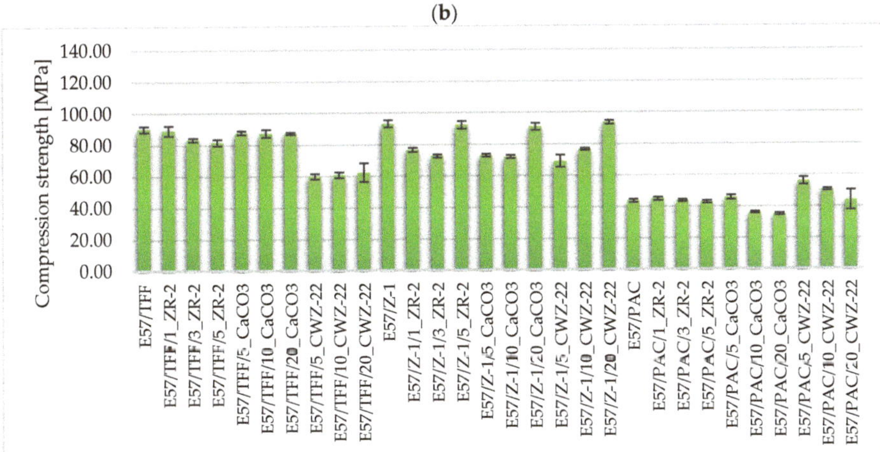

(**c**)

Figure 5. Compressive strength of epoxy compositions (**a**) based on Epidian 5 epoxy resin, (**b**) based on Epidian 53 epoxy resin, and (**c**) based on Epidian 57 epoxy resin.

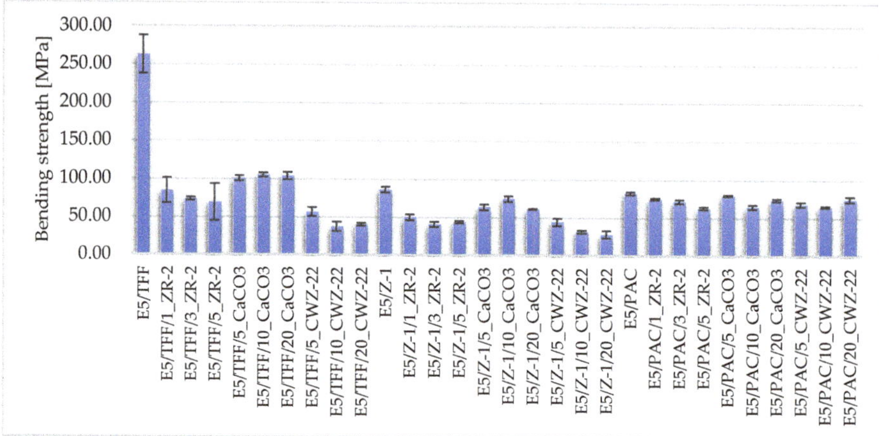

(**a**)

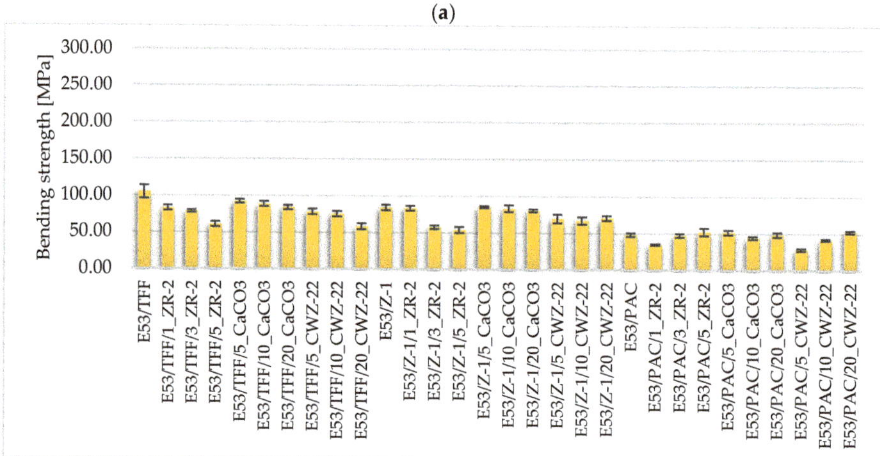

(**b**)

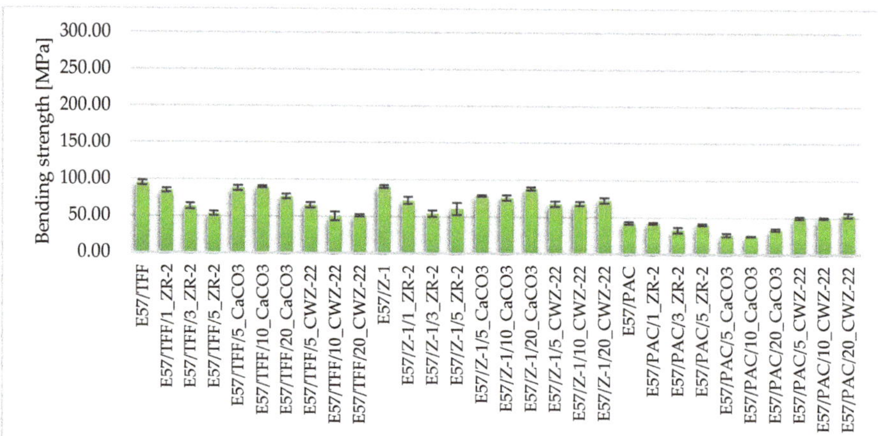

(**c**)

Figure 6. Bending strength of epoxy compositions (**a**) based on Epidian 5 epoxy resin, (**b**) based on Epidian 53 epoxy resin, and (**c**) based on Epidian 57 epoxy resin.

In the case of montmorillonite ZR-2, mixed results were evident: at low concentrations and in flexible compositions (E57/PAC), mechanical properties improved slightly, while in rigid matrices or at 5% addition, these effects were reversed. ZR-2 agglomeration and poor adhesion to the matrix were the main factors limiting the reinforcement efficiency, which correlates with data from Yu [45] and Yang et al. [42].

Bending, as a composite test, proved to be particularly sensitive to the quality of the interfacial bond and the uniformity of filler distribution (Figure 6). The highest bending values—exceeding 120 MPa—were obtained for compositions with $CaCO_3$ in E5/TFF/10% and E53/Z1/20% compositions. These results indicate that the propagation of microcracks in the tensile zone is effectively limited and the structural integrity of the entire coating is high. Fu and co-authors [46] note that well-distributed mineral particles can act as crack traps, delaying crack propagation.

Compositions containing CWZ-22, even with good wettability (Figure 7l), showed no increase in flexural strength. The presence of local agglomerates and micropores, also observed in SEM analysis, indicates that good dispersion does not guarantee mechanical strengthening—in line with the observations of Yang and Heo [42].

In compositions with montmorillonite, the improvement in flexural properties was limited to compositions with a flexible matrix (E57/PAC). In rigid compositions, no strengthening effect was observed—moreover, at higher ZR-2 concentrations, weakening effects related to phase discontinuity were evident.

4.2. SEM Analysis of Microstructure

In order to further analyze the influence of material composition on the functional properties of epoxy coatings, SEM microscopic observations of selected compositions were carried out. A set of twelve compositions (Figure 7) was selected so that the microstructure of materials differing in both the type of epoxy resin (E5, E53, E57) and the type of curing agent (TFF, Z1, PAC) and the filler used (ZR-2, $CaCO_3$, CWZ-22) could be compared. The obtained images made it possible to identify differences in the dispersion of the dispersed phase, the quality of interfacial adhesion, and the presence of defects, which can have a key impact on the durability and effectiveness of protective coatings made from the described compositions.

(**a**)

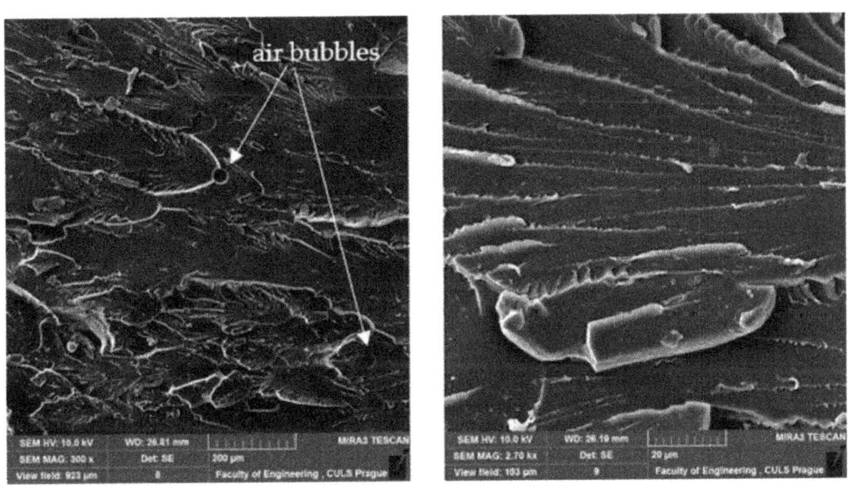

Figure 7. *Cont.*

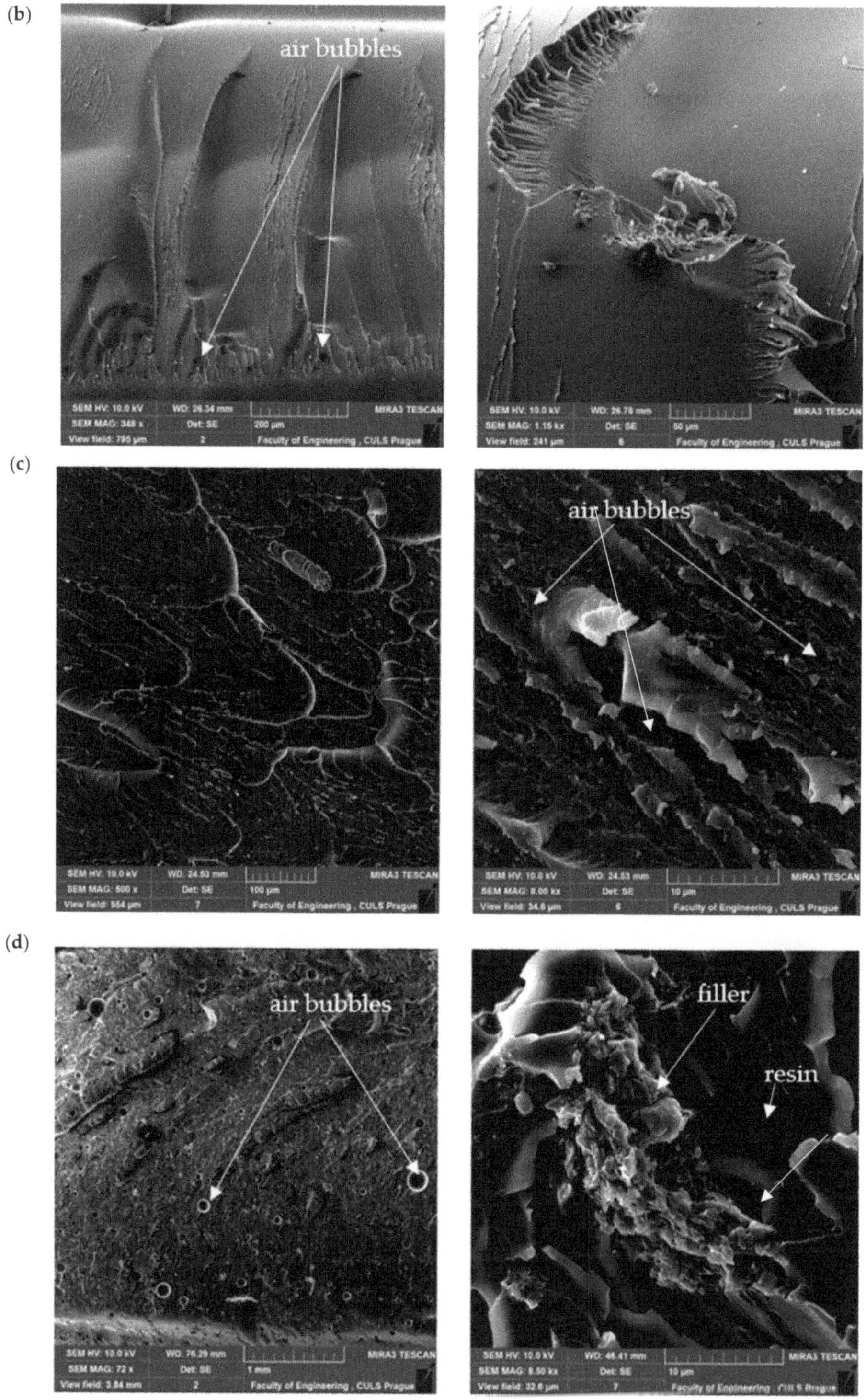

Figure 7. *Cont.*

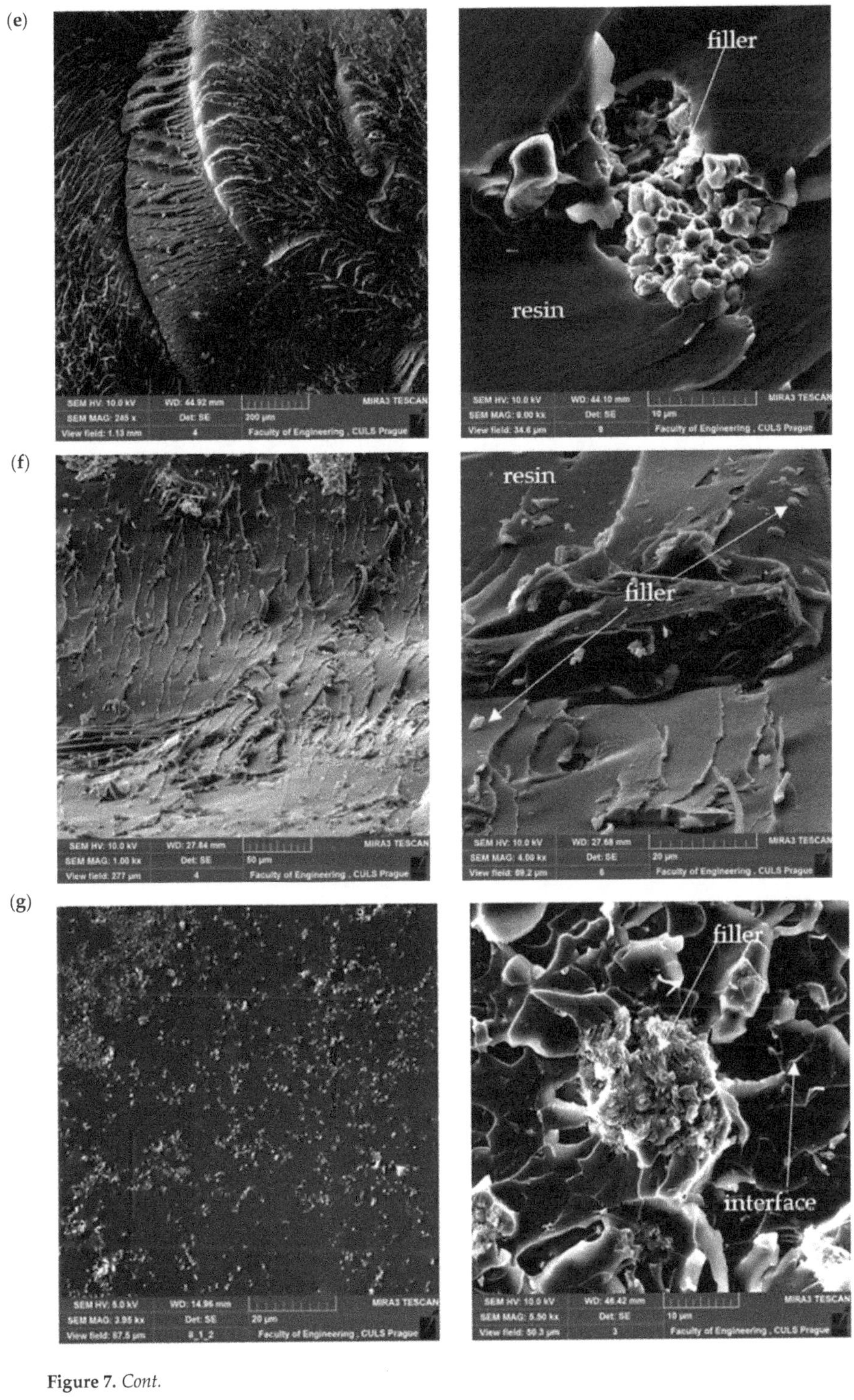

Figure 7. *Cont.*

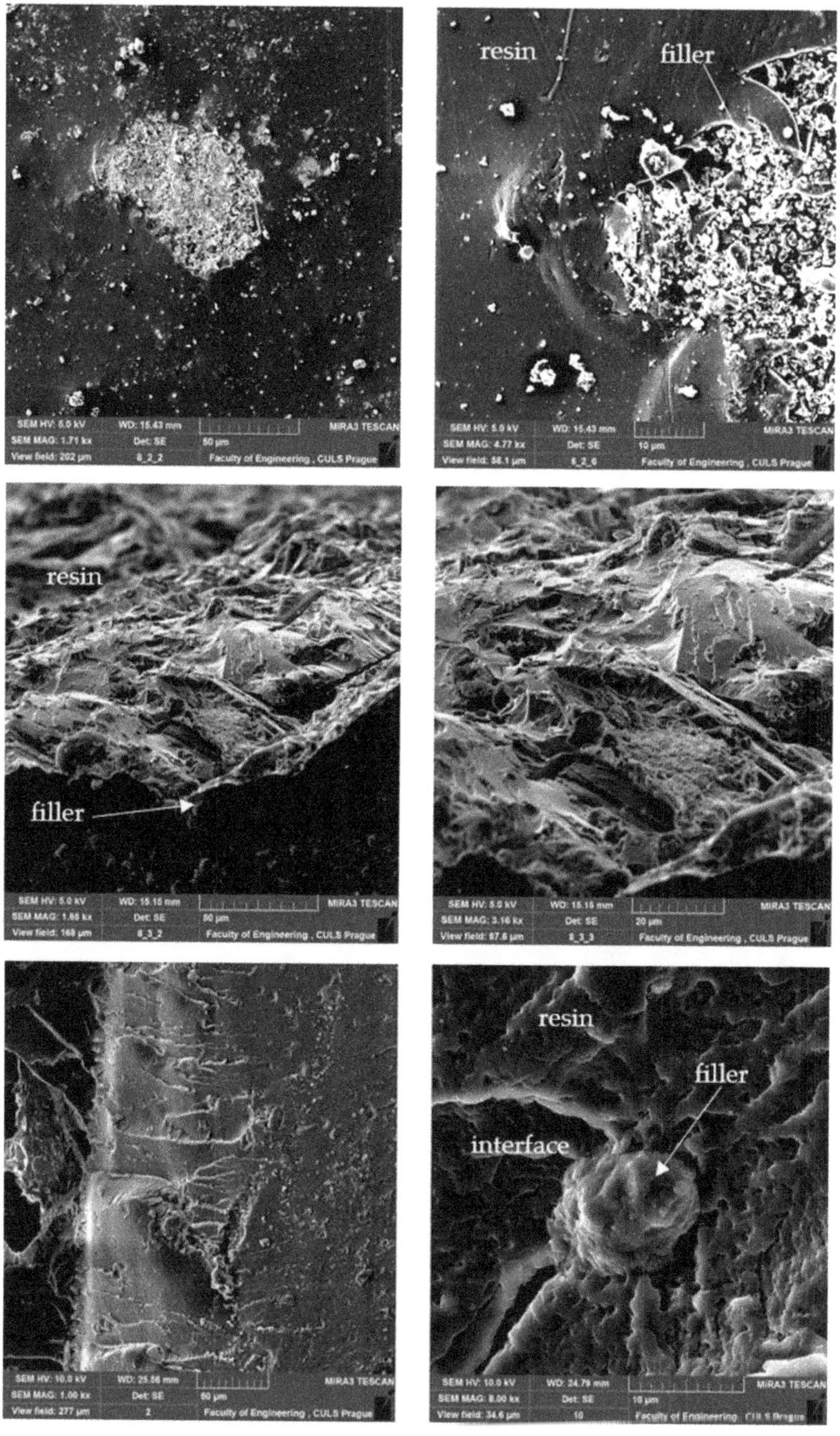

Figure 7. *Cont.*

(k)

(l)

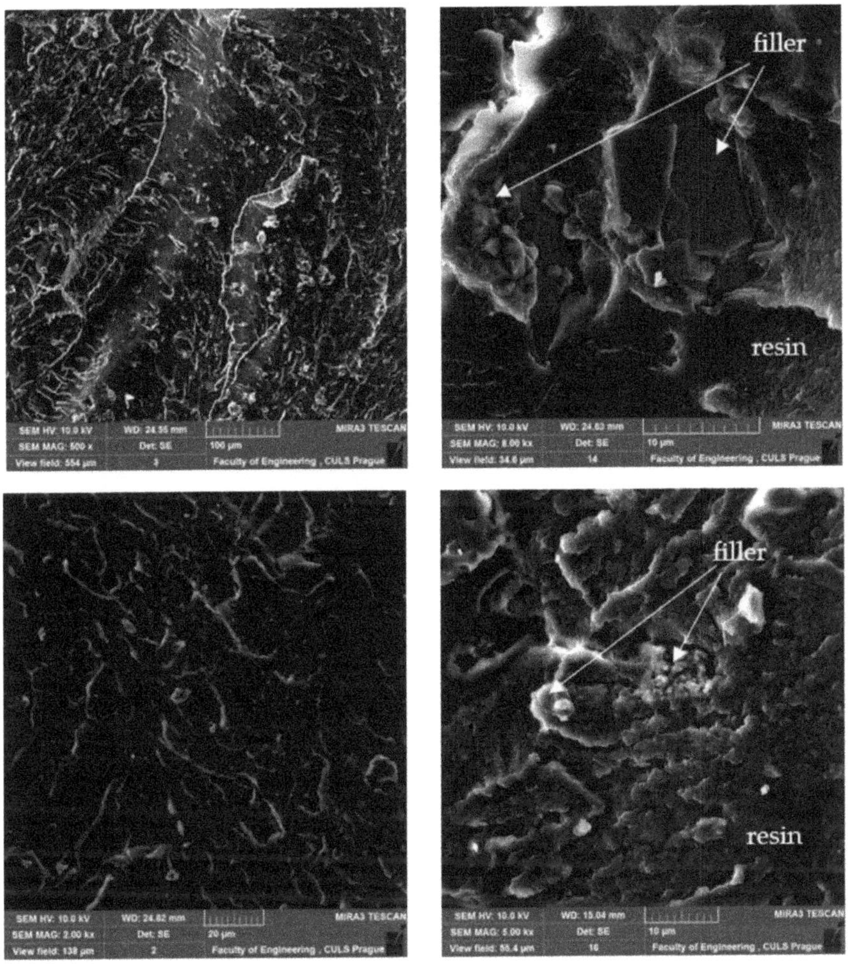

Figure 7. SEM images of selected epoxy compositions: (**a**) E57/TFF; (**b**) E57/Z1; (**c**) E57/PAC; (**d**) E5/TFF/5ZR-2; (**e**) E5/TFF/20CaCO$_3$; (**f**) E5/TFF/20CWZ-22; (**g**) E53/Z1/5ZR-2; (**h**) E53/Z1/20CaCO$_3$; (**i**) E53/Z1/20CWZ-22; (**j**) E57/PAC/5ZR-2; (**k**) E57/PAC/20CaCO$_3$; (**l**) E57/PAC/20CWZ-22.

An image of a reference sample of the E57/TFF composition (Figure 7a), containing only resin and curing agent, shows a homogeneous, continuous epoxy matrix, with visible individual pores (gas bubbles), without microcracks or delamination zones. This structure confirms the correct course of crosslinking and provides a reference point for evaluating the influence of modifying additives. In comparison, a sample of the E57/Z1 composition (Figure 7b), where the combination of the E57 resin with the stiff Z1 curing agent led to a morphology with significantly higher stiffness. The SEM microscopic image reveals the presence of microcracks and sharp phase transitions, typical of compositions with a high degree of cross-linking and less ability to relax stress. Such structures may be prone to brittleness, limiting their effectiveness as protective coatings in dynamic environments. Different structural properties were observed in a sample of the E57/PAC composition (Figure 7c), which uses the same type of resin but a more flexible curing agent, showing a slight waviness of the matrix surface, which may indicate a reduced degree of cross-linking and greater susceptibility to plastic deformation. The lack of visible defects confirms the good structural uniformity of the flexible compositions.

The addition of fillers introduces significant morphological changes. The E5/TFF/5ZR-2 composition sample (Figure 7d), which contains a low weight share of montmorillonite, shows the presence of small, lamellar ZR-2 particles. Their dispersion is moderate—locally there are agglomerates and small zones of stratification, which may limit the efficiency of stress transfer and affect local porosity. In addition, gas bubbles are visible, occurring on the surface of the tested breakthrough of the composition sample. In contrast, E57/PAC/5ZR-2 (Figure 7j), despite the identical filler concentration, presents a much more homogeneous structure with better particle embedding in the elastic matrix. The absence of defects and the correct dispersion testify to the synergy of the elastic matrix and the well-dispersed nano phase, which may explain the enhanced dynamic properties of this composition.

Significant differences are also observed when comparing $CaCO_3$ filled structures. The E5/TFF/20$CaCO_3$ sample (Figure 7e) is characterized by a compact structure with well-submerged filler particles, which is conducive to increased compressive strength. At the same time, the lack of local cracks and voids may indicate effective load transfer. However, agglomerations of filler particles are noticeable. In contrast, in the E53/Z1/20$CaCO_3$ sample (Figure 7h), a tighter packing of the filler in a rigid matrix was observed, with the presence of micropores and discontinuities at the phase boundaries. This configuration promotes stress concentration and may increase the coating's susceptibility to crack initiation.

In compositions containing CWZ-22 activated carbon in particulate form, such as E5/TFF/20CWZ-22 (Figure 7f), E53/Z1/20CWZ-22 (Figure 7i), and E57/PAC/CWZ-22 (Figure 7l), significant inhomogeneity in the distribution of the filling phase is noticeable. Activated carbon particles form irregularly shaped agglomerates, around which microcracks and porosity are often present. Particularly evident is poor adhesion between the filler and matrix, which limits structural integrity and can reduce barrier properties and resistance to service loads. In the case of sample E53/Z1/20CWZ-22, pronounced defects appear at the phase boundaries, indicating the possibility of delamination under varying loads. This situation may be due to the high proportion of filler in the composition.

The E53/Z1/5ZR-2 epoxy composition (Figure 7g), containing a low proportion of nanofiller in the solid resin, is characterized by the presence of zones of delamination and incomplete phase integration. Although particle aggregation is less severe than in the case of the CWZ-22 filler, the presence of voids and irregular phase transitions can negatively affect mechanical performance, according to the results of tensile and bending tests.

Finally, in the E57/PAC/20$CaCO_3$ sample (Figure 7k), a compact structure with clearly marked clusters of mineral filler is noted. Although agglomerates are not present, single cracks were observed, indicating correct compatibility. Local stiffening may limit deformability and adversely affect coating behavior in dynamic environments.

A compilation of SEM images proves that the microstructural properties of epoxy composites significantly depend on both the type of filler and flexibility of the matrix and the quality of interfacial interactions. Compositions characterized by structural homogeneity and good adhesion show better mechanical properties and protection potential. In contrast, the presence of agglomerates, microcracks, and delamination, typical of active fillers and rigid compositions, can significantly limit the functionality of the material in engineering applications.

5. Prediction Results Analysis

Different machine learning methods were applied, including neural networks and support vector machines, to develop models for three essential mechanical properties of the material: tensile strength (TS), compression strength (CS), and bending strength (BS). Each method was selected to ensure that the model architecture was appropriately suited to the specific characteristics and predictive challenges of the target property.

Table 10 shows the hyperparameters and structural details of the machine learning models that were most effective at predicting the TS, CS, and BS of the tested material. Each property was modeled using a different machine learning technique that was carefully selected based on its performance during training and validation. The table includes the following relevant configuration parameters: model architecture, activation function, regularization settings, training time, prediction speed, and data pre-processing methods.

Table 10. Hyperparameters and structural details of the machine learning models.

Model Hyperparameters	Neural Network Model for Tensile Strength [MPa]	SVM Model for Compression Strength [MPa]	Neural Network Model Bending Strength [MPa]
Prediction speed (obs/s)	73,000	23,000	67,000
Training time (s)	2.5936	2.6161	1.51
Layers	2	N/A	3
Layer sizes	10-10	N/A	10-10-10
Activation	ReLU	N/A	ReLU
Regularization	$\lambda = 0$	kernel scale (RBF kernel) = 0.75 ($\gamma \approx 0.889$); box constraint C $\approx$ 100 (auto); epsilon $\approx$ 2.5 MPa (auto)	$\lambda = 0$
Standardize data	Yes	Yes	Yes

For the TS, the bilayered NN model was the best. This model employed a simple feedforward architecture consisting of two hidden layers, each with ten neurons (layer sizes: 10–10). Both layers used the ReLU (rectified linear unit) activation function, which is commonly used due to its efficiency and ability to handle nonlinear relationships. No regularization was applied ($\lambda = 0$), and the input data were standardized prior to training. Training the model took approximately 2.59 s, and it offers a prediction speed of 73,000 observations per second. While the model is computationally efficient, its predictive accuracy for tensile strength was relatively low, suggesting that further refinement or architectural adjustments may be necessary.

The best results for CS were obtained using an SVM (support vector machine) model. As SVMs are not neural networks, parameters such as layers and activation functions are not applicable in this context. The regularization setting was configured as 'Auto', enabling the model to adaptively balance complexity and generalization. As with the other models, the input data were standardized. Training took approximately 2.62 s, with the model achieving a prediction speed of 23,000 observations per second. While this is slower than neural network models, it is still acceptable given the high level of accuracy. This model demonstrated excellent performance, proving highly effective in predicting compression strength.

The model for BS was also modeled using a neural network, which proved to be the most successful model in terms of predictive accuracy overall. This network was deeper, consisting of three hidden layers, each with ten neurons (layer sizes: 10–10–10). As with the tensile model, the ReLU activation function was used throughout. No regularization was applied ($\lambda = 0$), and the data were standardized prior to training. Training this model took only 1.51 s—the shortest time of all three models—and it achieved a prediction speed of 67,000 observations per second. This makes it suitable for fast, scalable applications. The combination of a deeper structure and an efficient activation function enabled this model to accurately capture the complex relationships governing bending strength.

Model performance was evaluated using standard indicators (Equations (1)–(5)) and the interpretation of feature importance via Shapley values. This enabled an assessment of both overall predictive accuracy and the contribution and influence of individual input variables in the prediction process. Model selection was based on a combination of multiple evaluation metrics, including R^2, RMSE, MAE, and MAPE. Although R^2 and RMSE were prioritized due to their intuitive interpretability and widespread use in materials research, all four metrics were considered in order to evaluate both absolute and relative accuracy. No formal weighting scheme was applied; instead, a comparative approach was used to select the model offering the most balanced and consistent performance across these indicators.

In the case of tensile strength (TS), the selected neural network model achieved the highest R^2 among all tested configurations, but its high MAPE indicated low reliability. This limitation was acknowledged, and the model was retained not for prediction in practice but to identify possible future directions for feature engineering and data refinement. As such, its feature importance results are interpreted cautiously and framed as exploratory insights.

Table 11 shows the results of the most successful predictive models developed to estimate three important mechanical properties of materials: tensile strength, compressive strength, and bending strength. Various machine learning techniques, including NN and SVM, were applied to each property during the modeling process. The method that achieved the highest accuracy during training and validation was selected for each property, and only these models are included in the table. The table summarizes the performance of these models based on the following key validation metrics: R-squared (R^2), root mean square error (RMSE), mean square error (MSE), mean absolute error (MAE), and mean absolute percentage error (MAPE).

Table 11. Model performance metrics.

Model Results	Neural Network Model for Tensile Strength [MPa]	SVM Model for Compression Strength [MPa]	Neural Network Model for Bending Strength [MPa]
R-Squared (Validation)	0.64	0.93	0.95
RMSE (Validation)	4483.5	6.4227	6.6236
MSE (Validation)	20,102,000.0	41.251	43.872
MAE (Validation)	694.66	3.8083	4.317
MAPE (Validation)	1494.2%	5.0%	7.1%

Based on the presented results, the NN model for predicting BS achieved the highest accuracy, with an R^2 value of 0.95. This indicates an excellent alignment between the predicted and actual values. Similarly, the SVM model for CS achieved an R^2 value of 0.93, also demonstrating strong predictive performance. These models also reported very low error values across all remaining metrics—for example, RMSE values of around 6.4–6.6 MPa, MAE below 5 MPa, and MAPE values below 7%—making them highly reliable for practical applications such as quality control, simulation, and materials optimization.

Conversely, the NN model for TS performed considerably worse. It achieved a relatively low R^2 value of 0.64, suggesting that a significant proportion of the variance in the target variable remains unexplained. It also produced substantially higher error values:

an RMSE of 4483.5, an MAE of 694.66, and an exceptionally high MAPE of 1494.2%. This indicates that its predictions are highly inaccurate and unreliable.

Several steps were undertaken in an attempt to improve the predictive performance of the TS model. These included optimization of ANN hyperparameters (e.g., number of layers and neurons, activation function types such as ReLU and Tanh, and regularization techniques), as well as testing of alternative algorithms such as support vector regression (SVR), decision trees, random forests, and ridge regression. In addition, various data preprocessing techniques, including standardization, min-max scaling, and logarithmic transformations, were evaluated. Despite these efforts, the TS model consistently underperformed in comparison to models for compression and bending strength. This suggests that the tensile strength data may be influenced by additional latent factors—such as filler–matrix interfacial variability, dispersion quality, or localized defects—which are not sufficiently captured by the current input features.

As a result, the interpretation of Shapley feature importance in the TS model should be regarded as indicative rather than conclusive. We caution against drawing definitive conclusions about the dominant predictors of tensile strength based on this model alone.

Due to the models' predictive performance in relation to TS, CS and BS, a deeper analysis was conducted to better understand the factors contributing to their accuracy. Advanced interpretability techniques based on Shapley values were employed to accomplish this, including Shapley importance and Shapley summary plots. These tools enable transparent evaluation of each model's decision-making process by quantifying the contribution of each input feature to the model's output.

The Shapley approach, which is based on cooperative game theory, calculates the Shapley value for each feature, representing the average marginal contribution of that feature across all possible subsets of input variables. This method ensures fair and balanced influence attribution, as it considers every possible feature combination and weighs each contribution equally. Applying this analysis to the best-performing models for each mechanical property provides a detailed picture of how different features influence predictions both globally (across the dataset) and locally (per observation).

The Shapley analysis not only reveals which variables have the greatest impact on model predictions but also the nature and direction of that impact. For instance, some features may consistently increase predicted values, while the effect of others depends on the input configuration, either increasing or decreasing predictions. This level of interpretability is crucial for validating model behavior, identifying key design parameters and supporting informed decisions in material optimization.

Figures 8 and 9 present a set of Shapley plots, providing a detailed explanation of how input features influence the prediction of three key mechanical properties: tensile strength, compression strength, and bending strength. These plots include both importance and summary plots.

The plots indicate the average absolute impact of each feature on model predictions. In the TS model, CWZ-22 had the highest importance with a maximum Shapley value of 4040.02, while Z-1 had the lowest at 51.16. For CS, PAC dominated with 8.57, and Z-2 contributed the least with 0.69. In the BS model, TFF had the strongest influence (41.17), while $CaCO_3$ showed the lowest (4.26). These values support the identification of the most influential formulation variables across each property.

The plots (Figure 9a–c) show both the magnitude and direction of the effect each feature has on individual model predictions. For TS, CWZ-22 exhibits a wide spread of Shapley values (both positive and negative), suggesting its contribution varies depending on specific formulation values. In contrast, Z-1 has consistently low influence.

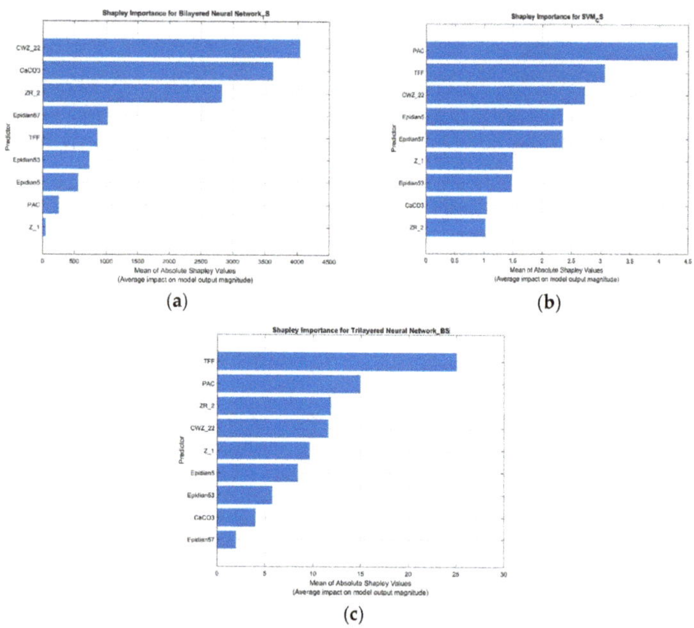

Figure 8. Shapley importance plots for the developed models: (**a**) NN model for TS; (**b**) SVM model for CS; (**c**) NN model for BS.

In the CS model, PAC shows strong and mostly positive influence, while Z-2 has limited effect. For BS, TFF consistently contributes positively to predicted strength, whereas $CaCO_3$ has a minor and slightly varying influence.

The Shapley importance plot in the TS model (Figure 8a) indicates that the most significant predictors are CWZ-22, $CaCO_3$, and ZR-2. These three variables clearly have the highest mean absolute Shapley values, confirming that they dominate the model's predictions. Their physical roles may relate to matrix stiffening (CWZ-22), filler content ($CaCO_3$), and reinforcement or fiber content (ZR-2), all of which typically have a significant impact in tensile loading conditions.

The Shapley summary plot (Figure 9a) further clarifies how these variables influence individual predictions. The widespread Shapley values for CWZ-22 suggest that, although it is generally impactful, its contribution can vary considerably from sample to sample. This indicates some level of model instability or nonlinearity in how this variable interacts with others. The same is true, albeit to a lesser extent, of $CaCO_3$ and ZR-2.

Secondary variables such as Epidian 57, TFF, and Epidian 53 demonstrate modest and relatively narrow distributions, with Shapley values centered around zero. Their role is more supportive and consistent. Variables such as PAC, Epidian 5, and, in particular, Z-1 demonstrate negligible importance and minimal variance, suggesting that they are not relevant for predicting tensile strength.

Despite identifying several key drivers, the model's overall predictive performance remains relatively low (R^2 = 0.64). This may be due to overreliance on a few strong predictors and the shallow depth of the network (two hidden layers).

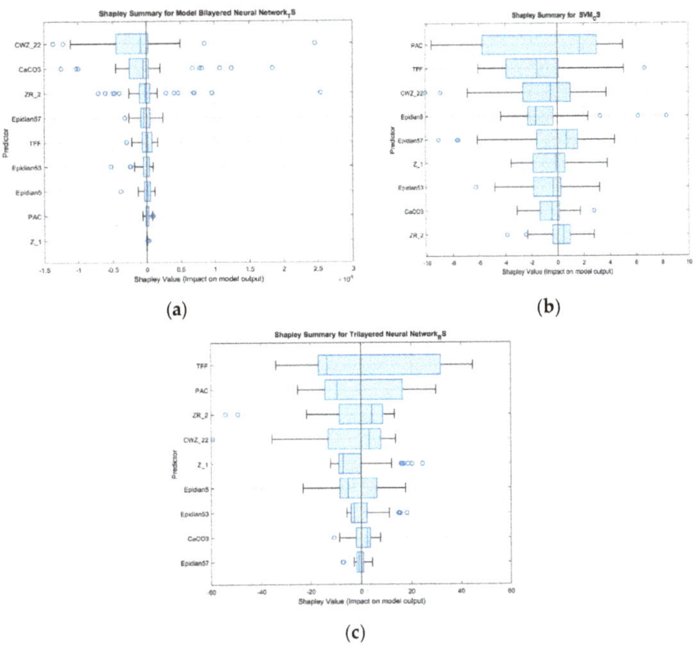

Figure 9. Shapley summary plots for the developed models: (**a**) NN model for TS; (**b**) SVM model for CS; (**c**) NN model for BS.

The model that predicts CS has a notably different importance profile (Figure 8b). The Shapley importance plot identifies PAC, TFF, and CWZ-22 as the top contributors. This is in contrast to the tensile strength model, where PAC was among the least relevant features. Here, however, PAC dominates, achieving the highest average Shapley value, closely followed by TFF.

The Shapley summary plot (Figure 9b) provides more nuance. The wide box plot for PAC, extending significantly in both positive and negative directions, shows that PAC can have both enhancing and suppressing effects on the model output, depending on the context. This suggests that the model has learned conditional relationships, for which SVMs operate well due to their ability to define non-linear boundaries. TFF shows similar bidirectional behavior, albeit with slightly less variability. Although influential, CWZ-22 has a more constrained and generally positive distribution.

Other variables, such as Epidian 5, Epidian 57, and Z1, demonstrate moderate importance with some variability. Conversely, $CaCO_3$ and ZR-2—highly important in the tensile model—have little influence here, highlighting the property-specific relevance of features.

This SVM model achieves excellent predictive accuracy (R^2 = 0.93) due to its ability to leverage complex, context-sensitive variables effectively, particularly PAC and TFF.

In the BS prediction task, the Shapley importance plot (Figure 8c) identifies TFF as the most critical feature, followed by PAC, ZR-2, and CWZ-22. These variables are key contributors to the model's high accuracy (R^2 = 0.95) and robust generalization.

The summary plot shows (Figure 9c) that TFF consistently has a positive influence across samples, establishing it as a stable and dominant predictor. PAC and ZR-2 also demonstrate a strong and varied influence, which is sometimes positive and sometimes negative. This suggests that their impact is more nuanced and sample-dependent. CWZ-22 retains moderate importance, but its range of influence is narrower than that of tensile strength predictions.

The lower-ranked variables—Z1, Epidian 5, Epidian 53, $CaCO_3$, and Epidian 57—contribute little to the model's decisions, with low and narrow distributions of Shapley values. Their roles are minor, and they do not appear essential in the bending context.

The model provides sufficient depth and flexibility to capture a broad range of interactions and nonlinear relationships. This is evident in how it balances multiple important features without depending excessively on a single variable.

The combined analysis of the Shapley importance and summary plots provides several key conclusions:

- TFF is the only variable that is consistently important across all three models. It is particularly dominant in the bending strength model, where it has the highest and most stable positive average impact. This suggests that TFF plays a versatile role, influencing a range of mechanical properties, which is likely due to its structural or adhesive characteristics in the composite.
- PAC is highly influential for compression and bending strength but almost irrelevant for tensile strength. Its behavior is variable and sometimes bidirectional, indicating that its role depends heavily on its interaction with the other components of the formulation.
- CWZ-22 has the greatest impact on tensile strength prediction, with reduced relevance in compression and bending. This suggests that its contribution may relate to stiffness or network formation under tensile loads.
- ZR-2 is important for tensile and bending strength but not compression, mirroring the pattern observed for CWZ-22.
- $CaCO_3$ is a key predictor only in the tensile model, highlighting its specific role in enhancing tensile resistance, likely through particle reinforcement or matrix modification.
- Z1, Epidian 53, Epidian 5, and Epidian 57 remain minor contributors across all tasks, though some exhibit small but variable influences, especially in specific samples.

Importantly, the Shapley summary plots emphasize the difference in impact directionality. The SVM model shows the greatest use of bidirectional contributions, especially for PAC and TFF, while the neural network models—particularly the deeper one used for bending—show more consistent, stable patterns of feature importance.

In addition to identifying key predictors, the Shapley analysis was interpreted within the context of materials science to determine why certain features dominate particular mechanical responses.

For example, PAC (porous activated carbon) was found to have the greatest influence on compression strength (CS). This is consistent with its fine particle size, rigidity, and ability to improve stress distribution and energy dissipation under compressive loading. In contrast, its impact on tensile strength (TS) was much lower, likely due to interfacial limitations—carbon-based fillers can act as stress concentrators if dispersion or adhesion is suboptimal.

Conversely, CWZ-22—a surface-modified filler—demonstrated the strongest influence on TS. Its enhanced surface activity likely improves interfacial bonding with the matrix, which is critical under tensile stress, as microcracks and failure typically initiate at the filler–matrix interface. However, its effect on CS and BS was lower, as these properties are more influenced by bulk behavior and particle morphology than by interfacial chemistry alone.

The widespread Shapley values observed for CWZ-22 also suggest that its contribution varies depending on other interacting features, such as concentration, resin type, or hardener system. This highlights the complexity of multi-phase composite behavior and the importance of analyzing not just main effects, but also interactions. This will be the subject of future work.

The trends identified through machine learning feature importance analysis were examined further in light of the microstructural observations obtained via SEM imaging (see Figure 7). Fillers exhibiting microporosity, weak interfacial bonding, or agglomeration, such as PAC and $CaCO_3$, were associated with limited improvements or even reductions in mechanical strength, particularly in tensile and bending tests. While these effects were not explicitly included as input features in the ML models, they are likely to be reflected implicitly in the 'filler type' variable, which serves as a proxy for the filler's physical behavior, dispersion quality, and interfacial characteristics.

For instance, the relatively low Shapley values of $CaCO_3$ across all strength properties are consistent with SEM findings of filler clustering and resin detachment. In contrast, CWZ-22 showed a more homogeneous distribution and fewer voids, which is consistent with its strong influence on tensile strength in the Shapley analysis. These correlations suggest that microstructural defects may explain why certain fillers underperform despite their chemical potential and highlight the importance of linking data-driven model interpretation with physical evidence from microscopy.

This extensive analysis confirms that each mechanical property is governed by a unique set of predictors, and the same variable can have entirely different roles depending on the target. It also underscores the need for dedicated model architectures and feature importance evaluation when working with complex materials data.

By combining global (mean) and local (distributional) interpretability tools like Shapley values, we gain not only performance insights but also practical knowledge for material design. This can support targeted optimization, formulation tuning, and data-driven decision-making in engineering applications.

Although the proposed machine learning (ML)-based approach shows strong potential for modeling and interpreting the effects of formulation variables, it also has inherent limitations. The models are trained on a finite dataset and may lose predictive power when applied to formulations outside the composition space that has been explored. Furthermore, complex microstructural interactions that are not captured by the input variables (e.g., filler distribution or interfacial defects) may reduce the accuracy of the models.

Nevertheless, this method is particularly well-suited to polymer composite systems where multiple variables influence multiple target properties and exhaustive experimental campaigns are impractical. The method's generalizability lies in its modular workflow: while specific models may require retraining for other systems, the combined use of targeted experimentation, machine learning modeling, and Shapley-based interpretation remains broadly transferable.

6. Conclusions

The conducted research clearly confirms the effectiveness of physical modification of epoxy adhesive materials using mineral, active, and nanofillers in the context of designing functional protective coatings. The results of mechanical tests and SEM microstructural analysis showed a significant influence of both the chemical composition and interfacial interactions on the behavior of the compositions under service loading conditions.

Among the tested systems, compositions based on Epidian 5 epoxy resin and TFF curing agent, modified with the addition of $CaCO_3$ calcium carbonate (10–20 wt%), showed the highest strength parameters (more than 60 MPa in tension, 145 MPa in compression, and 120 MPa in bending). This confirms that classical mineral fillers, when properly dispersed, can effectively strengthen the composite structure and increase its resistance to degradation under mechanical conditions.

Active fillers, such as CWZ-22 activated carbon in particulate form, despite favorable performance characteristics (vibration damping, conductivity), did not contribute to improving strength parameters, as confirmed by both mechanical strength and microstructure analysis (presence of pores, poor phase adhesion). On the other side, the use of montmorillonite ZR-2 allowed for obtaining favorable effects only in compositions with an elastic matrix (e.g., E57/PAC), where the particles were better dispersed and integrated into the continuous phase.

The novelty of this paper was the successful combination of classical experimental approaches with data analysis tools and machine learning methods. Machine learning models, based on input data on the type of resin, curing agent, and filler concentration, enabled effective prediction of mechanical properties of composites. This approach significantly improves the design process of epoxy compositions, which can also be used as functional protective coatings, enabling the selection of optimal formulations without the need to conduct costly, labor-intensive, and time-consuming tests each time. Predictive models were developed to forecast the mechanical performance of compositions based on input data such as the type of resin, hardener and filler, and their respective concentrations. Three high-performance models were selected:

- For tensile strength, a 2-layer neural network model achieved an $R^2 = 0.64$, an RMSE = 4483.5, and a MAPE = 1494.2%.
- For compression strength, an SVM model achieved an $R^2 = 0.93$, an RMSE = 6.42, and a MAPE = 5.0%.
- The most accurate model for bending strength was a 3-layer neural network, which obtained an $R^2 = 0.95$, an RMSE = 6.62, and a MAPE = 7.1%.

However, the prediction model for tensile strength demonstrated limited reliability, with a moderate R^2 value (0.64) and high error metrics. Multiple approaches were tested to improve this model—including algorithm switching, parameter tuning, and data transformations—but without significant success. This limitation suggests that tensile strength may be influenced by more complex or microstructural factors not included in the current dataset. Therefore, the feature importance results from the TS model should be treated with caution and not considered as definitive evidence of factor influence. Future work should consider expanding the dataset or integrating SEM-based descriptors to improve TS prediction accuracy.

Moreover, the results demonstrate that different modeling approaches are optimal depending on the nature of the mechanical property. Shapley value analysis enabled detailed interpretation of feature importance:

- TFF was the most consistent predictor across all models, particularly in bending.
- PAC significantly influenced compression and bending strength but had little effect on tensile strength.
- CWZ-22 and $CaCO_3$ were key drivers of tensile strength.

This interpretability is crucial for validating model predictions and directing future formulation efforts.

The analysis confirms that the integrated approach—combining microstructural observations, strength analysis, and predictive modeling—is an effective methodology for designing modern engineering materials. The results obtained can be successfully applied in industrial practice, especially in areas requiring coatings with high mechanical and chemical resistance, such as the construction, energy, marine, and automotive industries.

However, some limitations of the study should be noted. Although the dataset was carefully designed, it may not fully capture the variability in resin types, filler geometries, or environmental conditions affecting long-term performance. Furthermore, the current models are based on static laboratory data and do not consider real-world degradation mechanisms, such as UV exposure, moisture absorption, or thermal cycling. While the ML models applied were effective, they were based on relatively standard architectures and did not exploit deeper learning methods or ensemble approaches to their fullest potential.

Future research should focus on integrating advanced machine learning models, such as random forests, XGBoost, and deep neural networks, to more accurately capture the non-linear interactions between the input features and the predicted mechanical properties. Due to their greater complexity and flexibility, these algorithms are well suited to modeling high-dimensional, heterogeneous material data.

Another important area for research is to extend experimental datasets to include information on durability and the effects of aging, such as the long-term impact of environmental factors (e.g., humidity, UV radiation, and temperature fluctuations). Incorporating this data would enable long-term performance to be predicted and would enhance the applicability of the models in real-world operating conditions.

In parallel, future studies should place greater emphasis on analyzing full-scale coatings, considering parameters such as actual coating thickness, porosity, and interfacial adhesion. These characteristics significantly influence the performance of coatings in industrial applications and must be accounted for to improve the practical relevance and reliability of predictive models.

Although the developed machine learning models initially require a representative set of experimental data, they provide a valuable tool for future material design. Once trained, they facilitate the rapid in silico screening of new formulations, thereby reducing the need for extensive physical testing. This can shorten development time and reduce costs, particularly when only the top performers are selected for validation. Furthermore, the model architecture and learned relationships can be adapted to similar epoxy systems, facilitating knowledge transfer across related projects.

Finally, considerable potential exists in developing hybrid experimental–computational workflows that combine machine learning predictions with finite element simulations or micromechanical modeling. These multiscale approaches can bridge the gap between the structure of materials at a microscopic level and their behavior at a macroscopic level, enabling a more comprehensive and physics-informed process for designing materials.

Author Contributions: Conceptualization, I.M.-B. and K.A.; methodology, I.M.-B. and K.A.; software, I.M.-B. and K.A.; validation, I.M.-B. and K.A.; formal analysis, I.M.-B. and K.A.; investigation, I.M.-B. and K.A.; resources, I.M.-B. and K.A.; data curation, I.M.-B. and K.A.; writing—original draft preparation, I.M.-B. and K.A.; writing—review and editing, I.M.-B. and K.A.; visualization, I.M.-B. and K.A.; supervision, I.M.-B. and K.A.; project administration, I.M.-B. and K.A.; funding acquisition, I.M.-B. and K.A. All authors have read and agreed to the published version of the manuscript.

Funding: The research leading to these results has received funding from the commissioned task entitled "VIA CARPATIA Universities of Technology Network named after the President of the Republic of Poland Lech Kaczyński" under the special purpose grant from the Minister of Education and Science, contract no. MEiN/2022/DPI/2575, as part of the action "In the neighborhood—inter-university research internships and study visits".

Institutional Review Board Statement: Not applicable.

Informed Consent Statement: Not applicable.

Data Availability Statement: The original contributions presented in this study are included in the article. Further inquiries can be directed to the authors.

Conflicts of Interest: The authors declare no conflicts of interest.

References

1. Curtarolo, S.; Setyawan, W.; Hart, G.L.W.; Jahnatek, M.; Chepulskii, R.V.; Taylor, R.H.; Wang, S.; Xue, J.; Yang, K.; Levy, O.; et al. AFLOW: An Automatic Framework for High-Throughput Materials Discovery. *Comput. Mater. Sci.* **2012**, *58*, 218–226. [CrossRef]
2. Jain, A.; Ong, S.P.; Hautier, G.; Chen, W.; Richards, W.D.; Dacek, S.; Cholia, S.; Gunter, D.; Skinner, D.; Ceder, G.; et al. Commentary: The Materials Project: A Materials Genome Approach to Accelerating Materials Innovation. *APL Mater.* **2013**, *1*, 011002. [CrossRef]
3. Taub, A.I.; Luo, A.A. Advanced Lightweight Materials and Manufacturing Processes for Automotive Applications. *MRS Bull.* **2015**, *40*, 1045–1054. [CrossRef]
4. Karthikeyan, N.; Naveen, J. Progress in Adhesive-Bonded Composite Joints: A Comprehensive Review. *J. Reinf. Plast. Compos.* **2024**, 07316844241248236. [CrossRef]
5. Fekiač, J.J.; Krbata, M.; Kohutiar, M.; Janík, R.; Kakošová, L.; Breznická, A.; Eckert, M.; Mikuš, P. Comprehensive Review: Optimization of Epoxy Composites, Mechanical Properties, & Technological Trends. *Polymers* **2025**, *17*, 271. [CrossRef]
6. Szewczak, A. Impact of Epoxy Resin Modification on Their Strength Parameters. *Bud. Archit.* **2020**, *18*, 041–050. [CrossRef]
7. Kudžma, A.; Stonys, R.; Kilikevičius, A.; Višniakov, N.; Antonovič, V.; Škamat, J. Effect of Low Oxidation Graphene Oxide on Physical, Dynamic and Mechanical Properties of MCC Type Refractory Castable. In Proceedings of the 24th International Conference on Composite Structures (ICCS-24), Porto, Portugal, 14–16 June 2021.
8. Ahmad Temizi, M.A.; Meor Ahmad, M.I.; Khamis, N.K.; Ahmad Samsudin, M.Y. Merit Technologies Sdn. Bhd, Jalan PJU 3/45, Sunway Damansara, 47810 Petaling Jaya, Selangor Study of Mechanical and Thermal Properties for Epoxy Grouts Subjected to Seawater Conditioning at Elevated Temperature: Tensile Test and Compressive Test. *J. Kejuruter.* **2022**, *34*, 1017–1025. [CrossRef]
9. Yan, H.; Tan, J.; Chen, H.; He, T.; Zeng, D.; Zhang, L. Machine Learning-Based Prediction of Tribological Properties of Epoxy Composite Coating. *Polymers* **2025**, *17*, 282. [CrossRef]
10. Jang, H.; Ryu, D.; Lee, W.; Park, G.; Kim, J. Machine Learning-Based Epoxy Resin Property Prediction. *Mol. Syst. Des. Eng.* **2024**, *9*, 959–968. [CrossRef]
11. Berladir, K.; Antosz, K.; Ivanov, V.; Mitaľová, Z. Machine Learning-Driven Prediction of Composite Materials Properties Based on Experimental Testing Data. *Polymers* **2025**, *17*, 694. [CrossRef]
12. Yang, L.; Zhang, T.; Sun, W. Construction of Biocompatible Bilayered Light-driven Actuator Composed of rGO/PNIPAM and PEGDA Hydrogel. *J. Appl. Polym. Sci.* **2020**, *137*, 49375. [CrossRef]
13. Foroutani, K.; Ghasemi, S.M.; Pourabbas, B. Ordered Isoporous Membranes from Ionic Diblock Copolymers via SNIPS: Optimizing Effective Factors with a Structural Survey. *Prog. Org. Coat.* **2021**, *161*, 106554. [CrossRef]
14. Rudawska, A. Epoxy Adhesives. In *Handbook of Adhesive Technology*; CRS Press Taylor & Frances Group: Boca Raton, FL, USA, 2018; pp. 415–442.
15. Ma, X.; Zhang, Y.; Wang, X. Machine Learning for Composite Materials. *MRS Commun.* **2020**, *10*, 551–556. [CrossRef]
16. Raccuglia, P.; Elbert, K.C.; Adler, P.D.F.; Falk, C.; Wenny, M.B.; Mollo, A.; Zeller, M.; Friedler, S.A.; Schrier, J.; Norquist, A.J. Machine-Learning-Assisted Materials Discovery Using Failed Experiments. *Nature* **2016**, *533*, 73–76. [CrossRef]
17. Ramprasad, R.; Batra, R.; Pilania, G.; Mannodi-Kanakkithodi, A.; Kim, C. Machine Learning in Materials Informatics: Recent Applications and Prospects. *Npj Comput. Mater.* **2017**, *3*, 54. [CrossRef]
18. Xie, T.; Grossman, J.C. Crystal Graph Convolutional Neural Networks for an Accurate and Interpretable Prediction of Material Properties. *Phys. Rev. Lett.* **2018**, *120*, 145301. [CrossRef]
19. Czub, P.; Penczek, P. *Epoxy Resin Chemistry and Technology*; Wydawnictwa Naukowo-Techniczne: Warszawa, Poland, 2002; ISBN 978-83-204-2611-3.
20. Królikowski, W.; Rosłaniec, Z. Polymer Nanocomposites. *Kompozyty* **2004**, *9*, 3–16.
21. *BN-89 6376-02–Industry Standard*; Epoxy Resins Epidian 1, 2, 3, 4, 5, 6. Zakłady Chemiczne Organika-Sarzyna: Nowa Sarzyna, Poland, 2013; Access Location: Library of Lublin University of Technology. (In Polish)
22. *BN-73 6376-01–Industry Standard*; Epoxy Resins Epidian 51 i 53. Institution Developing the Standard—Zakłady Chemiczne in Oświęcim. Zakłady Chemiczne Organika-Sarzyna: Nowa Sarzyna, Poland, 1974. Available online: https://bcpw.bg.pw.edu.pl/dlibra/publication/12442/edition/11925?language=pl (accessed on 13 June 2025). (In Polish)
23. Information Catalogue of Ciech, S.A. Available online: https://www.zywicesarzyna.pl/ (accessed on 13 June 2025).

24. Bereska, B.; Iłowska, J.; Czaja, K.; Bereska, A. Curing Agents for Epoxy Resins. *Przem. Chem.* **2014**, *93*, 443–448.
25. Material Safety Data Sheet for Montmorillonite ZR-2 NanoBent. Available online: https://www.sorbentsystems.com/pdf/montmorillonitemsds.pdf?srsltid=AfmBOoqAlBCRt_AwRrrYWB-6NSwXF4a6tyuSBUSKVk-MzVitPNQECEwP (accessed on 13 June 2025).
26. Safety Data Sheet for CaCO3. Available online: https://www.carlroth.com/medias/SDB-P013-MT-EN.pdf?context=bWFzdGVyfHNlY3VyaXR5RGF0YXNoZWV0c3wyNDQ1NDh8YXBwbGljYXRpb24vcGRmfGFEWmhMMmhpTnk4NU1UYzJORGMwTlRVd016QXlMMU5FUWw5UU1ERXpYMDFVWDBWT0xuQmtaZ3xhMzc2MjBhZDhhMjgxMDExZDk2NDY2NDRmYTk5NjY5NGJmYmUxN2RiN2JkMWRlYWIwMWJmOWZkNGExMGFiZGY3 (accessed on 13 June 2025).
27. Characteristics Sheet of CWZ-22 Activated Carbon. Available online: https://mass-zone.eu/en/stanlab-activated-carbon-dust-cwz-22-1kg-p-8886.html (accessed on 13 June 2025).
28. Dickens, B. Model-Free Estimation of Outdoor Performance of a Model Epoxy Coating System Using Accelerated Test Laboratory Data. *J. Coat. Technol. Res.* **2009**, *6*, 419–428. [CrossRef]
29. Chen, Z.; Zhou, K.; Lu, X.; Lam, Y.C. A Review on the Mechanical Methods for Evaluating Coating Adhesion. *Acta Mech.* **2014**, *225*, 431–452. [CrossRef]
30. Ye, D.; Xu, Z.; Liu, H.; Zhang, Z.; Wang, P.; Wu, Y.; Yin, C. Terahertz Non-Destructive Testing of Porosity in Multi-Layer Thermal Barrier Coatings Based on Small-Sample Data. *Coatings* **2024**, *14*, 1357. [CrossRef]
31. *PN-EN ISO 527-1-Plastics*; Determination of Mechanical Properties in Static Tension. Europe. ISO: Geneva, Switzerland, 2012. (In Polish)
32. *ISO 604-Plastics*; Determination of Properties in Compression. Europe. ISO: Geneva, Switzerland, 2006.
33. *DIN-EN ISO 178-Plastics*; Determination of Flexural Properties. Europe. ISO: Geneva, Switzerland, 2019.
34. Hastie, T.; Tibsgirani, R.; Friedman, J. *The Elements of Statistical Learning*; Springer: New York, NY, USA, 2009.
35. James, G.; Witten, D.; Hastie, T.; Tibshirani, R.; Taylor, J. *An Introduction to Statistical Learning*; Springer Texts in Statistics; Springer: Cham, Switzerland, 2013; ISBN 978-3-031-38746-3.
36. Antosz, K.; Kozłowski, E.; Sęp, J.; Prucnal, S. Application of Machine Learning to the Prediction of Surface Roughness in the Milling Process on the Basis of Sensor Signals. *Materials* **2025**, *18*, 148. [CrossRef] [PubMed]
37. Thomson, W. The Shapley Value, a Crown Jewel of Cooperative Game Theory. In *Handbook of the Shapley Value*; Fragnelli, V., Sanchez-Soriano, J., Eds.; Series in Operations Research; Chapman and Hall/CRC: Boca Raton, FL, USA, 2019; pp. 1–15. ISBN 978-1-351-24141-0.
38. Nwoye, I.; Obelle, M.C.; Nwakpa, S.O.; Onyia, C.W.; Obuekwe, I.; Idenyi, N.E. Predictability of the Impact Strength of CaCO 3 -Epoxy Resin Composite Based on CaCO 3 Input Concentration and Sustained Stress at Impact. *Int. J. Mater. Lifetime* **2015**, *2*, 6–12.
39. Park, S.-J. *Interface Science and Composites*; Issn Ser; Elsevier Science & Technology: San Diego, CA, USA, 2011; ISBN 978-0-08-096348-8.
40. Miturska, I.; Rudawska, A.; Müller, M.; Valášek, P. The Influence of Modification with Natural Fillers on the Mechanical Properties of Epoxy Adhesive Compositions after Storage Time. *Materials* **2020**, *13*, 291. [CrossRef]
41. Miturska, I.; Rudawska, A.; Müller, M.; Hromasová, M. The Influence of Mixing Methods of Epoxy Composition Ingredients on Selected Mechanical Properties of Modified Epoxy Construction Materials. *Materials* **2021**, *14*, 411. [CrossRef]
42. Yang, G.; Heo, Y.-J.; Park, S.-J. Effect of Morphology of Calcium Carbonate on Toughness Behavior and Thermal Stability of Epoxy-Based Composites. *Processes* **2019**, *7*, 178. [CrossRef]
43. Rudawska, A.; Frigione, M. Aging Effects of Aqueous Environment on Mechanical Properties of Calcium Carbonate-Modified Epoxy Resin. *Polymers* **2020**, *12*, 2541. [CrossRef]
44. Shivamurthy, B.; Murthy, K.; Joseph, P.C.; Rishi, K.; Bhat, K.U.; Anandhan, S. Mechanical Properties and Sliding Wear Behavior of Jatropha Seed Cake Waste/Epoxy Composites. *J. Mater. Cycles Waste Manag.* **2015**, *17*, 144–156. [CrossRef]
45. Yu, D.; Wang, J.; Tian, J.; Xu, X.; Dai, J.; Wang, X. Preparation and Characterization of TiO2/ZnO Composite Coating on Carbon Steel Surface and Its Anticorrosive Behavior in Seawater. *Compos. Part. B Eng.* **2013**, *46*, 135–144. [CrossRef]
46. Fu, J.; Kamali-Bernard, S.; Bernard, F.; Cornen, M. Comparison of Mechanical Properties of C-S-H and Portlandite between Nano-Indentation Experiments and a Modeling Approach Using Various Simulation Techniques. *Compos. Part. B Eng.* **2018**, *151*, 127–138. [CrossRef]

materials

MDPI

Article

Computational Analysis of a Wind Turbine Blade for Different Advanced Materials

Panagiotis F. Fragkos and Efstathios E. Theotokoglou *

Laboratory of Testing and Materials, Department of Mechanics, School of Applied Mathematical and Physical Sciences, National Technical University of Athens, Zografou Campus, 15773 Athens, Greece; fragkospanos@mail.ntua.gr

* Correspondence: stathis@central.ntua.gr

Abstract: As wind turbine rotors grow in size and Greece advances its offshore wind energy initiatives, this study analyzes the structural behavior of offshore wind turbine blades using fluid–structure interaction (FSI) methods. The blade skin and shear webs of the International Energy Agency (IEA) 15 MW wind turbine, assumed to operate in the Aegean Sea, are examined. Computational fluid dynamics (CFD) simulations are conducted for two steady-state wind speeds based on local weather data, followed by finite element analysis (FEA) to assess advanced materials in terms of strength, cost, and carbon footprint. This is the first study to evaluate bamboo- and basalt-based composite materials under Greek offshore wind conditions using FSI methods. Bamboo composites are affordable and sustainable, but their limited durability reduces their viability in offshore environments. The simulation results indicate that using bamboo composites as blade skin may lead to damage due to the excessive loads on offshore wind turbine blades. In contrast, basalt fiber composites are also environmentally viable and offer superior strength, corrosion resistance, and long-term performance, making them a promising alternative. However, their naturally high density may impact the overall weight of the structure. This study concludes that offshore wind technology in the Aegean is feasible but remains costly and environmentally demanding. The further development and adoption of basalt fibers may serve as a gateway to more environmentally friendly offshore structures.

Keywords: computational analysis; advanced materials; fiber-reinforced composites; IEA 15 MW wind turbine; wind turbine blades; fluid–structure interaction; offshore structures

1. Introduction

Wind energy plays a crucial role in Greece's renewable energy strategy, contributing 23% of the country's total energy demand in 2024. To accelerate offshore wind deployment, the Greek government has set a target of at least 2 GW of offshore wind capacity by 2030, as outlined in the National Energy and Climate Plan (NECP). Harnessing Greece's substantial wind resources is essential for sustainable development, energy security, and climate change mitigation [1].

Beyond environmental imperatives, economic and regulatory factors influence Greece's renewable energy transition. The European Union (EU) and the United Nations (UN) impose climate targets through agreements such as the Paris Agreement [2], aiming for Europe to become the first climate-neutral continent by 2050. Meeting these objectives requires a shift from lignite-based energy production to renewable alternatives.

Despite ambitious targets, offshore wind energy adoption in Greece faces challenges. The deep waters of the Aegean Sea complicate fixed-foundation turbine installation, while high material costs, complex construction processes, and maintenance requirements further impact feasibility. Additionally, integrating offshore wind farms into the national grid presents logistical and financial constraints [3]. In their research, Ouria et al. [4] concluded that the capacity factor, hub height, and blade diameters are simultaneously the most decisive in the productivity of alternative wind turbines (Figure 1).

Figure 1. Offshore wind turbine with floating base [5].

An alternative to offshore wind turbines is the deployment of underwater tidal turbines, such as those featured in the NH1 Tidal Project [6]. However, unlike regions such as northern France, the Greek seas generally exhibit lower tidal energy potential, which reduces the efficiency and economic feasibility of tidal technologies in this context. In contrast, offshore wind energy benefits from more favorable wind conditions in Greece, as well as a more mature technological framework, stronger investment interest, and active development both regionally and in the literature.

Notably, wind energy development prioritizes two key design advancements: aerodynamic optimization and the scaling of wind turbines. Improved blade designs enhance aerodynamic efficiency, with a notable innovation being the integration of curved winglets, a feature commonly used in aircraft to reduce drag forces. However, despite their potential to slightly improve energy capture, winglets have not been widely adopted in wind turbines due to limited efficiency gains [7]. Additionally, increasing the rotor diameter enhances energy yield but introduces structural and logistical challenges. Onshore turbines face transportation constraints, particularly in mountainous regions, whereas offshore turbines allow for larger designs with fewer restrictions. While offshore wind turbines can theoretically scale without major constraints, practical challenges persist. Greece's first floating wind farm is expected to be operational east of Crete by 2028 [8], and large-scale development may introduce unforeseen environmental and engineering complexities, including potential interactions with marine ecosystems.

Wind turbine blades are subject to various forms of damage, particularly in regions experiencing high aerodynamic and mechanical loads. According to Katsaprakakis et al. [9], the most common locations for damage include the leading edge, trailing edge, blade tip, and root section. The leading edge serves as the primary impact zone for airborne particles, rain, and hail, making it highly susceptible to erosion. The trailing edge, where airflow from both blade surfaces converges, is prone to fatigue cracking. The blade tip, being the outermost section, experiences high stress and is frequently struck by lightning. The root section, which connects the blade to the hub, is particularly vulnerable to cyclic loading, stress concentration [10], and adhesive joint failures.

Offshore wind turbines face additional environmental challenges compared to their onshore counterparts [9]. High wind speeds and saline exposure accelerate leading edge erosion, increasing material degradation rates. Corrosion effects, particularly due to saltwater exposure, contribute to structural wear, especially at blade joints. Furthermore, biofouling occurs as marine organisms accumulate on blade surfaces, altering aerodynamics and increasing mechanical loads. These challenges necessitate advanced material selection and targeted maintenance strategies to ensure offshore turbine longevity.

Recent developments in computational engineering have advanced the analysis of offshore wind turbines. The International Energy Agency (IEA) introduced the IEA 15-megawatt (MW) Offshore Reference Wind Turbine (IEA-15-240-RWT) [11], a standardized model for offshore turbine research. Developed in collaboration with the Technical University of Denmark (DTU) and the National Renewable Energy Laboratory (NREL), this model supports industry-research cooperation and innovation in turbine design. The computer-aided design (CAD) model of the blade has been utilized in a wide variety of studies of offshore wind turbine blade behavior (Figure 2). Since its introduction in 2020, large-scale floating wind turbine studies have gained significant traction in both research and industry applications [12–14].

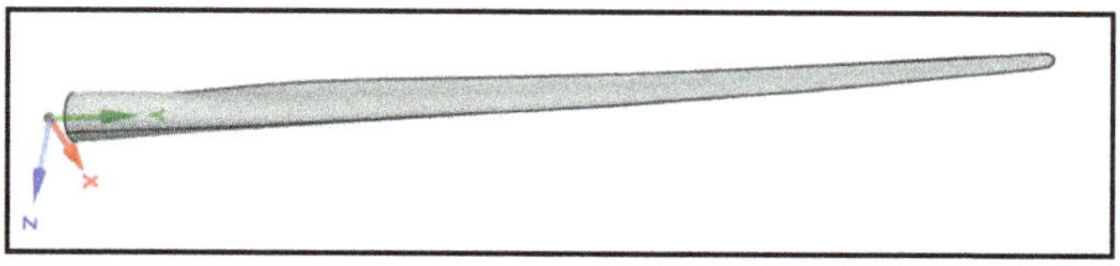

Figure 2. IEA 15 MW wind turbine blade CAD model.

This study builds upon methodologies established in previous research, including the work of Theotokoglou and Xenakis [15] and Nebiewa et al. [16]. Their research utilizes a geometric model of a wind turbine blade, incorporating periodic boundary conditions to simulate rotor performance while enabling localized material studies. Both studies use fluid–structural interaction (FSI) methods to determine the displacement and stresses of the blade for exposure to various wind speeds.

The present study focuses on the computational analysis of a theoretical candidate wind turbine blade for floating offshore wind farms. Given the limited availability of detailed internal geometry, emphasis is placed on the outer casing (blade shell or skin) as the primary structural component under investigation. The study examines the aerodynamic loads acting on the blade under nominal operating conditions ("rated conditions") and extreme conditions ("feathered" operation [17]). Additionally, a comparative assessment of different blade materials is conducted to evaluate their static performance.

Beyond structural analysis, this research also includes a preliminary economic and environmental evaluation of floating wind turbine deployment in Greek waters. This holistic approach aims to provide an initial assessment of the feasibility and performance of floating offshore wind turbines in the Greek seas, supporting future advancements in the country's offshore wind energy sector.

2. Materials and Methods

The cross-section of a standard wind turbine blade incorporates an internal support structure consisting of beams arranged in a box-spar configuration. Two of these beams, oriented perpendicular to the cross-section, are referred to as shear webs. The shear webs are structurally connected by two additional beams, known as spar caps [18].

Shear webs play a crucial role in enhancing the blade's mechanical performance, enabling it to behave like a solid beam while significantly reducing mass. Without them, the blade would lack the necessary structural integrity to withstand the substantial aerodynamic and gravitational forces imposed by wind loads (Figure 3).

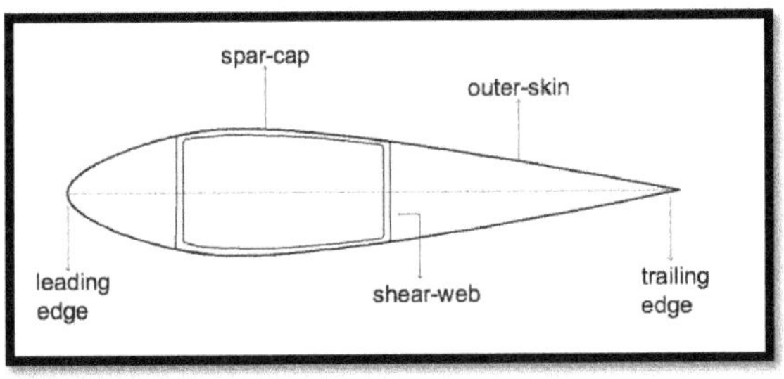

Figure 3. Internal structure of a typical wind turbine blade in section [18].

In this study, only the blade skin (casing) and shear webs are modeled. This approach ensures a balance between computational efficiency and accuracy, yielding realistic and reliable results for static analysis [14,18].

Modern wind turbine blades are predominantly constructed from composite materials, which consist of two or more constituent components that, when combined, exhibit superior mechanical properties not attainable individually. One critical property provided by these materials is resistance to aerodynamic forces, particularly in the flapwise direction (bending forces). The shear webs contribute significantly to this resistance (Figure 4). While resistance to flapwise aerodynamic forces is critical for wind turbine blades, this property is fundamentally governed by the material's modulus of elasticity E. A high modulus of elasticity E enhances structural stiffness, limits deflection, and contributes to favorable dynamic behavior by raising the blade's natural frequency. The latter is desirable because it prevents resonance with operating frequencies.

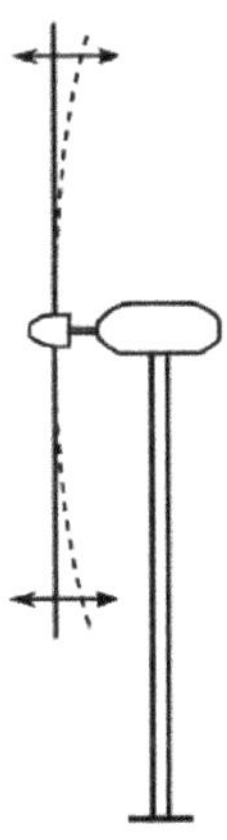

Figure 4. Flapwise bending moment on wind turbine blades [19].

The selection of suitable materials is a key consideration in wind turbine blade design. As turbine sizes increase, the mass of the blades rises rapidly, making material choice a crucial factor in maintaining structural efficiency. Additionally, excessive bending moments must be avoided to prevent blade-to-tower collisions. Furthermore, offshore wind turbine blades must endure harsh environmental conditions and sustain periodic loads due to continuous rotor rotation over extended operational periods [20].

The composite materials used in wind turbine blades are fiber-reinforced composites. These materials consist of a reinforcement component, which enhances mechanical properties, and a matrix component, which ensures optimal load transfer and structural integrity. Most fiber-reinforced composites used in wind energy applications employ a polymer matrix. The reinforcing fibers are selected for their high stiffness, high fracture toughness, and low density, making them ideal for large-scale wind turbine blades.

Fiberglass is a composite material consisting of glass fibers—either continuous or discontinuous—embedded in a polymeric matrix. This type of composite is widely produced due to its favorable mechanical properties, low cost, and compatibility with various manufacturing techniques for glass-reinforced plastics [15].

Carbon fibers are high-performance reinforcing agents extensively used in advanced polymer matrix composites beyond traditional fiberglass applications. Among all fibrous reinforcement materials, carbon fibers exhibit the highest specific modulus of elasticity and the highest specific strength [15]. They retain exceptional tensile strength and thermal resistance, making them particularly suitable for demanding environments. Additionally, carbon fibers are resistant to moisture [10], solvents, acids, and bases at room temperature, which is a significant advantage for offshore wind turbine applications.

Aramid fibers are known for their high strength and elasticity. They offer superior strength-to-weight ratios compared to metals, making them valuable for lightweight structural applications [15]. A notable drawback of aramid fibers is their susceptibility to moisture absorption [21], which may affect long-term performance in offshore conditions.

Basalt fiber composites are gaining attention as a promising alternative to conventional reinforcement materials in wind turbine blades. Derived from volcanic rock through a high-temperature melting and extrusion process, basalt fibers offer a unique combination of high strength, durability, and resistance to environmental degradation, making them particularly suitable for wind energy applications [22].

Table 1 presents the fundamental mechanical properties of various isotropic advanced composite materials that could be used in wind turbine blades. These materials exhibit homogeneity, meaning their density remains uniform throughout their volume. Additionally, they are assumed to be isotropic, possessing identical mechanical properties in all directions [15]. These isotropic properties were derived from homogenized composite behavior under multiaxial loading. This assumption is commonly used in preliminary finite element analysis (FEA) studies to reduce complexity and computational cost and is considered acceptable when directional dependencies are minor or when only global responses (e.g., deflection, stress distribution) are of interest [15]. For the basalt fiber composite, an equivalent plate theory was employed to represent the composite sections. This method homogenizes the laminate stack into an equivalent single-layer material with effective stiffness properties. This allows isotropic or quasi-isotropic modeling assumptions to be made [22,23].

However, in certain cases, anisotropic materials—specifically orthotropic materials—can also be considered. One such material of interest is bamboo, which has been formulated as a composite material (Table 2). Bamboo offers several key advantages, including low density, high sustainability, cost-effectiveness, rapid growth, and superior mechanical properties compared to many traditional woods. Its compatibility with composite manufacturing further enhances its potential as an alternative material for wind turbine blades [24,25]. A disadvantage is its tendency to absorb moisture, which impairs the fiber-matrix bonding and decreases the composite's strength and stiffness. An alternative is bamboo composites supplemented with thermosetting polymers, which may resist moisture, though such technology is still under development [26]. Given the limited research on bamboo composites for large-scale blade skins, its application in this context warrants further investigation.

Table 1. Mechanical properties of isotropic composites [15,22,23].

Material	E-Glass Fiber Composite	Aramid Fiber Composite	Carbon Fiber Composite	S-2 Glass Fiber Composite	Basalt Fiber Composite
Fiber type	Glass fiber	Aramid fiber	Carbon fiber	Glass fiber	Basalt fiber
Application	Blade skin	Blade skin	Shear webs	Blade skin	Blade skin
Density ρ [kg/m^3]	2568.70	1439.30	1320.37	2460.74	2650
Modulus of elasticity E [GPa]	72.4	112	138	86.9	86
Poisson ratio v	0.20	0.36	0.3	0.23	0.28
Shear Modulus G [GPa]	30	7	5.7	35	33.59

Table 2. Mechanical properties of anisotropic (orthotropic) material—bamboo composite [24,25].

Material	Bamboo
Fiber type	Natural fiber
Application	Blade skin
Density [kg/m^3]	1120
E_x [GPa]	22.37
E_y [GPa]	10
E_z [GPa]	10
G_{xy} [GPa]	4.30
G_{xz} [GPa]	4.30
G_{yz} [GPa]	3.30
v_{xy}	0.35
v_{xz}	0.35
v_{yz}	0.25

Table 2 presents the mechanical properties of a bamboo composite with a synthetic polymer matrix (epoxy)—an orthotropic material. The notations (x,y,z) in Table 2 correspond to the principal material axes of the Cartesian coordinate system of Figure 2. Unlike isotropic materials, orthotropic ones exhibit different mechanical behaviors depending on the direction of loading. For properties such as the Poisson's ratios and density, experiments may be performed on real bamboo specimens by using strain gauges [24]. However, determining the rest of the mechanical properties of natural fiber composites is challenging. Bora et al. [25] applied a micromechanical approach to estimate the elastic properties of a bamboo-based composite, using a square representative volume element (RVE) with periodic boundary conditions. This method accounts for the composite's repeating microstructure and enables a more accurate evaluation of its mechanical behavior compared to simpler analytical models.

3. IEA 15 MW Wind Turbine Blade Analysis

The IEA 15 MW wind turbine is designed for both fixed-bottom and floating offshore applications. In this study, the focus is on its floating base, as this configuration is particularly suitable for deployment in Greek waters. The key specifications of the wind turbine are presented in Table 3.

Table 3. IEA 15 MW wind turbine basic specifications [11,27].

Characteristic	Value	Units
Rated power	15	MW
Specific power	332	W/m^2
Number of blades	3	-
Cut-in wind speed	4	m/s
Cut-out wind speed	25	m/s
Rated wind speed	10.59	m/s
Rated rotational speed	0.79	rad/s
IEC Class 61400	IB	-

Figure 5 provides an engineering drawing illustrating the wind turbine's primary dimensions, along with its floating platform. As an IEC 61400 Class IB turbine, it is designed to withstand extreme wind speeds of up to 70 m/s [27].

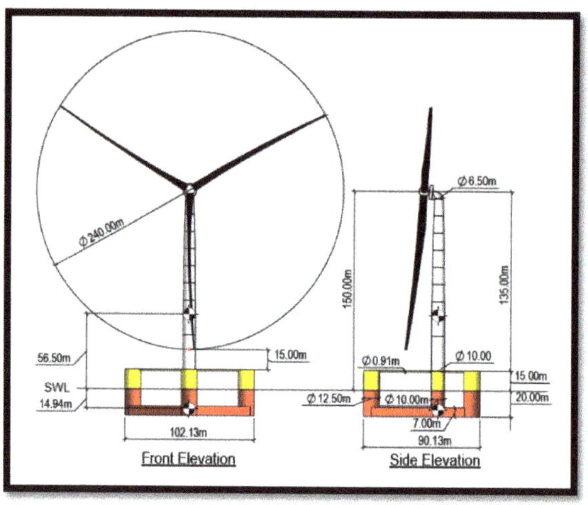

Figure 5. IEA 15 MW wind turbine dimensions [11].

The primary control mechanism of the wind turbine is pitch angle control, which regulates the blade pitch angle to optimize aerodynamic performance. Adjusting the pitch angle modifies the angle of attack, thereby controlling aerodynamic forces and influencing the power output. This regulation allows the torque to remain constant beyond the nominal operating point. Consequently, even as wind speeds increase, the electrical generator remains protected from overloading.

Figure 6 illustrates the relationship between the wind speed $|\vec{V}_\infty|$ [m/s] and the pitch angle θ [°].

The operation of a wind turbine can be categorized into three distinct regions based on wind speed conditions (Figure 6). In Region 1, which corresponds to low wind speeds (3–7 m/s), the available wind intensity is insufficient to reach the turbine's rated power. To maximize power capture, the pitch angle is adjusted to increase the generated torque.

Region 2 represents the rated power conditions, where the turbine operates close to its maximum designed power output. During this stage, the pitch angle remains at zero degrees to ensure optimal aerodynamic efficiency.

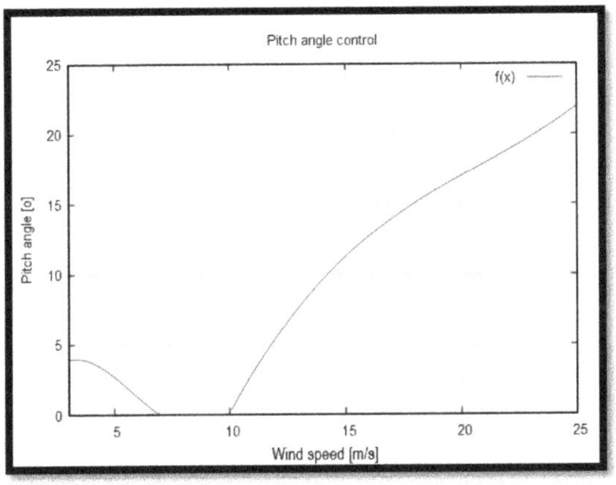

Figure 6. Wind speed $|\vec{V_\infty}|$ and pitch angle θ relationship.

In Region 3, also known as the power limitation mode, the turbine's control system actively adjusts the pitch angle to prevent overloading while maintaining maximum power output within safe operational limits. However, if wind speeds exceed an upper threshold, pitch angle adjustments alone are no longer sufficient to regulate power generation.

Beyond Region 3, when wind speeds exceed the cut-out limit, a dedicated braking system prevents the rotor from rotating. Simultaneously, the pitch angle is set to 90 degrees, fully engaging the feathered mode, ensuring minimal structural loading and protecting the turbine components from excessive stress.

Figure 7 illustrates the IEA 15 MW wind turbine blade in its CAD representation. The pronounced pre-bending curvature is clearly visible, a design feature that mitigates structural loads and reduces operational fatigue. Pre-bending is a critical consideration in offshore wind turbine blades, enhancing structural integrity and aerodynamic efficiency.

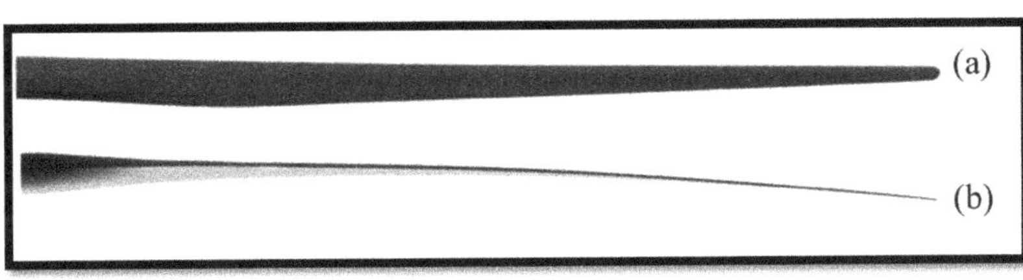

Figure 7. CAD representation of the (**a**) suction side and (**b**) trailing edge of the IEA 15 Wind Turbine blade [11].

The primary geometric and physical properties of the IEA 15 MW wind turbine blade are summarized in Table 4. Other geometric characteristics, such as airfoil types and thickness, are presented in Table 5.

Table 4. Primary geometric characteristics of the IEA 15 MW wind turbine blade [11].

Characteristic	Value	Units
Blade length	117	m
Root diameter	5.20	m
Length of root cylinder	3.54	m
Max chord	5.77	m
Blade mass	65,250	kg
Blade center of mass	26.80	m

Table 5. Geometry and blade thickness along length [28].

Airfoil	Radius Ratio (r/R) [%]	CAD Geometry Length [m]	Blade Skin Thickness [m]	Shear Web Thickness [m]
SNL-FFA-W3-500	15	20.55	0.056	0.040
FFA-W3-360	24.51	31.67	0.0485	0.036
FFA-W3-330blend	32.88	41.47	0.042	0.032
FFA-W3-301	43.91	54.37	0.033	0.027
FFA-W3-270blend	53.76	65.90	0.0256	0.023
FFA-W3-241	63.82	77.67	0.017	0.018
FFA-W3-211	80.11	96.73	0.007	0.012

4. Weather Conditions of the Site

One of the first offshore wind farms in Greece is expected to be installed near eastern Crete, in Sitia. Given the location of the wind turbine, but with limited available wind resource data, estimations must be made based on meteorological studies and Weibull distribution models.

The Weibull distribution is widely employed in modeling failure probabilities of physical systems due to external factors. It is particularly relevant to wind energy applications, as its flexible nature allows it to model wind speed frequency distributions over the operational lifespan of a wind farm. Compared to direct experimental wind measurements, the Weibull distribution provides more reliable long-term predictions of wind potential. The probability density function for wind speed is given by [29]:

$$f(u) = \frac{k}{c} \cdot \left(\frac{\left|\vec{V}_\infty\right|}{c}\right)^{k-1} \cdot e^{-\left(\frac{\left|\vec{V}_\infty\right|}{c}\right)^k} \tag{1}$$

where $\left|\vec{V}_\infty\right|$ is the wind speed, and k and c are parameters that form the final figure of the curves. Specifically, k is the shape parameter, which determines how variable the wind speed is. On the other hand, c is the scale parameter, which is related to the average wind speed. Determining these parameters requires statistical data, which is not fully available for the study site and thus must be estimated using empirical methods.

Wind speed measurements for meteorological or wind resource assessments are typically recorded over a year, at a height of 10 m, with the average wind speed calculated as:

$$\overline{u} = \sum_{i=1}^{v} \frac{u_i}{v}, i = 1, 2, 3, \ldots, v \tag{2}$$

For the Aegean Sea, the mean annual wind speed is reported to be 8.5–9.0 m/s, based on related studies conducted. Specifically, for the offshore region east of Sitia, the wind speed is estimated at 8.5 m/s. The site is also strategically advantageous, as it is free from major maritime traffic and has a relatively shallow seabed [30,31].

Since the IEA 15 MW wind turbine hub is located at 150 m (Figure 5), the wind speed at this height must be extrapolated using atmospheric boundary layer theory. The motion of air relative to the Earth's rotation forms a boundary layer, influenced by surface roughness (e.g., sea or land). For practical wind speed calculations, the power law equation is applied:

$$\frac{\left|\vec{V}_{\infty}\right|}{u_{ref}} = \left(\frac{z}{z_{ref}}\right)^{a} \tag{3}$$

where u_{ref} is the wind speed at reference height z_{ref}, and α is an empirical exponent dependent on terrain characteristics. For this study, a value of $\alpha = 0.14$, commonly used in wind energy literature, was selected [32].

The standard deviation of wind speed measurements is not available, but it is necessary for estimating the Weibull parameters k and c. Based on wind studies from the Aegean islands, the coefficient of variation (CV) for wind speed is estimated at 53.573%, derived from the average values of Mykonos and Lesvos. The CV, which indicates the variability of a dataset, is expressed as:

$$CV = \frac{s}{|\overline{u}|}\cdot 100 = \frac{s}{\overline{u}}\cdot 100[\%] \tag{4}$$

where s is the standard deviation of the wind speed measurements. Using $s = 4.55$ m/s for an average wind speed of 8.5 m/s, Bowden's empirical relationships can be applied to determine the Weibull parameters [33].

$$k \cong \left(\frac{s}{\overline{u}}\right)^{-1.086} \overset{(4)}{\widehat{=}} \left(\frac{CV}{100}\right)^{-1.086} \tag{5}$$

$$c = \frac{\overline{u}}{\Gamma\left(1+\frac{1}{k}\right)}[\mathrm{m/s}] \tag{6}$$

where Γ represents the Gamma function, which can be computed using numerical methods.

The calculated values are $k = 1.97$ and $c = 9.59$ m/s.

From the Weibull distribution analysis and Figure 8, two key operational regions are identified:

- Rated operating conditions: The highest wind speed occurrence is around 10 m/s, with a probability of 8.5%, aligning closely with the rated wind speed of 10.59 m/s for the IEA 15 MW wind turbine (Table 3).
- Cut-out wind speed: Wind speeds exceeding 35 m/s have an almost negligible probability. At this threshold, the turbine enters feathering mode, where rotation ceases to ensure structural integrity. This speed remains well below the turbine's design survival wind speed, confirming its operational safety (Table 3).

To simulate these two critical conditions, the wind velocity $\vec{V}_{\infty}$ is assumed to follow the +X-direction, and the turbine blade is positioned in a 3D computational environment in two configurations (Figure 9). In configuration (a), the blade is set in feathering mode at a 90-degree pitch angle relative to the wind flow, representing the extreme wind speed scenario (35 m/s wind speed). In configuration (b), the blade is positioned at a zero-degree pitch angle, corresponding to rated operating conditions (10.59 m/s wind speed).

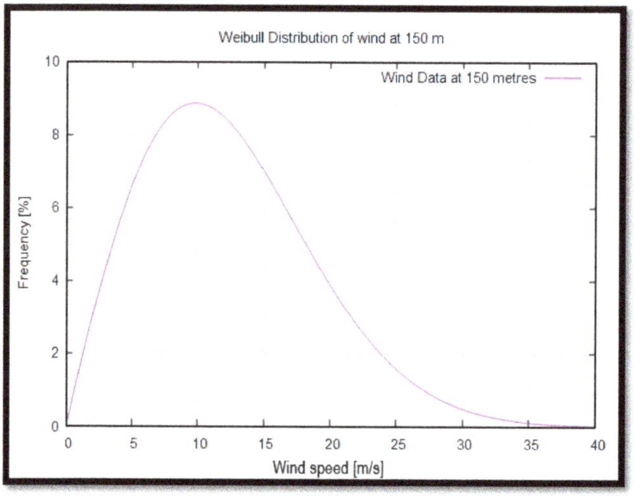

Figure 8. Weibull distribution of the wind velocity at 150 m above sea level.

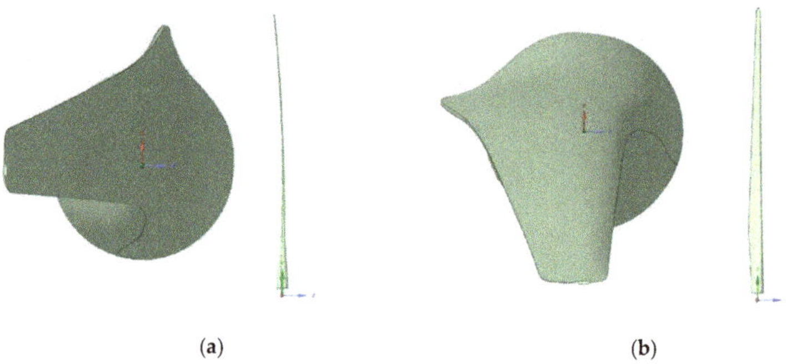

Figure 9. Configurations of the blade in the 3D environment. In configuration (**a**), the blade is in feathered mode. In configuration (**b**), the blade is operating at the rated conditions.

5. Computational Fluid Dynamics Analysis

For the analysis of complex fluid–structure interactions and aerodynamic behavior, computational fluid dynamics (CFD) methods are employed. These methods utilize specialized software to discretize the computational domain using the finite volume method (FVM). The Navier–Stokes equations, which govern fluid motion, are then solved numerically at each discretized control volume. The objective is to achieve solution convergence, with the degree of accuracy defined by the user.

The fundamental governing equations in this study are [34]:

$$\frac{\partial \rho}{\partial t} + div\left(\rho \vec{u}\right) = 0 \tag{7}$$

$$\rho \frac{D\vec{u}}{Dt} = -grad(p) + div\left(\overleftrightarrow{\tau}\right) \tag{8}$$

where ρ is the fluid density, p is the static pressure, $\vec{u}$ is the velocity vector, and $\overleftrightarrow{\tau}$ is the stress tensor. Equation (7) represents the continuity equation, ensuring mass conservation, while (8) corresponds to the momentum conservation equation, which describes the balance of forces in the fluid.

The simulations in this study were conducted using ANSYS Fluent© [35], a widely used CFD solver from the ANSYS© suite [35]. The analysis focuses on two steady-state flow conditions:rated operating conditions, for a wind speed of 10.59 m/s, and feathering mode, corresponding to extreme wind conditions at 35 m/s.

Given the relatively low wind velocity, the airflow is assumed to be incompressible. Table 6 summarizes the parameters selected for the flow modeling. Air is used as the working fluid, with standard ambient density and viscosity values applied. A constant temperature of 293.1 K (25 °C) is assumed, and thermal effects are neglected. As such, the influence of temperature on both the flow field and the composite material behavior is not considered in this study.

Table 6. Fluid mechanical analysis options.

Flow–viscosity model	K-omega (k-ω) (2 equations)
K-Omega model	Shear stress transport (SST)
Density [kg/m^3]	1.225
Viscosity [kg/ms]	1.7894×10^{-5}
Solution method	Pressure–velocity coupled
Gradient	Least squares-based
Pressure	Standard
Momentum	Second-order upwind
Turbulent Kinetic Energy—k	First-order upwind
Specific Dissipation Rate—ω	First-order upwind

Although air properties vary slightly with altitude, at the wind turbine hub height of 150 m, the density and viscosity decrease by approximately 2% compared to sea level. This variation is considered negligible for the present analysis.

The k-ω SST (shear stress transport) turbulence model was selected for this study as it provides improved accuracy over previous models. While alternative turbulence models were tested, the k-ω SST formulation yielded results that closely aligned with expected physical behavior. The k term represents the turbulent kinetic energy, while ω denotes the specific dissipation rate. Prior research indicates that the k-ω SST model effectively captures transitional flow phenomena at the fluid–wall interface, making it well-suited for wind turbine blade simulations [15,34,36].

The computational mesh of the fluid region consists of polyhedral elements, providing an efficient and precise representation of the flow field (Figure 10). To model transient phenomena, such as boundary layer formation on the blade, pentahedral elements are employed. Monitoring the boundary layer is crucial, as it influences flow transition from laminar to turbulent and can lead to flow separation, increasing drag and reducing aerodynamic efficiency. Pentahedral elements, as linear prismatic finite elements, form thin layers within the mesh, allowing for an accurate and efficient representation of boundary layer dynamics (Figure 10).

The flow domain is designed as an elongated cylindrical region, allowing for the full development of the airflow [14,15] and the prevention of reverse flow (Figure 11). The total length of the domain is approximately 2.7 km, similar to the strategy of Zhang et al. [14]. The inlet radius is 720 m, and the outlet radius is 1.5 km. At a distance of six lengths of the blade (720 m), the blade is positioned in relation to the entrance (inlet). The outlet is approximately over a length of sixteen blades (1940 m) from the blade model (Figure 12).

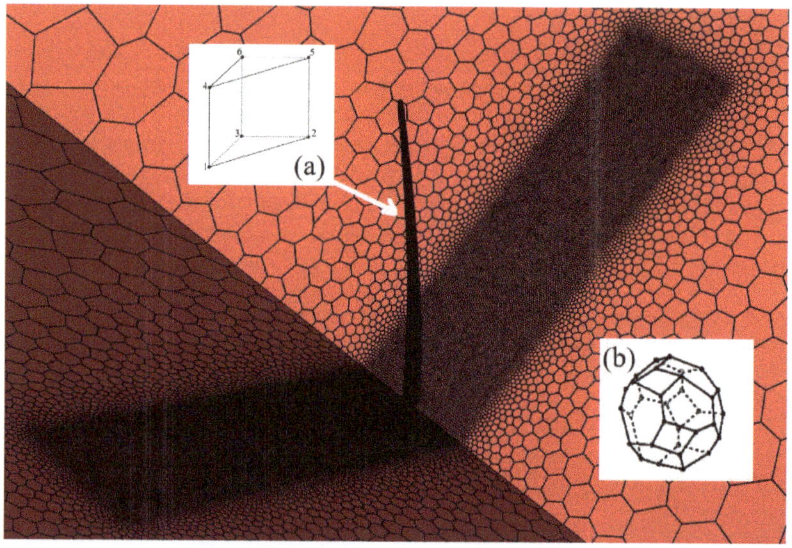

Figure 10. Fluid domain mesh. Prismatic elements (**a**) are used for the boundary layer along the blade. Polyhedral elements (**b**) are used for the rest of the flow domain.

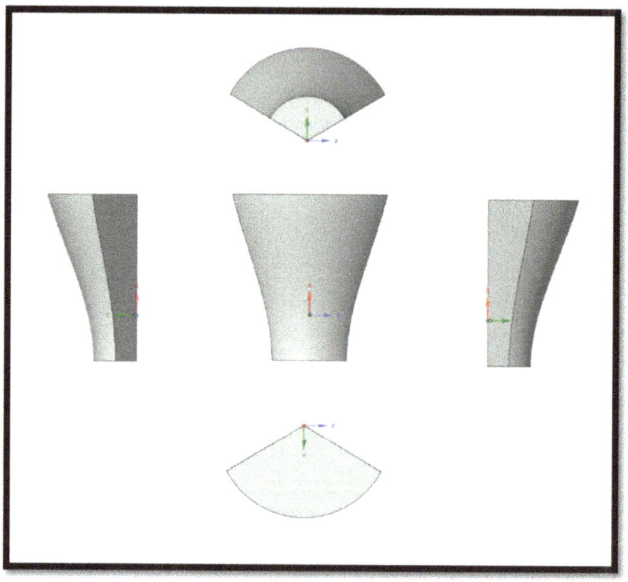

Figure 11. Views of the flow domain geometry.

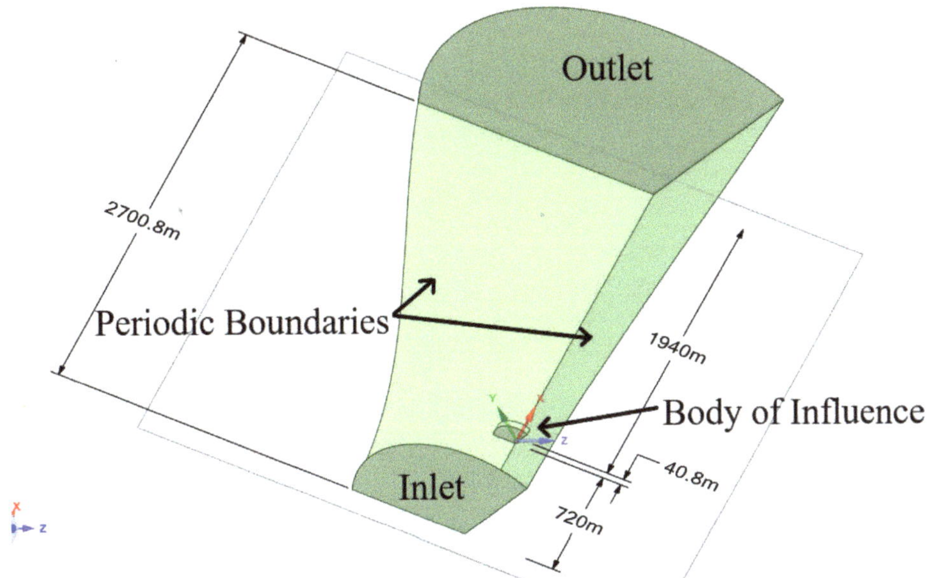

Figure 12. Size of the flow domain and its boundaries. The geometric model of the blade is located within the body of influence.

To optimize computational efficiency, the simulation domain represents a 120-degree sector of the rotor, corresponding to one-third of a full circle. These set of boundaries are called periodic, and they enforce the condition that the flow variables (e.g., velocity, pressure) on one periodic face are mapped to the opposite face with a rotational transformation. In cylindrical coordinates, this can be expressed as:

$$\vec{u}(r, \theta, z) = \vec{u}(r, \theta + \Delta\theta, z) \quad (9)$$

where $\vec{u}$ is the velocity vector, and $\Delta\theta = 120^\circ$ and (r, θ, z) are the cylindrical coordinates. By employing periodic boundary conditions, the entire rotor can be modeled using a single blade, significantly reducing computational cost. However, this simplification may overlook the potential effects of the interactions between the three blades and their influence on the wind flow dynamics.

The boundary conditions applied in the simulation are as follows:

- Inlet: Uniform velocity of 10.59 m/s or 35 m/s in the X+-direction, with a turbulence intensity of 5% and a turbulent viscosity ratio of 10.
- Outlet: Static pressure set to 1 atm.
- Wind turbine surface: No-slip condition applied to capture near-wall effects.
- Fluid region walls: Periodic boundary conditions to replicate full rotor operation. The axis of rotation is the global X axis.
- General conditions:
    - For rated operating conditions: rotational velocity of ω = 0.79 rad/s in the X+-direction.
    - For feathering mode: rotational velocity of ω = 0, corresponding to a stationary blade configuration.

Based on the data in Tables 7 and 8, the quality of the computational mesh is adequate for both cases.

Table 7. Attributes of the computational mesh of the flow domain and boundary layer in the case of the feathered blade ($\vec{V_\infty}$ = 35 m/s).

Mesh Attribute	Value
Number of elements	280,202
Surface size [m]	0.3
Inflation—first layer height [m]	0.3
Number of layers of prismatic elements (inflation—layers)	20
Orthogonal quality	0.95
Minimum orthogonal quality	0.383
Skewness average	0.0098
Maximum skewness	0.53

Table 8. Attributes of the computational mesh of the flow domain and boundary layer in the case of operation in the rated conditions ($\vec{V_\infty}$ = 10.59 m/s).

	Coarse Mesh	Fine Mesh
Mesh Attribute	**Value**	
Number of elements	307,741	3,106,827
Number of nodes	1,634,325	16,684,340
Surface element size [m]	0.6	0.3
Body of influence size [m]	1.5	0.8
Number of inflation layers	10	20
Orthogonal quality	0.956	0.966
Minimum rectangular quality (Minimum orthogonal quality)	0.150	0.237
Skewness	0.0200	0.0115
Maximum skewness	0.900	0.867
Thrust [MN] $\vec{T}$	0.66	0.73
Flapwise bending moment [MNm] $\vec{M}_{flap}$	−52.0	−58.7

For the case of the stationary blade modeled as a rigid wall ($\vec{V_\infty}$ = 35 m/s), a full mesh study was not conducted. However, incremental mesh refinement showed negligible changes in results, indicating mesh independence. This setup lacks the unsteady flow effects present in rotating simulations, reducing modeling complexity and sensitivity to mesh resolution, which justifies the lower element count.

The quality of the simulation results is primarily validated against reference data provided by the developers of the wind turbine model, as well as findings from other researchers [11–14]. For both steady-state cases, convergence was achieved after approximately 2000 iterations on average. Specifically, the simulations were considered converged when the residuals for continuity, velocity (in all three directions), and turbulence (k, ω) dropped below 10^{-6}. Due to differences in mesh density (as shown in Tables 7 and 8), computational time varied significantly between the two cases, with the finer mesh requiring substantially more time to converge.

A mesh sensitivity analysis was conducted for the rotating blade case under rated conditions (V_∞ = 10.59 m/s), comparing coarse (307,741 elements) and fine (3.1 million elements) meshes (Table 8). Key mesh metrics improved with refinement (orthogonal quality and skewness), and aerodynamic outputs, such as thrust and flapwise bending moment, differed by far less than 10% with available data. Beyond three million elements, the computational cost increased significantly, with minimal gains in accuracy.

A key measure of accuracy is the axial thrust generated by the wind turbine rotor during operation. In the fluent computing environment, this thrust can be approximated as the net force exerted on the blade surfaces in the direction of the wind (axial direction, X-direction, Figure 11). Exact data for the rotor thrust $\vec{T}_{rotor}$ at rated speed is unavailable. However, linear interpolation can be applied using data from other wind speeds, as shown in Table 9.

Table 9. Linear interpolation for the thrust [11].

Wind Velocity $\vec{V}_\infty$ [m/s]	Thrust of Rotor $\vec{T}_{rotor}$ [MN]
10.2096	2.2644
10.6584	2.4473

For a wind speed of 10.59 m/s, linear interpolation can be performed as follows:

$$\begin{gathered}\frac{10.6584-10.59}{10.6584-10.2096} = \frac{2.4473-\vec{T}_{rotor}}{2.4473-2.2644} \Leftrightarrow \\ \Leftrightarrow \vec{T}_{rotor} = 2.4473 - 0.0279 = 2.4194\ \mathbf{MN}\end{gathered} \tag{10}$$

Therefore, assuming that the thrust is evenly distributed over the three blades:

$$\vec{T} = \frac{\vec{T}_{rotor}}{3} = 0.806\ \mathbf{MN} \tag{11}$$

Another measure of evaluation of results is the flapwise bending moment exerted by the wind (Figure 4). Similar to before, for a wind speed of 10.59 m/s and the data from Table 10:

$$\vec{M}_{flap} = 64.4131\ \mathbf{MNm} \tag{12}$$

Table 10. Linear interpolation for the flapwise bending moment [11].

Wind Velocity $\vec{V}_\infty$ [m/s]	Flapwise Bending Moment $\vec{M}_{flap}$ [MNm]
10.2096	60.3222
10.6584	65.1486

By comparing the values obtained from Equations (11) and (12) with the simulation results presented in Table 8 and the findings of previous studies [13,14,37], it is determined that the calculated thrust and bending moment exhibit good agreement with established data. Specifically, the bending moment shows a maximum deviation of approximately 9.60%, while the thrust deviation reaches 8.75%. These discrepancies are considered acceptable given the limitations imposed by mesh resolution [14] and variations in computational solvers, which can introduce differences in numerical results [38]. In particular, the current study performed a mesh sensitivity analysis, demonstrating that no significant changes in the results occurred with an increase in mesh resolution. This suggests that mesh resolution was not a significant source of error. Differences in solver settings, including turbulence models, boundary conditions, and time-stepping schemes, can introduce variability in results. Additionally, variations in computational methods, such as turbulence model selection and numerical discretization, can further contribute to discrepancies in the results.

Since experimental data for stress distributions (e.g., strain gauge measurements from similar blades) were unavailable for direct comparison, the numerical results presented here are considered reasonable. However, validation of stress distributions with experimental data remains a valuable direction for future work, as it would help refine the accuracy of the simulations, especially for cases involving complex loading conditions.

Figures 13 and 14 illustrate the cumulative thrust and flapwise bending moment along the blade length for operation at the rated wind velocity ($\vec{V_\infty}$ = 10.59 m/s). Both quantities exhibit a generally proportional relationship with blade length. Notably, the flapwise bending moment reaches a maximum value approximately 100 times higher than the thrust force.

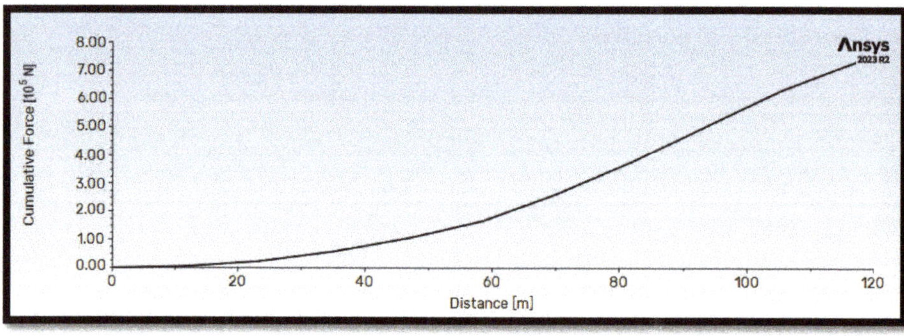

Figure 13. Cumulative thrust $\vec{T}$ along the length (distance from the center of rotation) of the blade. The wind velocity has a value of $\vec{V_\infty}$ = 10.59 m/s.

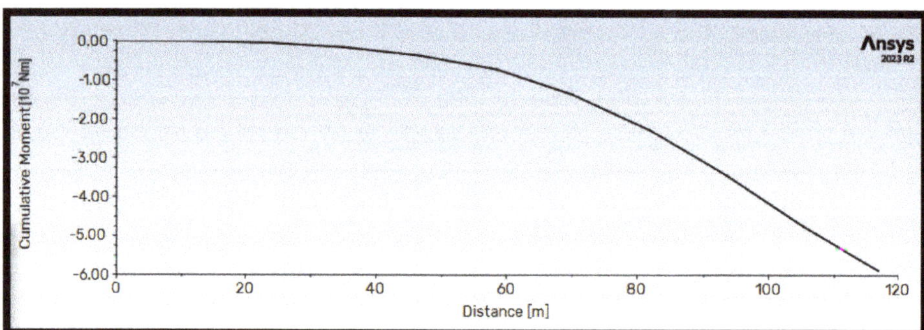

Figure 14. Cumulative flapwise bending moment $\vec{M}_{flap}$ along the length (distance from the center of rotation) of the blade. The wind velocity has a value of $\vec{V_\infty}$ = 10.59 m/s.

Figure 15 depicts the rotational velocity of the rotor. According to the no-slip condition, the fluid velocity matches the blade velocity at the point of contact. The observed range of rotational velocity aligns with expected values [11].

The pressure distribution on the blades in Zhang et al. [14] ranges from −10 kPa to 4.82 kPa, while the results of this study (Figure 16) exhibit a comparable range, specifically −7.3 kPa to 5.28 kPa for the dense mesh, corresponding to a 9.9% deviation. As expected, the leading edge and pressure side experience higher pressures compared to the trailing edge and suction side.

The wind velocity results under rated conditions align with those reported by Zhang, Y. et al. [14], with a maximum deviation of 5% at peak wind speed (Figure 17).

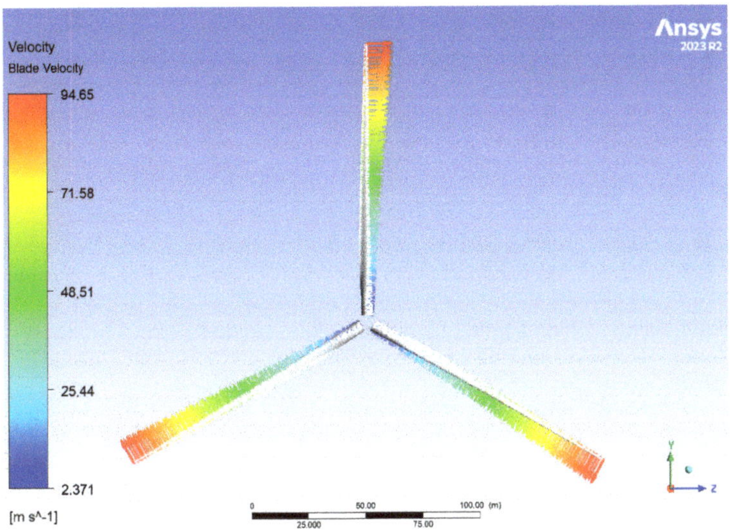

Figure 15. Rotational velocity of the rotor for operation at the rated wind speed $\left|\vec{V}_{\infty}\right| = 10.59$ m/s.

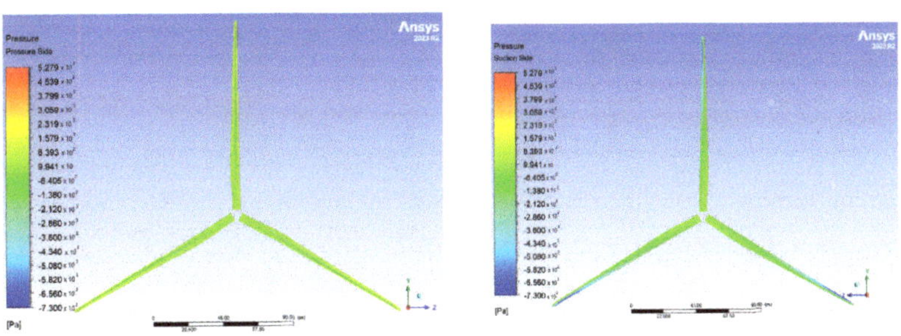

(**a**) (**b**)

Figure 16. Pressure contours on the blade surface for $\vec{V}_{\infty} = 10.59$ m/s. The pressure side is depicted in (**a**), while the suction side is depicted in (**b**).

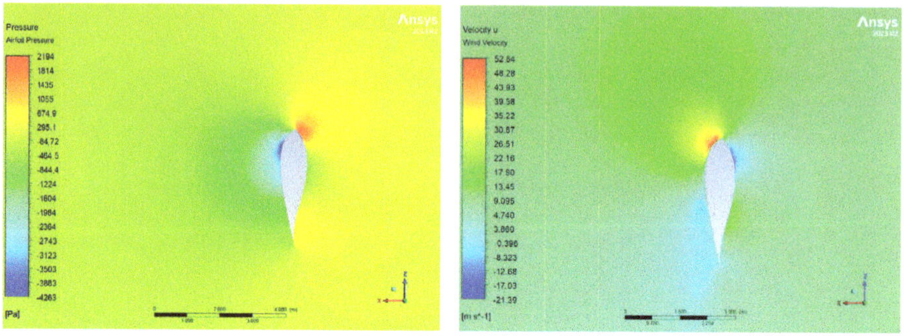

Figure 17. (**a**) Pressure contour of the airflow at r/R = 0.67 for $\vec{V_\infty}$ = 10.59 m/s. (**b**) Flow velocity contour at r/R = 0.67 for $\vec{V_\infty}$ = 10.59 m/s.

Figure 18 illustrates the effect of placing the blade in feathered mode. Despite an increase in wind speed, the pressure decreases by a factor of 10, demonstrating the effectiveness of this configuration in reducing aerodynamic loads. Given the adequacy of the CFD results, FSI analysis was subsequently employed for computational structural evaluation.

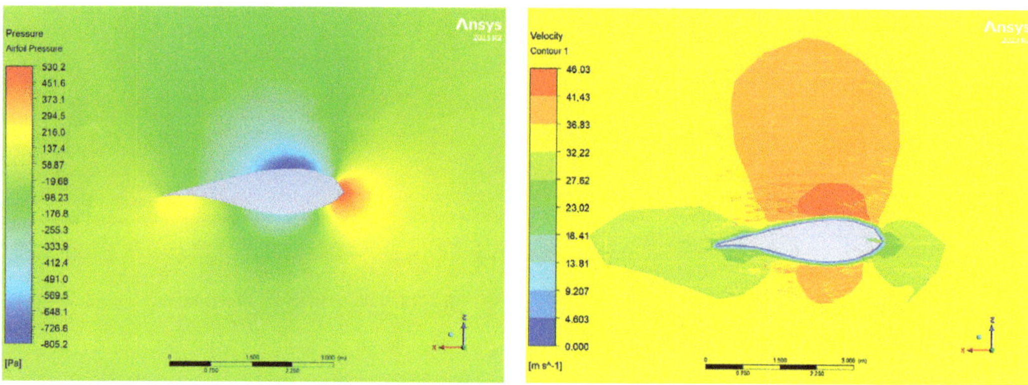

Figure 18. (**a**) Pressure contour of the airflow for $\vec{V_\infty}$ = 35 m/s. (**b**) Flow velocity contour for $\vec{V_\infty}$ = 35 m/s.

6. Computational Structural Analysis

For the analysis of complex static problems, the finite element analysis (FEA) method is employed. Similar to CFD methods, FEA utilizes computational software to discretize the examined structure into a mesh of finite elements. The governing equations are then numerically approximated and solved for each element using appropriate numerical analysis techniques.

In ANSYS© Mechanical [39], the displacement-based finite element formulation is applied, meaning that the equations are solved with respect to nodal displacements. For static analysis, the fundamental equations to be solved are as follows [40]:

$$\nabla \cdot \vec{\sigma} + \vec{F} = 0 \; or \tag{13a}$$

$$\overset{\sim e}{F} = \overset{\sim e}{k} \cdot \overset{\sim e}{q} \tag{13b}$$

$$\tilde{q} = \tilde{N} \cdot \overset{\sim e}{q} \tag{14}$$

$$\tilde{\varepsilon} = \tilde{B} \cdot \overset{\sim e}{q} \tag{15}$$

$$\tilde{\sigma} = \tilde{D} \cdot \tilde{\varepsilon} \tag{16}$$

Equation (13a) represents the equilibrium condition of the system, while Equation (13b) is obtained after discretization into finite elements, where $\overset{\sim e}{q}$ is the element displacement matrix and $\overset{\sim e}{k}$ is the element stiffness matrix. Equation (14) defines the relationship between the overall body displacement $\tilde{q}$ and the element displacements using the shape functions $\tilde{N}$, which depend on the selected finite element type. Equation (15) links strain $\tilde{\varepsilon}$ to nodal displacements through the strain–displacement matrix $\tilde{B}$. Finally, Equation (16) represents Hooke's law, where stress $\tilde{\sigma}$ is related to strain $\tilde{\varepsilon}$ via the material stiffness matrix $\tilde{D}$.

For an isotropic material, the stiffness matrix $\tilde{D}$ takes the form:

$$\tilde{D} = \frac{E}{(1+v)(1-2v)} \begin{bmatrix} 1-v & v & v & 0 & 0 & 0 \\ v & 1-v & v & 0 & 0 & 0 \\ v & v & 1-v & 0 & 0 & 0 \\ 0 & 0 & 0 & \frac{1-2v}{2} & 0 & 0 \\ 0 & 0 & 0 & 0 & \frac{1-2v}{2} & 0 \\ 0 & 0 & 0 & 0 & 0 & \frac{1-2v}{2} \end{bmatrix} \tag{17}$$

where E is the modulus of elasticity, and v is Poisson's ratio. For orthotropic materials, the stiffness matrix $\tilde{D}$ differs due to direction-dependent material properties and is no longer symmetric.

In this study, small displacements are assumed. The materials used are considered homogeneous and to remain within the linear elastic region. The solver utilized is APDL© (ANSYS Parametric Design Language) [39].

At the time of this study, an original CAD model suitable for structural analysis of the IEA 15 MW blade was not available [11]. Instead, the analysis utilized the parametric CAD model developed by Mendoza, ASE et al. [41], which is publicly accessible. This model incorporates surface elements as internal shear webs. It includes only the blade skin and shear webs, omitting other important structural components typically present in wind turbine blades, such as the spar caps. Most notably, spar caps are responsible for carrying a significant portion of the bending loads in large blades, and their absence likely leads to an underestimation of blade stiffness and an overestimation of stress in the remaining components. Additionally, the simplified modeling of shear webs as surface elements may alter the load transfer mechanisms within the blade. As a result, the stress distribution and deformation patterns observed in the simulations may not fully represent the actual behavior of the complete blade structure, limiting the accuracy of the structural performance predictions under operational loads.

The procedure for properly orienting the blade, considering both operational cases, follows the approach illustrated in Figure 9. This step is crucial for the accurate transfer of CFD data to the FEA model using the fluid–structure interaction (FSI) method. For each surface type, including the blade skin and shear webs, the appropriate thickness is assigned based on the values provided in Table 5.

Second-order shell elements (ANSYS© SHELL281 [39]) are chosen for the static structural analysis (Figure 19). These elements are well-suited for modeling thin-walled structures, making them appropriate for the blade's thickness, as specified in Table 5. Additionally, they ensure high mesh quality while maintaining computational efficiency, as indicated by the element count in Table 11.

As with the CFD analysis, only one blade is examined, thus excluding the hub and tower. This approach disregards dynamic effects such as oscillations from the floating base. Their inclusion was deemed unnecessary given the already simplified nature of the model. The CAD model used for the structural analysis differs from the model used in the CFD analysis, and this discrepancy could introduce minor deviations in the results due to the FSI method.

Environmental effects such as solar radiation and marine conditions were also not considered. While these factors significantly impact material degradation and blade longevity, this study focuses solely on two steady-state flow conditions, which do not capture the full complexity of real-world environmental influences.

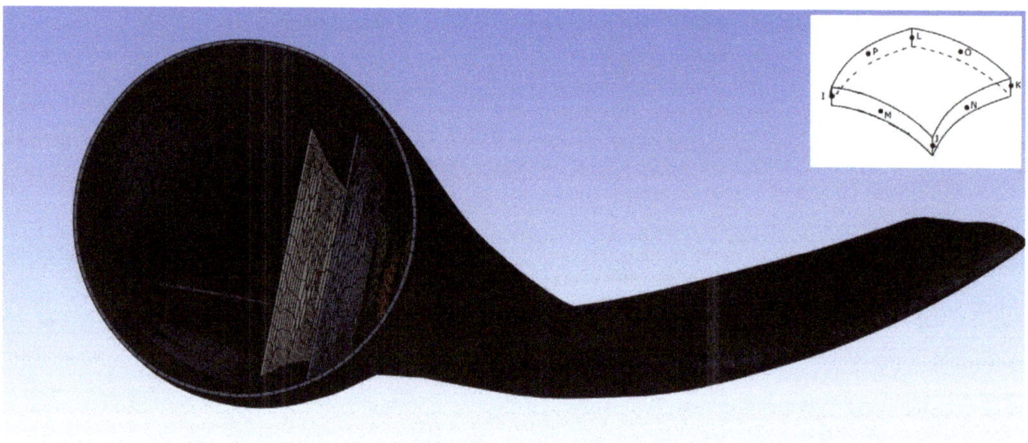

Figure 19. Mesh of the geometric blade model for static structural analysis. In the upper right corner, the SHELL281 element is depicted. It is a second-order element with 8 nodes.

Table 11. Structural analysis mesh properties.

Attribute	Value
Number of elements	40,156
Nodes	122,929
Area size [m]	0.2
Average skewness	0.06
Element quality	0.947

In this one-way FSI analysis, the steady-state aerodynamic loads obtained from the CFD simulations were applied as input forces for the static structural analysis, thus serving as boundary conditions. For higher accuracy, a two-way FSI analysis—where structural deformation affects the aerodynamic loading—would be required. However, this approach is often avoided due to the significant computational resources it demands [15–17].

The blade is fixed at the root, with reaction forces and moments applied at a reference point 3 m from the root cross-section. This setup simulates the presence of the hub, which is not explicitly included in the CAD model. The root boundary condition prevents unintended displacements, ensuring accurate solver calculations. The hub is modeled as a rigid body, serving as the reference point for displacement constraints.

For rated operating conditions, the rotation of the blade is simulated by applying an angular velocity of +0.79 rad/s in the direction of the wind.

To assess the structural integrity of the blade under aerodynamic loading, the equivalent von Mises stresses were calculated and compared to the ultimate tensile strength of the materials. The von Mises stress is determined using the following equation:

$$\sigma_{eq} = \sqrt{\sigma_x^2 + \sigma_y^2 + \sigma_z^2 - \sigma_x\sigma_y - \sigma_y\sigma_z - \sigma_z\sigma_x + 3\left(\tau_{xy}^2 + \tau_{yz}^2 + \tau_{zx}^2\right)}\ [\mathbf{Pa}] \quad (18)$$

where σ represents the normal stresses and τ represents the shear stresses. The maximum tensile strength values for the materials are provided in Table 12.

Table 12. Maximum tensile strength for materials [42–44].

Material/Fiber	Ultimate Tensile Strength (UTS) [MPa]
Carbon Fiber	1800–2400
E-Glass	500–800
S-2-Glass	1200–1500
Basalt	700–1200
Bamboo	350–500
Aramid	2000–2400

The ultimate tensile strength values were used in the absence of data on yield strength, as the materials were considered isotropic. Thus, the von Mises criterion was applied to assess the risk of material failure. For bamboo, an orthotropic material, shear strength values ranging from 40 MPa to 80 MPa were considered. For other conventional materials, such as glass fiber reinforced polymer (GFRP), shear strength values typically exceed 95 MPa [44,45].

Five material configurations are analyzed, as detailed in Tables 1 and 2. In all cases, the shear webs are composed of carbon fiber composites, while five different materials are evaluated for the blade skin.

The bamboo-based composite exhibits the highest displacement and stress values (Figure 20). When the blade is feathered, the root section experiences the greatest loading (Figures 21 and 22). But under rated operating conditions, the highest stress concentrations occur at the blade tip and leading edge (Figures 23–26). The normal stresses are lower in magnitude compared to the equivalent von Mises stresses, making the latter more suitable for assessing material performance and structural integrity (Figure 23). Shear stresses reach their peak on the XZ plane, likely due to the rotational motion of the blade, with maximum values concentrated near the leading edge (Figure 26).

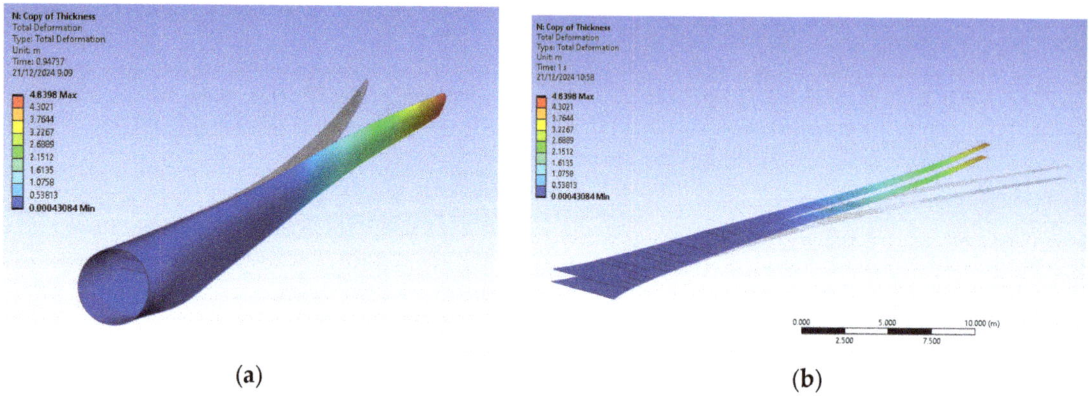

Figure 20. Displacement of (**a**) the blade skin and (**b**) the shear webs. $\vec{V_\infty}$ = 10.59 m/s, for bamboo composite blade skin.

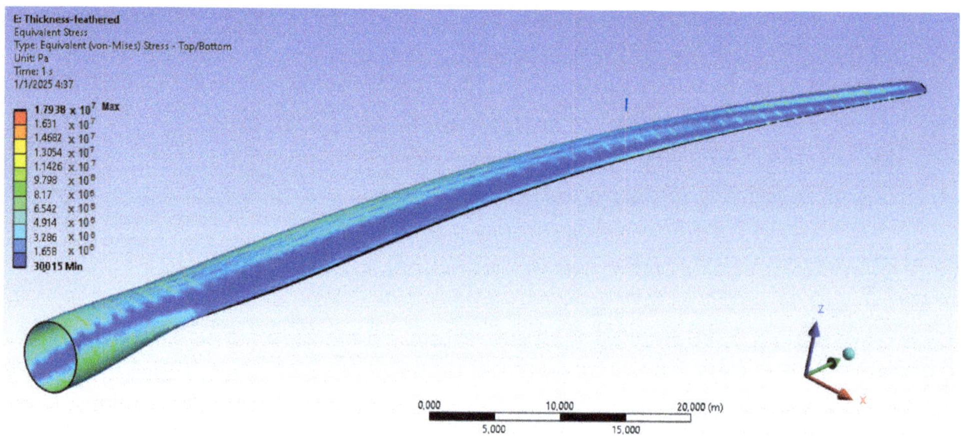

Figure 21. Equivalent von Mises stresses in the case of the bamboo-based blade skin. Higher von Mises stresses are present near root of the blade. $\vec{V_\infty}$ = 35 m/s.

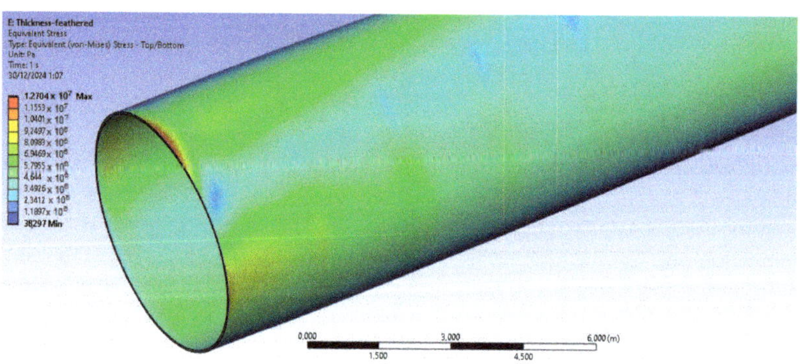

Figure 22. Close up of the blade, for the equivalent von Mises stresses in the case of the basalt fiber-based blade skin. $\vec{V_\infty}$ = 35 m/s.

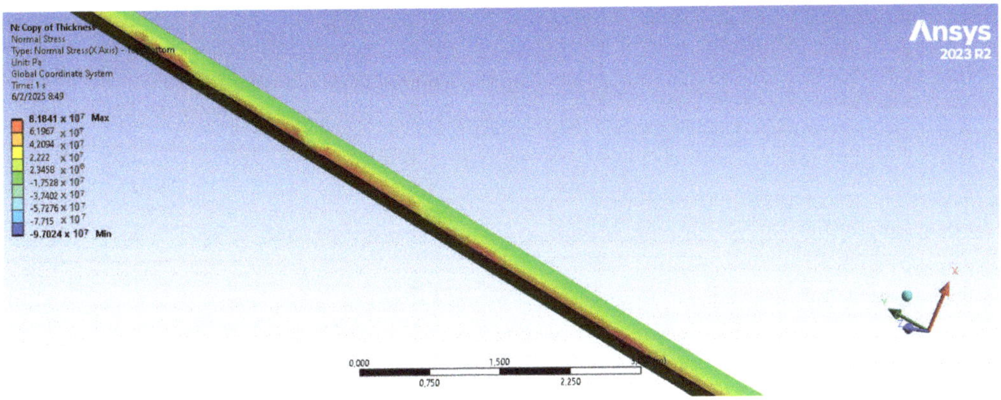

Figure 23. Normal stresses (X-axis) for bamboo-based blade skin. Higher normal stresses are present near the leading edge of the blade. $\vec{V}_\infty = 35$ m/s.

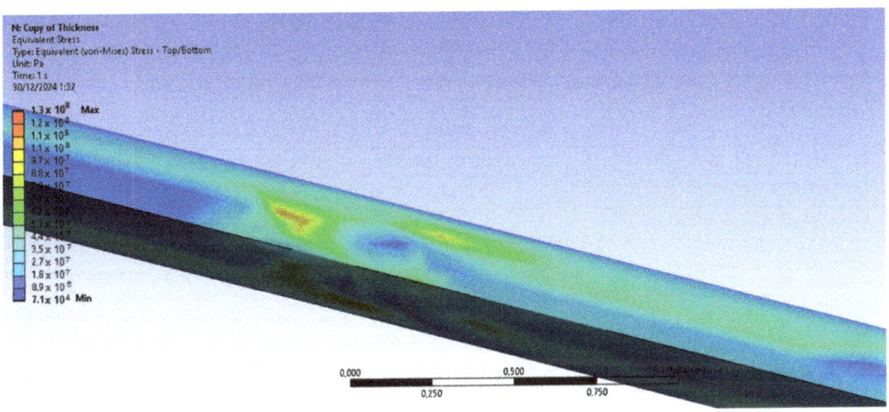

Figure 24. Equivalent von Mises stresses in the case of the bamboo-based blade skin. Higher von Mises stresses are present near the leading edge of the blade. $\vec{V}_\infty = 10.59$ m/s.

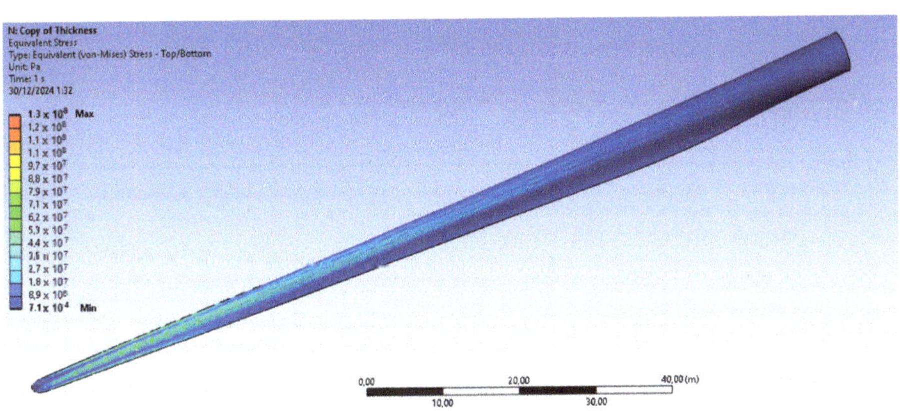

Figure 25. Equivalent von Mises stresses in the case of the bamboo-based blade skin. Pressure side of the blade. Higher von Mises stresses are present close to the blade tip. $\vec{V}_\infty = 10.59$ m/s.

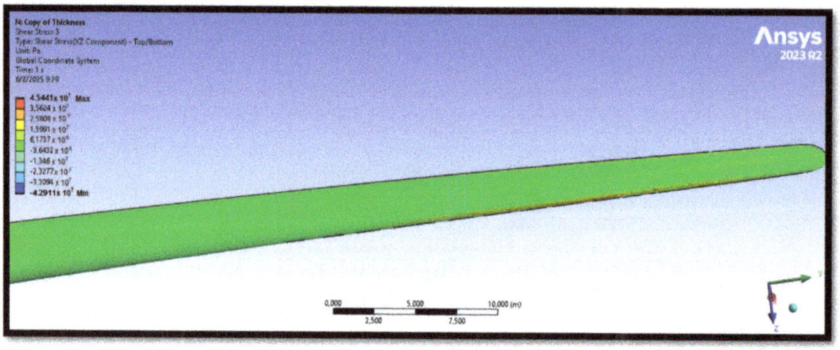

Figure 26. Shear stresses (XZ plane) in the case of the bamboo-based blade skin. The XZ plane corresponds to the leading edge, as the blade is spinning ($\vec{V_\infty}$ = 10.59 m/s).

7. Results and Discussion

We have already mentioned, that bamboo-based composite exhibits the highest displacement and stress values (Figure 20). However, all blade skin materials display similar overall stress and displacement distributions, meaning that Figures 20–26 are representative of all configurations. Maximum equivalent von Mises stresses, displacements, and shear stresses for each scenario are summarized in Table 13.

Table 13. Maximum equivalent von Mises stresses and maximum total displacement for each blade skin scenario.

Blade Skin Material	E-Glass		S-2-Glass		Aramid		Basalt		Bamboo	
Wind Velocity $\vec{V_\infty}$ [m/s]	10.59	35	10.59	35	10.59	35	10.59	35	10.59	35
Equivalent von Mises stresses (maximum) [MPa]	93.76	12.63	94.01	12.67	95.88	12.65	94.42	12.7	133	17.94
Maximum total displacement [m]	2.467	0.2374	2.098	0.2035	1.333	0.1571	2.190	0.2053	4.834	0.8078
Maximum shear stress [MPa]	45.29	7.00	45.43	7.02	46.26	7.06	45.44	7.04	45.44	9.62

To assess the validity of the results, a comparison is made with existing literature. A study by Theotokoglou and Xenakis [15] reports equivalent von Mises stress distributions of a similar order of magnitude, as well as comparable locations for stress maxima. Given that their analysis was conducted for smaller blades under similar environmental conditions, it is reasonable that the present study yields higher stress and deformation values. Notably, under rated conditions, the maximum equivalent von Mises stresses in their study are approximately 30% lower than those obtained in this work.

The magnitude of the computed stress values is consistent with those presented by Mendoza N et al. [13] and Zhang et al. [14]. In their investigations, the maximum von Mises stresses reached values in the range of hundreds of MPa, aligning with the findings of this study. Furthermore, the total blade displacement was found to be significant, approaching 5 m. According to Mendoza N. et al. [13], such displacements do not pose a risk of tower strikes. However, their studies focus on transient phenomena, which can lead to variations in the magnitude and location of maximum stress and displacement. Additionally, the materials analyzed in their work were predominantly orthotropic, whereas the present study assumes isotropic behavior for most materials. Another distinction is that the shear webs in this work are composed of carbon fiber, whereas previous studies [11] utilized glass fiber-reinforced composites. The selection of carbon fiber enhances structural rigidity, reducing total displacement but increasing mass.

The equivalent von Mises stresses exhibit a direct correlation with the modulus of elasticity E of most materials, with the exception of the bamboo composite, which is orthotropic. While the maximum equivalent stresses generally follow expected trends, total displacement is not strictly proportional to the elastic modulus or stress levels. Additionally, all materials demonstrate similar maximum shear stress values.

Based on the ultimate tensile strength data (Table 12), most materials can be deemed structurally viable for offshore wind turbine blades. However, bamboo-based composites present a higher risk of fracture and permanent deformation, as the maximum stresses approach their tensile strength limits (Table 13). Furthermore, bamboo exhibits relatively low shear strength compared to alternative materials, which may compromise long-term structural integrity.

The data in Table 13 are visually represented in Figures 27 and 28. Notably, transitioning the blade into feathered mode significantly reduces the load, even as the intensity of the wind increases.

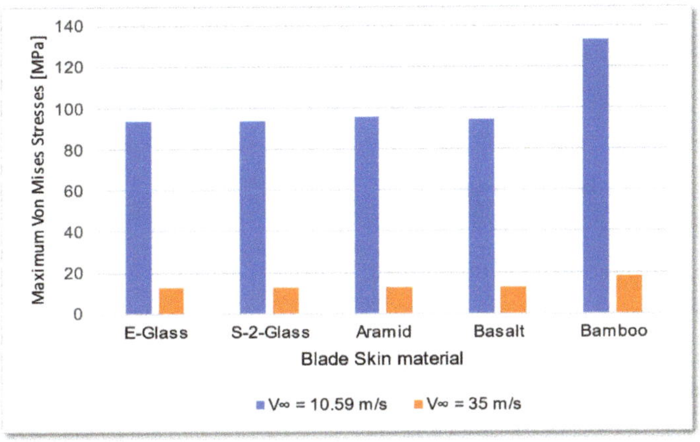

Figure 27. Maximum equivalent von Mises stresses for each blade skin material scenario.

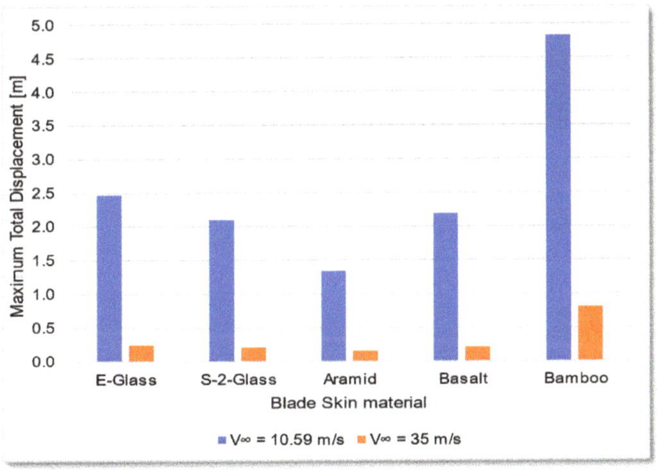

Figure 28. Maximum total displacement for each blade skin material scenario.

Carbon fiber was included in the study, specifically as the material used for the shear webs across all blade skin configurations, consistent with the IEA 15 MW reference design. For this reason, carbon fiber was not separately evaluated as a skin material, as doing so would have introduced redundancy without significant added insight. The von Mises stress results for the shear webs made of carbon fiber are implicitly reflected in the overall structural analysis but were not individually tabulated, as the focus was on the comparative performance of the blade skin materials.

Based on the simulation results, the blade mass was estimated for each material scenario. By integrating approximate emission factors [45–47] and cost per unit mass data [46,48,49] for each material, a comparative assessment was conducted in terms of both carbon footprint and economic feasibility. To account for variability in the data, an uncertainty margin of 20% was applied to the carbon footprint estimates (Figure 29) and 10% to the cost estimates (Figure 30), based on the range of values reported in the literature.

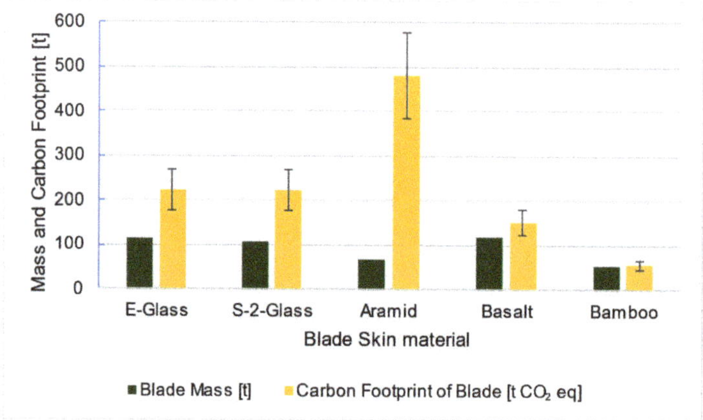

Figure 29. Comparison of all blade skin material scenarios in regard to the mass and carbon footprint of each wind turbine blade. The error bars correspond to an uncertainty of 20%. CO_2 refers to carbon dioxide emissions.

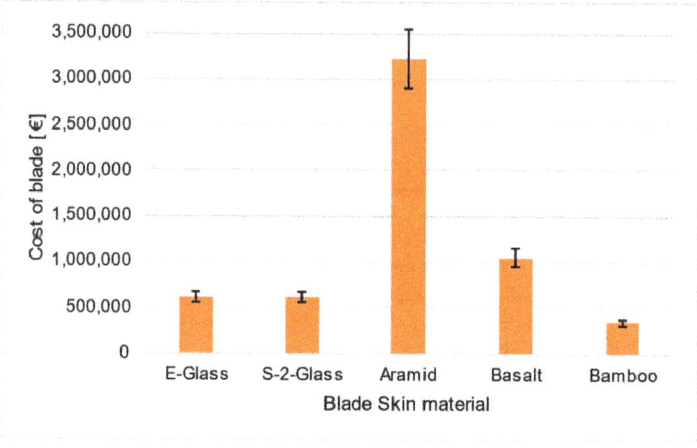

Figure 30. Comparison of all blade skin material scenarios in regard to the material cost of each wind turbine blade. The error bars correspond to an uncertainty of 10%.

As shown in Figure 29, basalt fiber-based composites emerge as a promising ecological alternative, though density reduction or material redistribution may be necessary to optimize performance. Aramid fiber composites offer excellent mechanical properties but are costly, as shown in Figure 30. However, a hybrid composite combining aramid and basalt fibers could mitigate both financial and environmental impact. Glass fiber composites remain a widely used and reliable choice, with various formulations available to meet specific design requirements—such as low-density orthotropic materials used in the IEA 15 MW blade.

Natural fiber composites, however, are unsuitable for offshore wind turbines due to excessive deformation and structural failure risks. Specifically, the von Mises stress (133 MPa) reaches the maximum tensile strength, while the presented shear stress surpasses the reported shear strength limit (40 MPa), indicating a high probability of failure. Additionally, marine durability remains uncertain, making them impractical for large-scale offshore applications. Nevertheless, natural fiber composites may still be viable for smaller, domestic wind turbines, where their recyclability, low cost, and reduced environmental impact provide advantages.

8. Conclusions

This study focuses on the computational analysis of the IEA 15 MW wind turbine blade, with an emphasis on evaluating various composite materials. The research aligns with national plans for the development of floating offshore wind turbines, which feature large-diameter rotors requiring meticulous material selection to optimize performance, durability, and cost-effectiveness.

The material selection was categorized into two distinct groups. For the shear webs, carbon fiber composite was utilized, while the blade skin was examined using multiple options, including E-glass, S-2 glass, Aramid fiber-based composite, basalt fiber-based composite, and natural fiber composite (bamboo). A computational study was conducted using ANSYS© [35,39] to assess the viability of these materials, particularly focusing on the application of emerging and environmentally sustainable options. The aerodynamic geometry of the blade was derived from original sources, while additional publications provided data for static structural modeling.

Following aerodynamic analysis, a static structural evaluation was conducted, incorporating the derived aerodynamic loads into the stress analysis model. Equivalent von Mises stresses, shear stresses, and displacements were determined. The material combinations were assessed based on production cost, environmental impact, and mechanical performance. The findings revealed a trade-off between economic and environmental benefits and structural integrity. Natural fiber composites, while cost-effective and environmentally sustainable, exhibited performance limitations, reaching the threshold of permanent deformation and failure. Other materials increased the overall blade mass, negatively affecting aerodynamic efficiency. Basalt fiber composites emerged as a promising alternative due to their high resistance to corrosion from seawater and ultraviolet exposure, as well as their strong mechanical properties.

The findings support the integration of alternative composite materials, such as basalt and bamboo fibers, in wind turbine manufacturing to address cost and sustainability concerns.

The implementation of a two-way FSI model would enable a more dynamic assessment of blade deformation under aerodynamic loads. Additionally, modeling the tower, hub, and floating base would provide a more comprehensive understanding of wave-induced effects on structural behavior. The study of transient phenomena, such as fluctuating wind speeds and wave interactions, could further enhance simulation accuracy. Expanding material comparisons and testing various material combinations would offer deeper insights into optimal design strategies. Future work could also explore multi-material configurations or optimization studies to evaluate the strategic placement of high-performance fibers while maintaining cost-effectiveness. Finally, a fatigue analysis based on Weibull probability distributions could provide critical data on blade lifespan under specific environmental conditions, such as those prevalent in the Aegean Sea.

Author Contributions: Conceptualization, P.F.F. and E.E.T.; methodology, P.F.F.; software, P.F.F.; validation, P.F.F. and E.E.T.; formal analysis, P.F.F.; investigation, P.F.F.; resources, P.F.F.; data curation, P.F.F.; writing—original draft preparation, P.F.F.; writing—review and editing, E.E.T.; visualization, P.F.F. and E.E.T.; supervision, E.E.T. All authors have read and agreed to the published version of the manuscript.

Funding: This research received no external funding.

Institutional Review Board Statement: Not applicable.

Informed Consent Statement: Not applicable.

Data Availability Statement: The original contributions presented in this study are included in the article. Further inquiries can be directed to the corresponding author.

Conflicts of Interest: The authors declare no conflicts of interest.

References

1. Ministry of the Environment and Energy. Renewable Energy Sources—Wind Energy. Available online: https://ypen.gov.gr/energeia/ape/technologies/aioliki/ (accessed on 10 October 2023).
2. United Nations. Paris Agreement. Available online: https://unfccc.int/process-and-meetings/the-paris-agreement (accessed on 11 March 2025).
3. Stehly, T.; Duffy, P.; Mulas Hermando, D. *Cost of Wind Energy Review: 2024 Edition*; National Renewable Energy Laboratory: Golden, CO, USA, 2024. [CrossRef]
4. Ouria, M.; Moura, P.S.; de Almeida, A.T. Will Wind Power be Cost-Effective for Decarbonizing the City of Tabriz-Iran, Whose Economy is Oil-Based, and How? *Wind Eng.* **2024**, *48*, 740–764. [CrossRef]
5. Untrakdrover. Agucadoura WindFloat Prototype. 2012. Available online: https://commons.wikimedia.org/wiki/File:Agucadoura_WindFloat_Prototype.jpg (accessed on 13 March 2025).
6. Euronews Green. Inside the Underwater Turbine Farm Set to Generate Clean Energy from France's Tides. 2025. Available online: https://www.euronews.com/green/2025/03/18/inside-the-underwater-turbine-farm-set-to-generate-clean-energy-from-frances-tides (accessed on 29 April 2025).
7. Koragappa, P.; Verdin, P.G. Design and Optimisation of a 20 MW Offshore Wind Turbine Blade. *Ocean Eng.* **2024**, *305*, 117975. [CrossRef]
8. Euronews. A National Priority: Greece Is Powering Ahead with Plans for First Offshore Wind Farms. *Euronews*. 1 November 2023. Available online: https://www.euronews.com/green/2023/11/01/a-national-priority-greece-is-powering-ahead-with-plans-for-first-offshore-wind-farms (accessed on 11 March 2025).
9. Katsaprakakis, D.A.; Papadakis, N.; Ntintakis, I.A. Comprehensive Analysis of Wind Turbine Blade Damage. *Energies* **2021**, *14*, 5974. [CrossRef]
10. Lagdani, O.; Tarfaoui, M.; Nachtane, M.; Trihi, M.; Laaouidi, H. A Numerical Investigation of the Effects of Ice Accretion on the Aerodynamic and Structural Behavior of Offshore Wind Turbine Blade. *Wind Eng.* **2021**, *45*, 1433–1446. [CrossRef]
11. Gaertner, E.; Rinker, J.; Sethuraman, L.; Zahle, F.; Anderson, B.; Barter, G.; Abbas, N.; Meng, F.; Bortolotti, P.; Skrzypinski, W.; et al. *Definition of the IEA 15-Megawatt Offshore Reference Wind Turbine*; NREL/TP-75698; National Renewable Energy Laboratory: Golden, CO, USA, 2020. Available online: https://docs.nrel.gov/docs/fy20osti/75698.pdf (accessed on 11 March 2025).
12. Chang, Z.; Zheng, Y.; Qu, M.; Gao, X.; Tian, X.; Liu, G. Motion Analysis of International Energy Agency Wind 15 MW Floating Offshore Wind Turbine under Extreme Conditions. *J. Mar. Sci. Eng.* **2024**, *12*, 1166. [CrossRef]
13. Mendoza, N.; Robertson, A.; Wright, A.; Jonkman, J.; Wang, L.; Bergua, R.; Ngo, T.; Das, T.; Odeh, M.; Mohsin, K.; et al. Verification and Validation of Model-Scale Turbine Performance and Control Strategies for the IEA Wind 15 MW Reference Wind Turbine. *Energies* **2022**, *15*, 7649. [CrossRef]

14. Zhang, Y.; Song, Y.; Shen, C.; Chen, N.-Z. Aerodynamic and Structural Analysis for Blades of a 15 MW Floating Offshore Wind Turbine. *Ocean Eng.* **2023**, *287*, 115785. [CrossRef]
15. Theotokoglou, E.E.; Xenakis, G. GE 1.5 XLE Wind Turbine Blade Analysis with Computational Methods for Various Composite Materials. *WSEAS Trans. Appl. Theor. Mech.* **2023**, *18*, 56–67. [CrossRef]
16. Nebiewa, A.M.; Abdelsalam, A.M.; Sakr, I.M.; El-Askary, W.A.; Abdalla, H.A.; Ibrahim, K. Studying and Analysis of Deformation and Stresses in Horizontal-Axis Wind Turbine Using Fluid-Structure Interaction Method. *Wind Eng.* **2024**. [CrossRef]
17. Stewart, G.M.; Lackner, M.A.; Arwade, S.R.; Hallowell, S.; Myers, A.T. Statistical Estimation of Extreme Loads for the Design of Offshore Wind Turbines During Non-Operational Conditions. *Wind Eng.* **2015**, *39*, 629–640. [CrossRef]
18. Theotokoglou, E.E.; Balokas, G.A. Computational Analysis and Material Selection in Cross-Section of a Composite Wind Turbine Blade. *J. Reinf. Plast. Compos.* **2014**, *34*, 101–115. [CrossRef]
19. Morim, R.B.; Carnielutti, F.M.; da Rosa, L.D.; Hubner, G.R.; Franchi, C.M.; de Souza, C.E.; Pinheiro, H. Analysis of Wind Turbine Power Generation with Individual Pitch Control. In Proceedings of the 2019 IEEE PES Innovative Smart Grid Technologies Conference—Latin America (ISGT Latin America), Gramado, Brazil, 15–18 September 2019. [CrossRef]
20. Mishnaevsky, L., Jr.; Branner, K.; Petersen, H.N.; Beauson, J.; McGugan, M.; Sørensen, B.F. Materials for Wind Turbine Blades: An Overview. *Materials* **2017**, *10*, 1285. [CrossRef] [PubMed]
21. DuPont. *Kevlar® Aramid Fiber: Technical Guide*; DuPont: Wilmington, DE, USA, 2019.
22. Zheng, T.; Zhao, C.; He, J. Research on Fatigue Performance of Offshore Wind Turbine Blade with Basalt Fiber Bionic Plate. *Structures* **2023**, *47*, 466–481. [CrossRef]
23. Reddy, S.P.; Suresh, R.; Hanamantraygouda, M.B.; Shivakumar, B.P. Use of Composite Materials and Hybrid Composites in Wind Turbine Blades. *Mater. Today Proc.* **2021**, *46*, 2827–2830. [CrossRef]
24. Lu, H.; Hongyang, L.; Xu, J.; Ma, N.; Zhou, Z.; Song, Y.; Yu, Y.; Zhang, X. Study on the Variation Pattern and Influencing Factors of Poisson's Ratio of Bamboo. *Front. Mater.* **2022**, *9*, 896756. [CrossRef]
25. Bora, M.; Sharma, A.V.; Kumari, P.; Sahoo, N. Investigation of Bamboo-Based Vertical Axis Wind Turbine Blade under Static Loading. *Ocean Eng.* **2023**, *285*, 115317. [CrossRef]
26. Hasan, K.M.F.; Al Hasan, K.M.N.; Ahmed, T.; György, S.-T.; Pervez, M.N.; Bejó, L.; Sándor, B.; Alpár, T. Sustainable Bamboo Fiber Reinforced Polymeric Composites for Structural Applications: A Mini Review of Recent Advances and Future Prospects. *Case Stud. Chem. Environ. Eng.* **2023**, *8*, 100362. [CrossRef]
27. Mahfouz, M.Y.; Salari, M.; Hernández, S.; Vigara, F.; Molins, C.; Trubat, P.; Bredmose, H.; Pegalajar-Jurado, A. Public Design and FAST Models of the Two 15 MW Floater-Turbine Concepts. Available online: https://corewind.eu/publications/ (accessed on 11 March 2025).
28. Keprate, A.; Bagalkot, N.; Siddiqui, M.S.; Sen, S. Reliability Analysis of 15 MW Horizontal Axis Wind Turbine Rotor Blades Using Fluid-Structure Interaction Simulation and Adaptive Kriging Model. *Ocean Eng.* **2023**, *288*, 116138. [CrossRef]
29. Montgomery, D.C.; Runger, G.C. Weibull Distribution. In *Applied Statistics and Probability for Engineers*; Tziola Publications: Thessaloniki, Greece, 2021; pp. 143–144.
30. Kaldellis, J.K. What is the Wind Energy Progress in Greece? Prospects and Problems. In *The Age of Wind Energy*; Springer: Cham, Switzerland, 2020; pp. 23–40. [CrossRef]
31. Tsarknias, N.; Gkeka-Serpetsidaki, P.; Tsoutsos, T. Exploring the Sustainable Siting of Floating Wind Farms in the Cretan Coastline. *Sustain. Energy Technol. Assess.* **2022**, *54*, 102841. [CrossRef]
32. Tieo, J.J.; Skote, M.; Srikanth, N. Suitability of Power-Law Extrapolation for Wind Speed Estimation on a Tropical Island. *J. Wind Eng. Ind. Aerodyn.* **2020**, *205*, 104317. [CrossRef]
33. Teimourian, H.; Abubakar, M.; Yildiz, M.; Teimourian, A. A Comparative Study on Wind Energy Assessment Distribution Models: A Case Study on Weibull Distribution. *Energies* **2022**, *15*, 5684. [CrossRef]
34. Tsangaris, S. *Fluid Mechanics—Theory and Exercises*, 2nd ed.; Tsotras Publications: Athens, Greece, 2016.
35. ANSYS Inc. Ansys Fluent. Available online: https://www.ansys.com/products/fluids/ansys-fluent (accessed on 11 March 2025).
36. Adanta, D.; Fattah, I.M.R.; Muhammad, N.M. Comparison of Standard k-ε and SST k-ω Turbulence Model for Breastshot Waterwheel Simulation. *J. Mech. Sci. Eng.* **2020**, *7*, 44. [CrossRef]
37. Maximiano, A. Benchmark of PivotBuoy Compared to Other Floating Systems; Horizon H2020, European Union. 2021. Available online: https://pivotbuoy.eu/wp-content/uploads/2021/06/D5.4-Benchmark-of-PivotBuoy-versus-other-floating-systems-v1.1.pdf (accessed on 11 March 2025).
38. Thomas, M.; Boettcher, N.; Huish, C.; Shakya, P.; Seibi, A.C.; Arias, D.; Shekaramiz, M.; Masoum, M.A.S. Assessing Wind Turbine Blade Durability Longevity Utilizing National Renewable Energy Laboratory Tools. *Wind Eng.* **2024**, *48*, 123–135. [CrossRef]
39. ANSYS Inc. ANSYS Mechanical. Available online: https://www.ansys.com/products/structures/ansys-mechanical (accessed on 11 March 2025).

40. Tsamasfyros, I.G.; Theotokoglou, E.E. *Finite Element Method I*; Symmetria: Athens, Greece, 2015.
41. Mendoza, A.S.E.; Mishra, I.; Griffith, D.T. An Open-Source NuMAD Model for the IEA 15 MW Blade with Baseline Structural Analysis. In Proceedings of the AIAA SCITECH 2023 Forum, National Harbor, MD, USA, 23–27 January 2023; p. 2093. [CrossRef]
42. Callister, W.D.; Rethwisch, D.G. *Materials Science and Engineering*, 9th ed.; Tziola: Thessaloniki, Greece, 2017.
43. Elahi, A.H.M.; Hossain, M.M.; Afrin, S.; Khan, M.A. Study on the Mechanical Properties of Glass Fiber Reinforced Polyester Composites. In Proceedings of the International Conference on Mechanical, Industrial and Energy Engineering, Khulna, Bangladesh, 26–27 December 2014.
44. Mousavi, S.R.; Zamani, M.H.; Estaji, S.; Tayouri, M.I.; Arjmand, M.; Jafari, S.H.; Nouranian, S.; Khonakdar, H.A. Mechanical Properties of Bamboo Fiber-Reinforced Polymer Composites: A Review of Recent Case Studies. *J. Mater. Sci.* **2022**, *57*, 3143–3167. [CrossRef]
45. Ecofys. *Methodology for the Free Allocation of Emission Allowances in the EU ETS Post-2012—Sector Report for the Glass Industry*; European Commission Report: Brussels, Belgium, 2009. Available online: https://climate.ec.europa.eu/system/files/2016-11/bm_study-glass_en.pdf (accessed on 11 March 2025).
46. Krauklis, A.E.; Karl, C.W.; Gagani, A.I.; Jørgensen, J.K. Composite Material Recycling Technology—State-of-the-Art and Sustainable Development for the 2020s. *J. Compos. Sci.* **2021**, *5*, 28. [CrossRef]
47. Maiti, S.; Islam, M.R.; Uddin, M.A.; Afroj, S.; Eichhorn, S.J.; Karim, N. Sustainable Fiber-Reinforced Composites: A Review. *Adv. Sustain. Syst.* **2022**, *6*, 2200258. [CrossRef]
48. Bortolotti, P.; Berry, D.; Murray, R.; Gaertner, E.; Jenne, D.; Damiani, R.; Barter, G.; Dykes, K.A. *Detailed Wind Turbine Blade Cost Model*; National Renewable Energy Laboratory: Golden, CO, USA, 2019. Available online: https://www.nrel.gov/docs/fy19osti/73585.pdf (accessed on 11 March 2025).
49. Mohammed, K.; Zulkifli, R.; Tahir, M.F.M.; Gaaz, T.S. A Study of Mechanical Properties and Performance of Bamboo Fiber/Polymer Composites. *Results Eng.* **2024**, *23*, 102396. [CrossRef]

materials

MDPI

Article

Optimization of SiC–TiC Composite Manufacturing by Electroconsolidation Method

Vyacheslav Ivzhenko [1], Jolanta Natalia Latosińska [2,*], Edvin Hevorkian [3], Miroslaw Rucki [4], Tamara Kosenchuk [1], Natalia Shamsutdinova [5], Tadeusz Szumiata [3], Volodymyr Chishkala [6] and Arturas Kilikevicius [4]

1 V.M. Bakul Institute of Superhard Materials, National Academy of Sciences of Ukraine, Avtozavodska Str. 2, 04074 Kyiv, Ukraine; ivv@ism.kiev.ua (V.I.); avl@ism.kiev.ua (T.K.)
2 Faculty of Physics and Astronomy, Adam Mickiewicz University, Uniwersytetu Poznańskiego 2, 61-614 Poznań, Poland
3 Faculty of Mechanical Engineering, Casimir Pulaski Radom University, Stasieckiego 54-B1, 26-600 Radom, Poland; e.gevorkian@urad.edu.pl (E.H.); t.szumiata@urad.edu.pl (T.S.)
4 Institute of Mechanical Science, Vilnius Gediminas Technical University, Sauletekio al. 11, LT-10223 Vilnius, Lithuania; m.rucki@urad.edu.pl (M.R.); arturas.kilikevicius@vilniustech.lt (A.K.)
5 Plasma Tech LLC, Chmelnitske Sh. 13, 21036 Vinnytsia, Ukraine; shamsut.natalka49@gmail.com
6 Department of Reactor Engineering Materials and Physical Technologies, V.N. Karazin Kharkiv National University, m. Nezalezhnisti 4, 61022 Kharkiv, Ukraine; vchishkala@ukr.net
* Correspondence: jolanta.latosinska@amu.edu.pl

Abstract: Modern SiC-based materials are of paramount importance in that they serve as wear-resistant and thermal protectors and as next-generation single-photon sources for photonic and quantum solutions. Efforts are underway to identify more efficient methods of manufacturing SiC-based ceramic materials. The objective of this paper is to provide a description of the optimization of sintering SiC–TiC composites by the electroconsolidation method. The influence of titanium carbide content on the physical and mechanical properties of SiC–TiC composites obtained by spark plasma sintering (SPS) at a pressure of 45 MPa was studied. It was found that compared to sintered silicon carbide, the porosity of composites with 40 mol% TiC decreased from ~30% to 0%, the crack resistance increased from 2.9 to 6.1 MPa × $m^{0.5}$, and the hardness increased from 2.9 to 21.5 GPa. The influence of sintering temperature and holding time on SiC–TiC composites' physical and mechanical properties during sintering at a pressure of 45 MP was also investigated. An increase in temperature from 1900 °C to 2000 °C resulted in an approximately 30% rise in the composite hardness. An extension of the time allotted for the sintering process from 30 to 45 min resulted in a decrease in both the fracture toughness and hardness of the material. The utilization of two- and three-dimensional vector spaces of material features was proposed as a novel methodology for the description of manufacturing process optimization.

Keywords: silicon carbide; titanium carbide; electric spark sintering; temperature; pressure; structure; porosity; manufacturing optimization

1. Introduction

Functional materials employed in modern sensing devices exhibit a wide range of physical properties, including piezoelectricity, pyroelectricity, and sensitivity to external stimuli such as temperature, pressure, and chemical environments. Materials such as barium titanate, zinc oxide, and various compositions based on ferroelectric and piezoelectric compounds have drawn significant research attention due to their high sensitivity and operational stability under diverse conditions [1–3]. Their performance—including dielectric response, thermal conductivity, and structural stability—plays a crucial role in defining sensor output characteristics, sensitivity, and accuracy [4,5]. Furthermore, the surface morphology and nanostructural features of these materials greatly influence their interaction with the surrounding environment. These findings have significant implications for the development of next-generation sensors, which show considerable potential in terms of enhanced selectivity and miniaturization potential [6].

In recent years, silicon carbide (SiC, carborundum) has emerged as a material of choice for a wide range of contemporary applications due to the unique properties it exhibits [7]. It has been utilized as a fundamental component in the fabrication of bulk materials and coatings. SiC has become a prevalent material in the manufacturing of power electronics, electric vehicles, and solar and wind turbine systems [8], and its primary function is enhancing the efficiency of components. Its heat resistance, high resistance to radiation, and low neutron absorption rate render it an optimal solution as a thermal protector, especially for the construction of nuclear reactors. SiC-based photodetectors have been shown to exhibit large gains in, as well as high sensitivity and a rapid and consistent response to, α, β, neutron, X/γ, and UV radiation [9,10]. Furthermore, in the domain of abrasive, grinding, and cutting tools, SiC has been demonstrated to offer enhanced chemical and mechanical durability. Recently, SiC was tested for its suitability in the production of prototypes of implants and prostheses [11,12], which, in addition to durability, must demonstrate both biocompatibility and corrosion resistance. Recent interest in SiC has seen a marked escalation following the discovery of point defects in its crystal lattice (the so-called color centers), which can serve as next-generation single-photon sources [13–15]. Single-photon sources are a key element in the practical implementation of photonic and quantum computing solutions. In this extensive range of applications, the efficient manufacturing of novel materials based on SiC is of particular importance.

Silicon carbide appears in more than 250 polymorphic forms, including various amorphous phases (thin films and fibers) as well as crystalline polytypes (differing in so-called A, B, and C stacking) [16]. There are two crystalline polymorphs, namely, the high-temperature stable α-form (hexagonal/rhombohedral; 2H-SiC, 4H-SiC, 6H-SiC, nH-SiC, 15R, and 21R), which is most common, and the low-temperature stable β-form (cubic; 3CSiC) [17]. The structural distinctions between both forms can be attributed to the primitive unit cell type, the order of alternation and periodicity of layers along one crystallographic axis, and the degree of hexagonality.

The main advantage of SiC is its high hardness, surpassed only by diamond, cubic boron nitride, and boron carbide [18], but its low fracture toughness of 3–4 $MPa \cdot m^{0.5}$ makes it less reliable in critical applications. However, SiC exhibits a low self-diffusion coefficient due to the dominating C—Si covalent bond, which affects its densification process and comprises 88% of its bonds, with ionic bonds comprising the remaining 12%. Thus, it is necessary to select an appropriate sintering method and thermodynamic simulation by adding certain elements that do not promote the decomposition of SiC at elevated temperatures [19]. In particular, alumina, yttria, and zirconia may serve as additives for sintered silicon carbide-based composites with high mechanical properties [20,21].

Titanium carbide (TiC) is a metallic mineral known to become plastic at elevated temperatures. TiC fcc crystals have 12 slip systems and can undergo slip deformation at high temperatures. TiC exhibits excellent properties, including a high melting point (3140 °C) and hardness (HV = 30.3 GPa) [22], which collectively serve to secure its classification as a covalently bonded ceramic. Nevertheless, the low level of toughness exhibited by this material results in its resemblance to other non-oxide ceramics. Elevated temperatures have been shown to cause a substantial increase in the size of the grains, which has a deleterious effect on the fracturing and bending strength of this material. The inferior mechanical property of TiC, i.e., its intrinsic brittleness, exerts a considerable restriction on its utilization. However, the use of TiC in the production of wear-resistant tools as a conductivity-improving additive and nucleating agent has been well documented. In recent years, considerable research activity has been dedicated to the synthesis of titanium carbide (TiC)-based nanoceramics.

It has long been established that the addition of certain amounts of TiC powder of varying fineness has a positive effect on the mechanical and oxidation properties of α-SiC [23]. A high-toughness SiC–TiC particulate composite material was produced through a combination of SiC and TiC, both of which are recognized as having low levels of toughness when utilized individually. Recent findings have indicated that the addition of titanium carbides improves wear resistance by ca. 25% compared to that of 90 wt.% SiC–10 wt.% B_4C [24]. Research on high-density SiC–$(Ti_xZr_{1-x})B_2$ and SiC–TiB_2 composite ceramics fabricated by solid-state spark plasma sintering demonstrated that materials of relative density 99.97% exhibited higher flexure strength and fracture toughness by 242.6% and 53.6%, respectively, than single-phase SiC ceramics [25]. The authors of the abovementioned study concluded that the toughening effect could be attributed to the shell–core structure.

Even though additive manufacturing (AM) techniques for the fabrication of silicon carbide ceramics are attracting increasing interest [26], the sintering methods of various modifications remain dominant in this area [27,28]. The parameters of the sintering process enable control over the densification process and hence the resulting structure and related mechanical properties of SiC-based composites.

In previous research, the kinetics of the densification process was investigated in relation to SiC–TiC and SiC–VC systems [29]. It was found that the interface area between phases plays an important role in the densification process. Liu et al. [30] used molecular dynamic simulations to investigate the interaction of submicron particles of SiC during the sintering process. Particular attention was paid to the evolution of surface morphology and microstructure, radial distribution function, atomic displacement distribution, mean square displacement, and structural transformation, as well as the diffusion coefficient. The authors concluded that when the sintering time was increased, the dominant mechanism shifted from particle surface diffusion to interface diffusion, increasing atomic migration and interface energy consumption. Lei and co-authors [31] used the finite element method to predict and control the sintering processes of porous SiC cylindrical samples—particularly, temperature distribution, but also electrical and thermal fields. They found that the temperature increase in the core of the samples was proportional to their porosity, increasing from 1960 °C up to 2010 °C when porosity increased from 1% up to 60%.

Conventionally, composites are fabricated using spark plasma sintering (SPS) or hot-pressing (HP) methodologies. SPS has been shown to result in accelerated densification and, in many cases, reduced grain growth in comparison with conventional HP methods. Electroconsolidation, a process that employs an electric current to consolidate powder into a dense material, offers several significant advantages over both aforementioned densification methods. The advantages of this approach are numerous, including enhanced densification quality (obviating the necessity for coating), the capacity to produce fine-grained structures, the augmented mechanical properties of the ceramic (e.g., density, hardness, and fracture toughness), and energy efficiency in comparison to classical methods.

The present study is aimed at investigating the influence of different parameters of the sintering process on the mechanical properties of SiC–TiC ceramics fabricated using electroconsolidation. In particular, the effect of temperature and holding time on hardness and fracture toughness was investigated in relation to the titanium carbide percentage in the composites. The utilization of two- and three-dimensional vector spaces of material features was proposed as a novel methodology for the description of manufacturing process optimization.

2. Materials and Methods

2.1. Experimental Section

In the course of the experiments, α–SiC powder grade M2 produced by Zaporizhzhia Abrasive Plant (Zaporizhzhia, Ukraine) was used. The powder, with an average particle size of 2 μm, contained ca. 98% of SiC and no more than 0.1 wt.% Fe, 1.5 wt.% O, and 0.4 wt.% C. Figure 1 presents an SEM image of the powder and an SEM–EDX (scanning electron microscope with energy-dispersive X-ray) analysis of the powders. Table 1 presents the composition of the analyzed SiC powder.

Table 1. The composition of SiC.

	Spectrum 1	Spectrum 2	Spectrum 3	Average	Standard Deviation
C	59.94	59.30	54.99	58.08	2.69
O	4.17	4.58	4.17	5.03	1.15
Si	35.89	36.12	35.89	36.90	1.54

The additive was TiC powder with an average particle size of 4 μm. Its X-ray absorption spectroscopy (XAS) diffractogram is shown in Figure 2. Table 2 presents the respective composition of the TiC powder.

Table 2. The composition of the TiC additive.

	Spectrum 11	Spectrum 10	Spectrum 9	Spectrum 8	Average	Standard Deviation
C	30.67	23.07	26.28	23.42	25.86	3.51
O	—	15.02	—	—	—	—
Al	0.36	0.38	0.33	0.27	0.34	0.05
Ti	68.96	56.64	73.39	76.30	68.82	8.66
Fe	—	4.88	—	—	—	—

TiC powder was added to SiC in proportions of 5, 20, 30, and 40% by mass. The two components were blended together for 24 h with a ball mill in a wet environment, using balls made of hot-pressed silicon carbide. After this, the samples were sintered in graphite molds using the electroconsolidation apparatus described in [32].

The high compaction of sintered silicon carbide requires high temperatures, high pressure, and a longer holding time than most refractory powders. In the case of the applied SPS apparatus, the strength of the graphite molds did not allow for an increase in the uniaxial pressure above 50 MPa. In addition, the powders were of a microscale and exhibited low sinterability compared to nanopowders, so that the holding time was prolonged to 5, 15, 30, and 45 min for two different sintering temperatures: 1900 °C and 2000 °C. For comparison, using submicron powders, it was possible to sinter the composition 80 wt.% SiC + 20 wt.% (Al_2O_3 + ZrO_2) at a temperature of 1860 °C, under a uniaxial pressure of 30 MPa, with a holding time of 3 min [21]. The sintering process took place under a current of 5000 A and voltage 5 V, in a vacuum of 1.5 Pa, ensuring a heating rate of 300 °C/min. A uniaxial mechanical pressure of 45 MPa was applied after the temperature reached 1000 °C, and it was released 3 min after turning off the current. Specimens obtained in this process were of a cylindrical shape with a diameter ∅ 11 mm and a height of 5 mm.

Density and related porosity were determined according to the standard EN ISO 3369:2014. The Vickers hardness HV was measured under the load 150 N using the Matsuzawa MXT70 tester (Matsuzawa Co. Ltd., Akita, Japan). Indentation was examined with the NU-2E microscope produced by Carl Zeiss (Jena, Germany) with 750× zoom. From the indentation and the length of radial cracks, fracture toughness K_{1c} was calculated according to the well-known Evans and Charles equation [33].

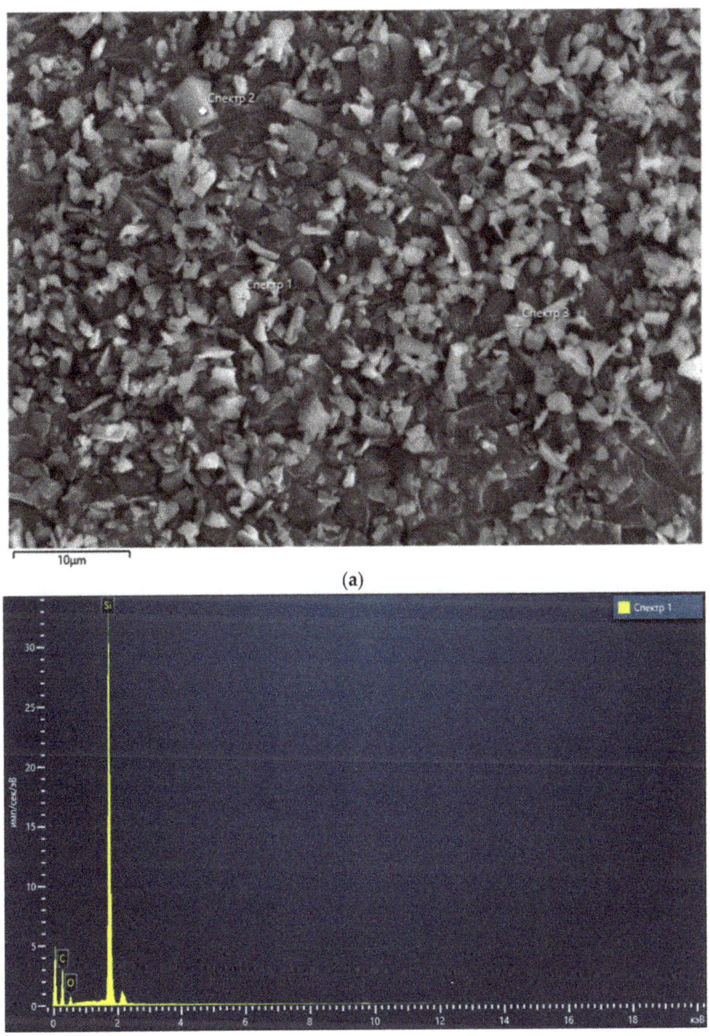

Figure 1. SEM-EDX analysis of the SiC powder: (**a**) SEM image; location of samples taken for the further analysis (Table 1) is marked as using numbers 1, 2, and 3; (**b**) composition.

(**a**)

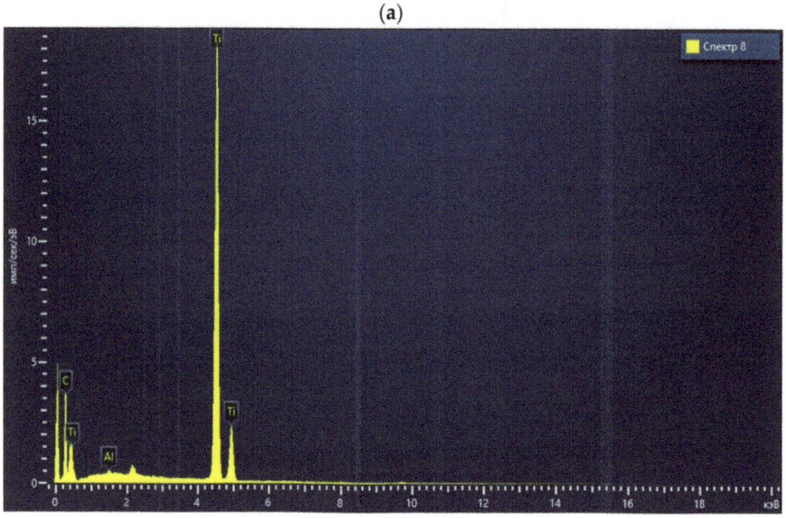

(**b**)

Figure 2. SEM-EDX analysis of the TiC powder: (**a**) SEM image; location of samples taken for the further analysis (Table 2) is marked as using numbers 11, 10, 9 and 8; (**b**) composition.

The sintered samples were examined with the Tescan Vega 3 SBH EP (TESCAN, Brno, Czech Republic) scanning electron microscope (SEM). During measurement, the accelerating voltage was 20–30 kV in backscattered electron (BSE) and secondary electron (SE) modes at different magnifications. The resolution of this device is as high as 3.5 nm. The chemical composition of the samples was determined by element mapping, point analysis, and the scanning of a surface of 0.09–0.25 mm^2. The application of the Bruker Quantax 610 M energy-dispersive X-ray spectroscope (EDS) (Bruker, Billerica, MA, USA) permitted the identification and quantification of elements in the range from B (z = 5) to Am (z = 95).

2.2. Material Feature Vector Approach

Conventionally, the evaluation of material quality is undertaken via manual analysis of a given set of properties, with these properties being determined experimentally. A more computationally oriented approach, which is an alternative to the aforementioned method, involves the use of dedicated n-dimensional feature vectors (vectors of features) in an n-dimensional feature vector space. The n-dimensional feature vectors constitute a distinct category of vectors, designated as such due to their capacity to articulate a set of features through a single mathematical quantity. The n-dimensional material feature vector can be defined as a numerical representation of material features, which is based on an ordered list of numerical values that characterize the given material. Thus, the material feature vector is constructed based on a tuple describing the set of properties for a given type of material. From a mathematical perspective, a vector is preferable to a tuple for the purposes of subsequent analysis. From a physical perspective, the material features are described in a uniform manner using material feature vectors combining grain size, hardness, and fracture toughness, i.e., the parameters which constitute the basis for the assessment of material quality. The angles and magnitudes describing these vectors are thus related directly to the material's properties.

In classical data analysis, n-dimensional vectors that possess both direction and magnitude are employed as a convenient tool for numerical representation, thereby enabling a wide range of analyses. A plethora of methodologies exists for the comparison of vectors based on diverse mathematical metrics, including cosine or Euclidean ones.

The cosine similarity (CS) between the material feature vectors can be defined using the cosine metric as follows:

$$d_c(p,q) = \frac{\sum_i p_i q_i}{\left(\sum_i p_i^2 \sum_i q_i^2\right)^{1/2}} \tag{1}$$

where <p> = $\{p_i\}$ and <q> = $\{q_i\}$ are the material feature vectors. The numerical value of $d_c(p, q)$ ranges from 0 to 1, and $d_c(p, q) = 1$ indicates the identity of vectors. The error value of $d_c(p, q)$ can be derived using the absolute value of the total differential:

$$|\Delta d_c(p,q)| = \sum_i \frac{q_i \sum_{j\neq i} p_j^2 - p_i \sum_{j\neq i} p_j q_j}{\left(\sum_i p_i^2\right)^{3/2}\left(\sum_i q_i^2\right)^{1/2}}|\Delta p_i| + \sum_i \frac{p_i \sum_{j\neq i} q_j^2 - q_i \sum_{j\neq i} p_j q_j}{\left(\sum_i p_i^2\right)^{3/2}\left(\sum_i q_i^2\right)^{1/2}}|\Delta q_i| \tag{2}$$

where <Δp> = $\{\Delta p_i\}$ and <Δq> = $\{\Delta q_i\}$ are the error vectors.

The cosine metric is a useful tool for the purpose of comparing vectors based on their directional similarity. However, since the vectors are normalized, any differences in magnitude between them are rendered negligible.

The Euclidean distance (ED) between the material feature vectors can be defined using the Euclidean metric, also known as the L2 norm, as follows:

$$d_E(p,q) = \left(\sum_i (p_i - q_i)^2\right)^{1/2} \tag{3}$$

where <p> = $\{p_i\}$ and <q> = $\{q_i\}$ are the material feature vectors. The numerical value of $d_E(p, q)$ ranges from 0 to infinity, and $d_E(p, q) = 0$ indicates the identity of vectors. The error value of $d_E(p, q)$ can be derived using the absolute value of the total differential:

$$|\Delta d_E(p,q)| = 2\left|\sum_i (p_i - q_i)\right|(|\Delta p_i| + |\Delta q_i|) \tag{4}$$

where <Δp> = $\{\Delta p_i\}$ and <Δq>$\{\Delta q_i\}$ are the error vectors.

The Euclidean metric facilitates the identification of more intricate relationships and consequently enables a more precise delineation of the similarity between vectors and the materials' features. Unlike the cosine metric, it measures the absolute difference in vector values.

There is no universal rule for choosing between the Euclidean and cosine metrics, as their suitability largely depends on the nature of the data. From a mathematical standpoint, the utilization of hybrid approaches, incorporating both metrics, appears optimal. This approach is underpinned by the principle that each metric enables the evaluation of distinct aspects. Irrespective of the metric under consideration, the vector approach has the advantageous capability of capturing regularities, thus facilitating the optimization of material features. Nevertheless, cosine metrics, which do not capture deep relationships between materials, may be less accurate.

The application of a heat map technique provides a visual representation of features. It facilitates the capture of the most salient data by representing the magnitude of individual values within a dataset by color [34]. In this study, grid color-coded heat maps were used to facilitate the visualization of the similarity between material feature vectors.

The material feature approach described above was combined with use of mathematical metrics as similarity measures and heat maps for the visualization of the distance between vectors. It shares distant semantic and methodological connotations with the method used for the in-depth analysis and prediction of binding modes in protein–ligand complexes [34,35].

The validity of the material feature vector approach was verified through an additional analysis that utilized distinct sets of experimental data for a different material. The verification focused primarily on composites which have been the subject of recent research but constituted a set separate from SiC–TiC composites. For the abovementioned set of data, the classical analysis in conjunction with the results obtained based on material feature vectors yielded convergent results.

3. Results and Discussion

3.1. Structure of the Composite

The structure of the sintered material consisted of dark gray grains of the matrix SiC material and light gray inclusions of the titanium carbide. An example structure of the composition 60 wt.% SiC–40 wt.% TiC sintered at 2000 °C under a uniaxial pressure of 45 MPa for 30 min is shown in Figure 3a. A visual representation of the magnitude of individual values as colors is provided in Figure 3b. A section of each image has been cut out and enlarged to show the boundaries between grains in Figure 3c,d.

The heat map visualization shown in Figure 3d reveals boundaries between grains of SiC, depicted in orange, and inclusions of TiC, depicted in cyan. The scale bars are shown in the images. The same color scale was used for both figures.

A micro-X-ray fluorescent spectroscopy (M-XRF) analysis of the grain boundaries between phases revealed that under a uniaxial pressure of 45 MPa, the interaction zone covered ca. 1.5 µm. Figure 4a shows the diagram corresponding to the radiation registered along the green arrow shown in Figure 4b. This feature can be attributed to a certain non-stoichiometry of TiC [36] and the appearance of very small amounts of a liquid phase in the Si–Ti–C system [37,38].

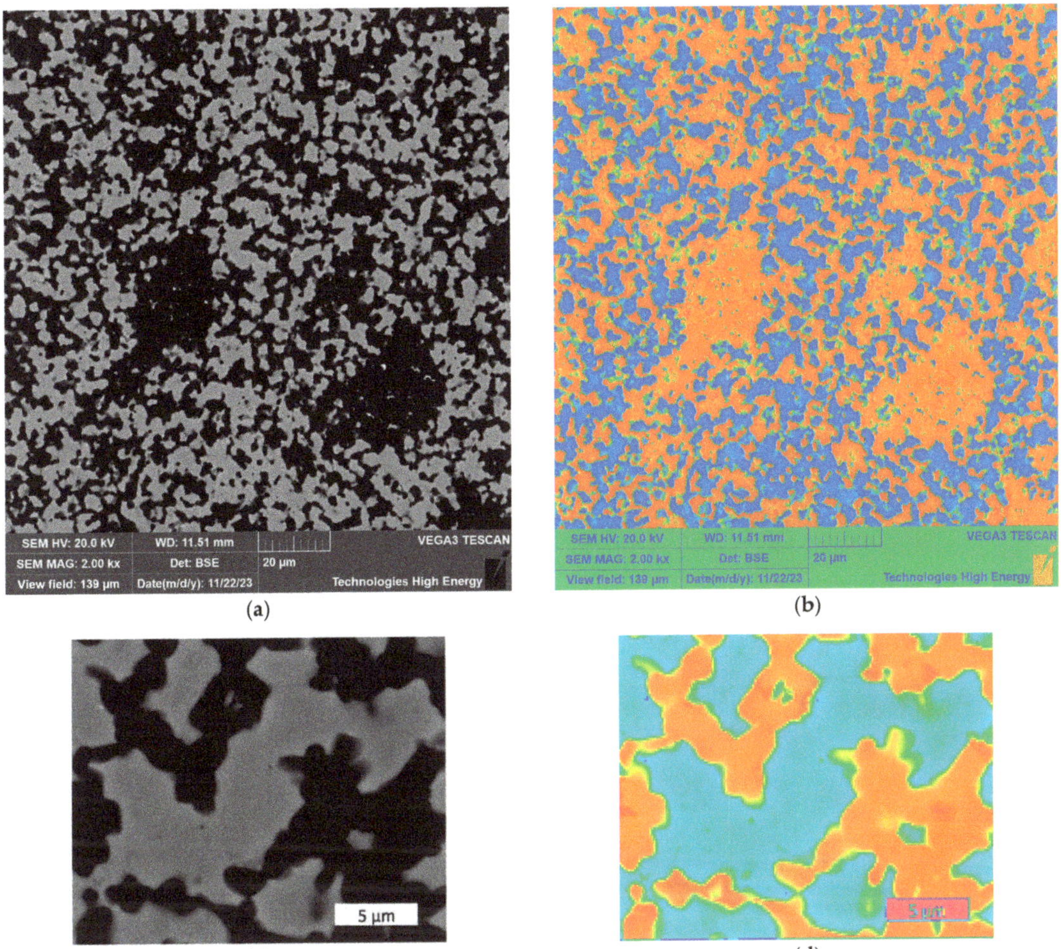

Figure 3. Microstructure of 60%SiC–40%TiC material sintered at 2000 °C under uniaxial pressure of 45 MPa for 30 min (**a**); the boundaries are better resolved using a heat map (SiC grains and TiC inclusions are depicted in orange and blue/cyan, respectively (**b**)). A section of each image has been cut out and enlarged to show the boundaries between grains (**c**,**d**). The color scale ranges from red to blue, with thresholds set at every 10% increment in image brightness.

The proportion of titanium carbide directly affected the porosity of the SiC–TiC composites. It was found that an increase in TiC content in the powder mixture resulted in a decreased porosity of the sintered material when all other parameters of the process were kept unchanged. The diagram shown in Figure 5 reveals the dependence of porosity on the TiC content for composites sintered at 2000 °C under a uniaxial pressure of 45 MPa for 30 min. It is noteworthy that when the proportion of titanium carbide reached 40 wt.%, the porosity was reduced to almost 0%.

Conversely, the porosity of the sintered material is directly dependent on the holding time t and sintering temperature T. The corresponding diagrams for the composition of 60 wt.% SiC–40 wt.% TiC are presented in Figure 6. After 30 min, porosity was reduced below 1%, and for longer holding times, the sintering temperature made no difference.

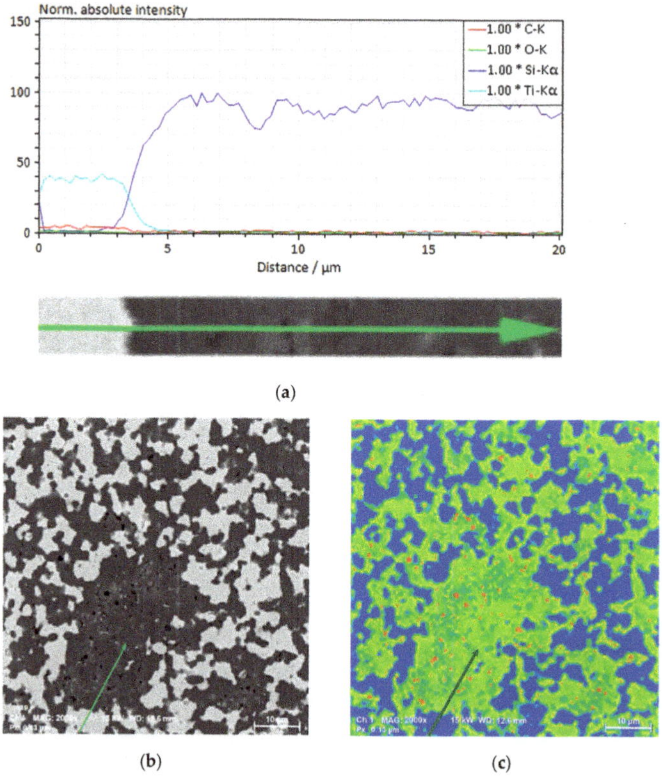

Figure 4. The interaction zone between SiC and TiC phases in the 60 SiC–40 TiC composite: (**a**) M-XRF diagram, (**b**) the analyzed area with the sampling direction marked with a green arrow, and (**c**) a heat map facilitating the more precise delineation of the sample granularity and boundaries. The color scale ranges from red to blue with thresholds set at every 10% increment in image brightness.

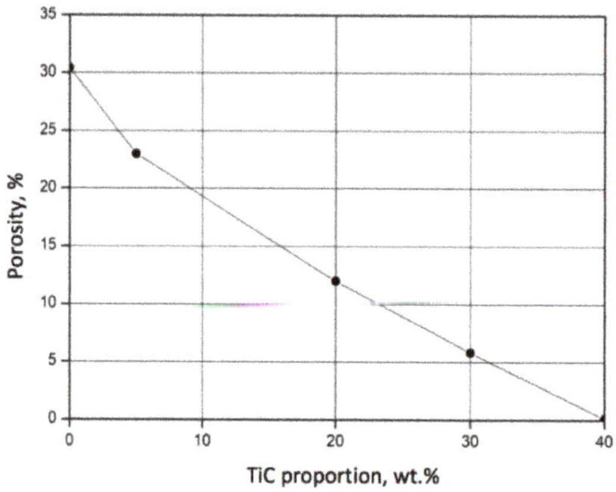

Figure 5. Effect of TiC content on porosity of material sintered at 2000 °C under uniaxial pressure of 45 MPa for 30 min.

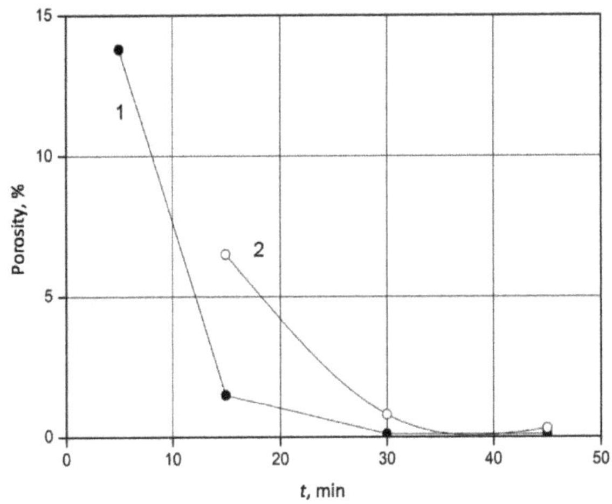

Figure 6. Effect of the holding time on the porosity of the 60 wt.% SiC–40 wt.% TiC material sintered under a uniaxial pressure of 45 MPa at the following temperatures: 1–T = 2000 °C (•); 2–T = 1900 °C (○).

3.2. Mechanical Properties of the Composite

The complex interaction between SiC and TiC influenced not only the porosity but also the mechanical properties of the composite. An increase in both fracture toughness K_{1c} and hardness HV was registered for a higher proportion of TiC, as shown in Figure 7.

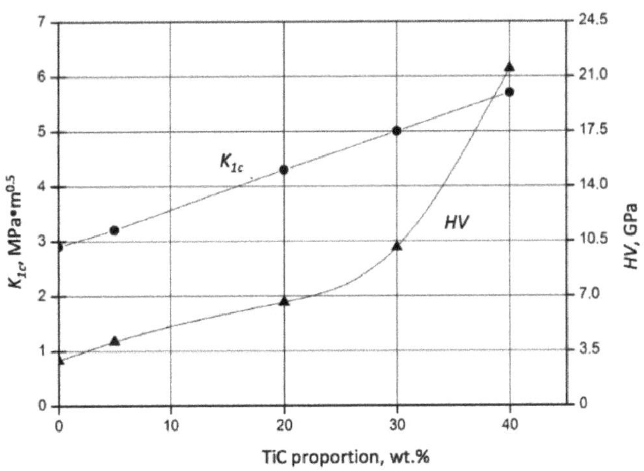

Figure 7. Effect of TiC content on the fracture toughness K_{1c} (marked •) and hardness HV (marked ▲) of SiC–TiC composites sintered at 2000 °C under a uniaxial pressure of 45 MPa for 30 min. (The standard deviation of K_{1c} and HV varied from 0.3 to 0.5 MPa·m$^{0.5}$ and from 0.8 to 1.0 GPa, respectively).

The improvement in fracture toughness with increased TiC addition can be attributed to the mismatch between thermal expansion coefficients α–SiC = 4.8 × 10^{-6}/°K and α–TiC = 7.4 × 10^{-6}/°K and to the related residual stresses [39]. It was found that the addition of TiC in a proportion of 20 wt.% to the SiC powder of a 2 μm particle size resulted in an increase in the fracture toughness of the sintered composite from 2.9 MPa·m$^{0.5}$ up to

4.3 MPa·$m^{0.5}$, while hardness HV increased from 2.9 GPa up to 6.6 GPa. A further increase in TiC content caused an almost proportional enhancement in fracture toughness, which reached K_{1c} = 5.7· MPa·$m^{0.5}$ for 40 wt.% of TiC. On the other hand, a twofold increase in the TiC content from 20 wt.% up to 40 vol.% resulted in a threefold increase in hardness HV up to 21.5 GPa. When the composite was sintered at lower temperatures (1900 °C), the respective values of toughness and hardness were reduced by ca. 25%. However, a further increase in the sintering temperature caused the significant worsening of the mechanical properties of the composites. This phenomenon can be explained by extensive SiC grain growth at higher temperatures.

It should be noted that an improvement in the fracture toughness of a composite is accompanied by an increase in hardness. The composite 60 wt.% SiC–40 wt.% TiC–40 TiC exhibited better properties compared to the results for pressureless sintering reported in [40], where the highest hardness of 20 GPa was found in samples containing 5 vol.% TiC, and a fracture toughness of 5.5 MPa·$m^{0.5}$ was reached in samples containing 10 vol.% TiC, sintered at 1885 °C for 1 h. Despite the larger grain sizes compared to nanoscale initial powders, as well as the lower temperatures used in [41], the fracture toughness reached in this study for the 60 wt.% SiC–40 wt.% TiC material was 15% higher than that of SiC sintered under 25 GPa at 1500 °C. Simultaneously, a hardness above 21 GPa was reached, typical for titanium carbide matrix composites [42]. Further analysis revealed a dependence of the composite's toughness and hardness on the holding time at different temperatures. In particular, the diagrams of K_{1c} and HV for the composition 60 wt.% SiC–40 wt.% TiC sintered at 1900 °C and 2000 °C are presented in Figures 8 and 9, respectively.

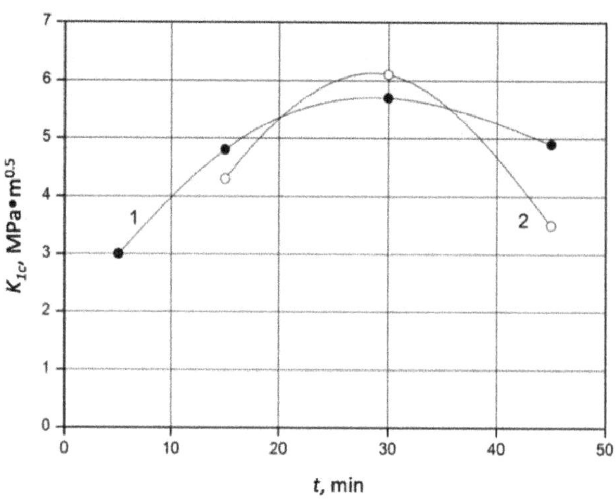

Figure 8. Effect of the holding time on the fracture toughness K_{1c} of the 60 wt.% SiC–40 wt.% TiC material sintered at the following temperatures: 1–T = 2000 °C (•); 2–T = 1900 °C (○). (The standard deviation of K_{1c} varied from 0.3 to 0.5 MPa·$m^{0.5}$).

The experimental results show that after the sintering of the composite 60 wt.% SiC–40 wt.% TiC at a temperature of 1900 °C for 15 min, fracture toughness K_{1c} increased up to 3.5 MPa·$m^{0.5}$, while hardness HV increased up to 9.9 GPa. When the holding time was prolonged up to 30 min at a sintering temperature of 1900 °C, K_{1c} increased up to 6.1 MPa·$m^{0.5}$, while HV increased up to 16.6 GPa. The further prolongation of the holding time up to 45 min caused a decrease in fracture toughness down to 3.5 MPa·$m^{0.5}$ and a decrease in hardness down to 9.9 GPa. Apparently, this was a consequence of the increased sizes of silicon carbide grains. The sintering of the composite 60 wt.% SiC–40 wt.% TiC at a temperature of 2000 °C after 15 min provided a fracture toughness of 4.8 MPa·m$^{0.5}$ and a hardness of 15.1 GPa. An increase in the holding time up to 30 min caused significant enhancements in crack resistance to 5.7 MPa·m$^{0.5}$ and in hardness to 21.5 GPa. However, after the sintering of this composite for 45 min, a substantial drop in the toughness and hardness of the composite was observed, caused by the growth of silicon carbide grains.

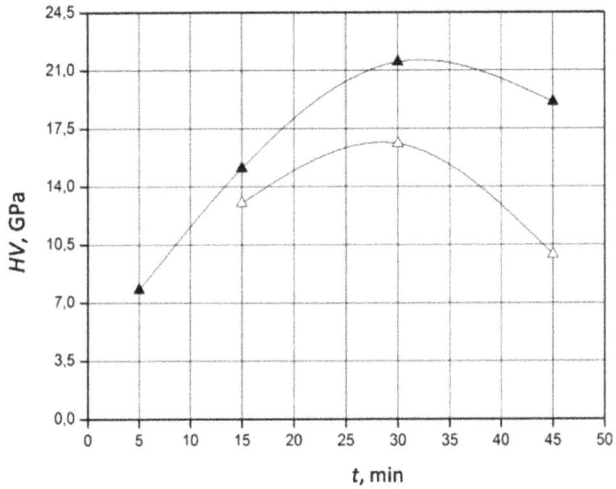

Figure 9. Effect of the holding time on the hardness HV of the 60 wt.% SiC–40 wt.% TiC material sintered under a uniaxial pressure of 45 MPa at the following temperatures: 1–T = 2000 °C (▲); 2–T = 1900 °C (△). (The standard deviation of HV varied from 0.8 to 1.0 GPa).

The literature review indicated that systematic studies of both HV and K_{1c} in relation to changes in TiC percentage in SiC–TiC composites had only been conducted by Ueki et al. [43]. In the aforementioned study, the researchers used a different technique—hot pressing in an Ar inert gas atmosphere at a temperature of 2150 °C and a pressure of 40 MPa for 2 h. The starting material was β–SiC powder, which was transformed in this process to a high-temperature phase. The authors obtained a composite that exhibited higher HV but lower K_{1c}. Furthermore, the composite displayed a higher level of porosity in comparison with the material produced in this research. This outcome is most likely attributable to elevated temperatures exceeding 2000 °C, a condition that has been demonstrated to be deleterious. In the time-consuming process proposed by Ueki et al. [43], the optimum doping level was ascertained to be 50%. However, the lower doping level proposed in this study is more advantageous because higher TiC admixture results in low strength (internal brittleness). It is evident that the properties of composites obtained by Chang et al. [44] and Cabrero et al. [45] are also inferior.

3.3. Material Feature Vectors Describing the Mechanical Properties of the Composite

As follows from the analysis of the experimental data, the combination of porosity, fracture toughness, and hardness is responsible for the unique and highly desirable properties of the composites. However, the process of comparing individual material features to identify or predict the optimal set of features may be time-consuming. To improve its efficiency, the use of a material feature vector, characterizing the structure and mechanical properties of composites, was proposed. The process of evaluating and quantifying the similarity between material feature vectors describing different SiC–TiC composites constitutes a pivotal element in both the formulation of recommendations and the identification of anomalies.

Each value in the 2D and 3D material feature vectors corresponds to a specific property of the sample under analysis: K_{1c} and HV or porosity, K_{1c}, and HV, respectively.

The magnitude of the material feature vector and the angle between vectors, presented in Table 3, increased with the increase in TiC content. The sole exception is the TiC proportion 30 wt.%, which serves to demonstrate that both length and angle must be considered.

Table 3. The 2D and 3D material feature vectors and angular distance (in degrees) between TiC 0% and four different TiC contents, namely, 5, 20, 30, and 40%. (The margin of error is no greater than 5%.)

TiC Content [%]	2D [K_{1c}, HV]		3D [Porosity, K_{1c}, HV]	
	Magnitude	Angle	Magnitude	Angle
0	4.40	—	32.30	—
5	5.03	−1.50	23.54	4.51
20	8.16	8.51	14.51	26.50
30	11.63	14.01	13.09	55.12
40	22.00	24.99	22.00	82.90

The vector spaces associated with these feature vectors, so-called 2D and 3D material feature spaces, are shown in Figures 10a and 10b, respectively.

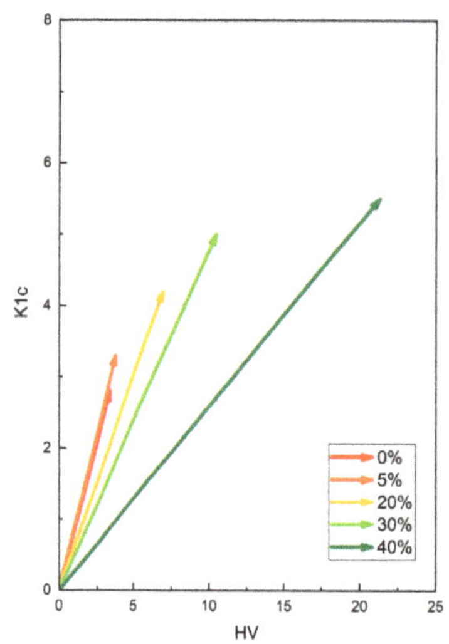

(a)

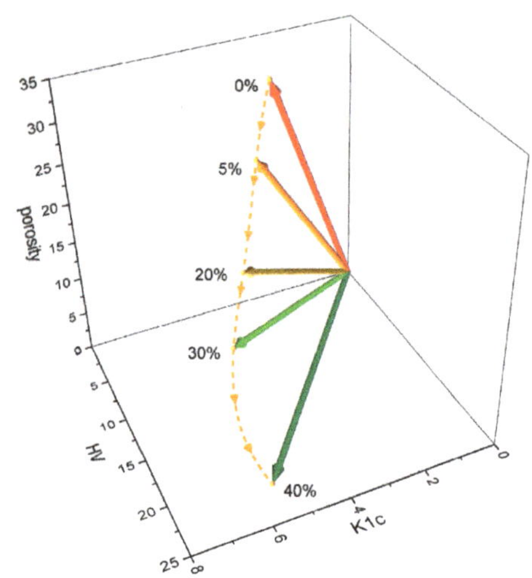

(b)

Figure 10. A visualization of the feature vector in the material feature space: (**a**) 2D and (**b**) 3D.

The path that follows the vector at any given point (trajectory), as shown in Figure 10b, describes the change in the physical properties of the material upon a change in TiC content in the composite. Based on this trajectory, the physical values at intermediate points, i.e., for other percentages of TiC, can be easily interpolated. The higher the angle, the better the material—i.e., the more the vector deviates from the K_{1c}/HV plane, the worse the properties the composite exhibits will be.

The variation in the feature vector orientation with increasing TiC content is shown in Figure 11.

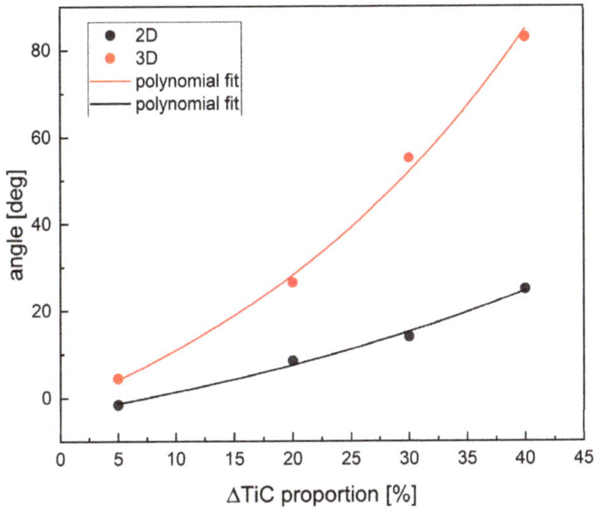

Figure 11. The angle between the feature vector describing pure TiC and the 5, 20, 30, and 40% percentage admixtures of TiC, respectively.

The alteration in the orientation of the feature vector in space upon an increase in TiC content is characterized by a monotonic progression. Porosity exerts a considerably more pronounced effect on the angle changes.

The changes in the orientation of the material feature vector in space upon an increase in TiC percentage (visualized based on data from [43,45]) are non-monotonic (Figure 12).

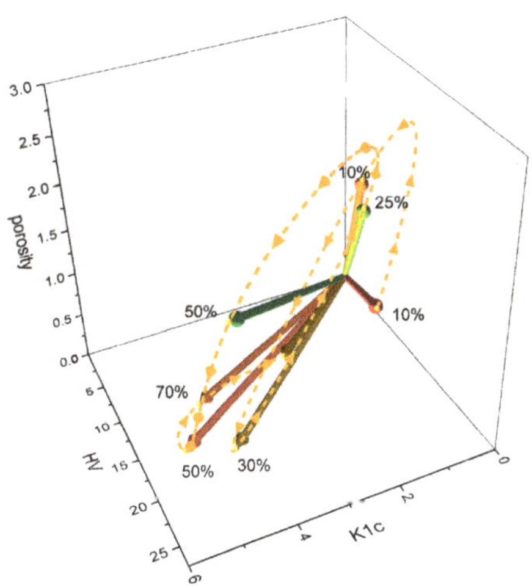

Figure 12. A visualization of the material feature vector in the 3D material feature space, with data from Refs. [43,45].

The trajectories (marked in yellow) illustrate the differences in the physical properties of the material resulting from the variation in the TiC content within the composite, as observed in both studies. They demonstrate the challenges associated with obtaining composites that exhibit a continuous transition in properties upon TiC percentage increase. However, the more deviated the vector was from the K_{1c}/HV plane, the worse the obtained composite properties were. This simple principle remains applicable.

In the context of material recommendation, two feature vectors describing preferences should be similar if they share common features or lead to the same choices. Comparisons of their similarity estimated using metrics that consider magnitude and direction (Euclidean distance, ED) or direction (cosine similarity, CS) can be helpful. As the ED value increases, the discrepancy between vectors becomes more pronounced. Conversely, as the CS decreases, the similarity between vectors becomes more apparent. A value of CS closer to 1 indicates a greater level of similarity, while a value closer to 0 indicates no similarity. (The margin of error pertaining to distances does not exceed 5%.)

The ED and CS distances between each pair of the 2D and 3D feature vectors are summarized in a tabular form and colored using a heat mapping technique in Tables 4 and 5.

Table 4. Euclidean distance between material feature vectors: 0, 5, 20, 30, and 40% percentage admixtures of TiC. A red–yellow–green scheme, where green indicates a maximum value and red indicates a minimum value, is used.

2D ED	0	5	20	30	40	3D ED	0	5	20	30	40
0	0	0.64	3.86	7.43	18.1	0	0	9.02	20.37	27.04	36.77
5	0.64	0	3.32	6.91	17.64	5	9.02	0	11.49	18.35	28.98
20	3.86	3.32	0	3.59	14.36	20	20.37	11.49	0	6.99	18.71
30	7.43	6.91	3.59	0	10.81	30	27.04	18.35	6.99	0	13.09
40	18.1	17.64	14.36	10.81	0	40	36.77	28.98	18.71	13.09	0

Table 5. Cosine similarity between material feature vectors: 0, 5, 20, 30, and 40% percentage admixtures of TiC. A red–yellow–green scheme, where green indicates a maximum value and red indicates a minimum value, is used.

2D CS	0	5	20	30	40	3D CS	0	5	20	30	40
0	1	0.9997	0.989	0.9703	0.9064	0	1	0.9969	0.8949	0.5718	0.1236
5	0.9997	1	0.9848	0.9636	0.895	5	0.9969	1	0.9261	0.631	0.1913
20	0.989	0.9848	1	0.9954	0.9589	20	0.8949	0.9261	1	0.8767	0.5393
30	0.9703	0.9636	0.9954	1	0.9817	30	0.5718	0.631	0.8767	1	0.8724
40	0.9064	0.895	0.9589	0.9817	1	40	0.1236	0.1913	0.5393	0.8724	1

Optimization requires a different approach from material feature vector variability, as the proximity of vectors is no longer a desirable feature. Conversely, it has been demonstrated that an increase in distance is directly correlated with an enhancement in material properties. As TiC content increases, the discrepancy between the ED similarity values becomes more pronounced (Figure 13a), but the CS similarity becomes more apparent (Figure 13b).

As follows from the above data, the 3D material feature vector is much more sensitive to the TiC proportion than the 2D material feature vector. Thus, the effect of porosity is significant and well reproduced by the 3D feature vector. The changes in the orientation of the feature vector in space and the similarity measures of material feature vectors indicate changes of a monotonic nature caused by the increase in the percentage of TiC.

A range of techniques exists for the manufacturing of composite materials, which in turn leads to significant fluctuations in material features. Consequently, it is challenging to predict the optimal percentage of admixtures for the manufacturing of composite materials. (The repeatability of materials obtained using such techniques is often poor.) In contrast to other methods, the behavior of the 2D and 3D material feature vectors for the SiC–TiC composites obtained by electroconsolidation is monotonic. This finding indicates that the material properties of SiC–TiC exhibit a smooth transition in response to TiC percentage.

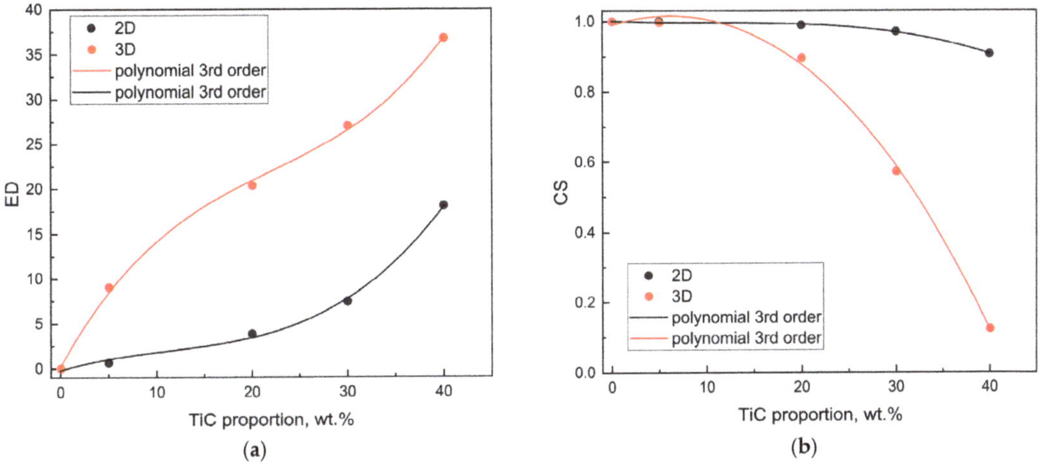

Figure 13. The similarity of the material feature vectors versus TiC proportion: (**a**) Euclidean distance; (**b**) cosine similarity.

Similar analysis was performed based on the data collected with different holding times and temperatures (Table 6).

Table 6. The influence of the holding time on the 2D and 3D material feature vectors and their angular distances (in degrees). (The margin of error is no greater than 5%.)

Holding Time, Min	2000 °C				1900 °C			
	2D $[K_{1c}, HV]$		3D $[Porosity, K_{1c}, HV]$		2D $[K_{1c}, HV]$		3D $[Porosity, K_{1c}, HV]$	
	Magnitude	Angle	Magnitude	Angle	Magnitude	Angle	Magnitude	Angle
5	7.80	—	16.03	—	—	—	—	—
15	15.08	4.06	15.38	49.72	13.66	—	14.92	—
30	21.86	7.50	21.86	59.84	17.31	2.73	17.38	18.94
45	18.66	7.39	18.66	61.14	10.41	1.75	10.41	23.77

A visualization of the material feature vectors in 3D material feature space is shown in Figure 14.

The highest magnitude of 2D and 3D material feature vectors indicates the best option. The clustered nature of vectors at 1900 °C and their scattered nature at 2000 °C illustrate the role that holding time plays in composite features.

The Euclidean distance and cosine similarity between material feature vectors at different holding times (T = 2000 °C and 1900 °C) are summarized in tabular form and colored using a heat mapping technique in Tables 7–10.

The highest ED (Tables 7 and 9) indicates the optimal holding time. Thus, the effect of holding time is significant and well described by both the 2D and 3D feature vectors. CS distance is not as good a measure of holding time changes as ED, which can be seen in Tables 8 and 10. The smallest CS should indicate the best option, but the 3D CS points to 45 min instead of 30 min, which is a result of the magnitude of the vectors being neglected by cosine metrics. Consequently, the CS is of limited use in this context.

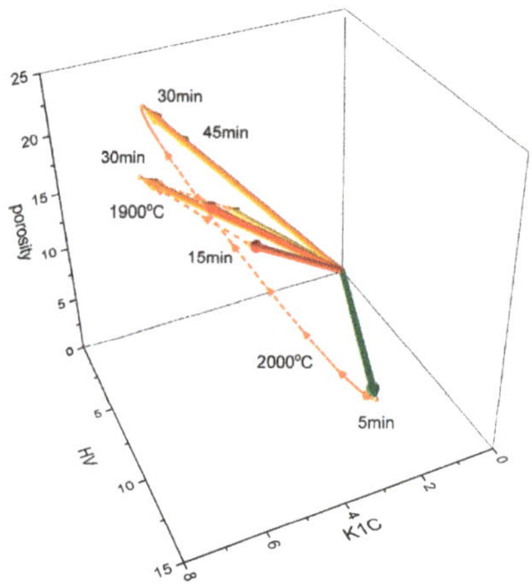

Figure 14. A visualization of the influence of holding time on the 3D material feature vectors.

Table 7. Euclidean distance between material feature vectors at different holding times (cells are colored to denote varying degrees of magnitude; T = 2000 °C). A red–yellow–green scheme, where green indicates a maximum value and red indicates a minimum value, is used.

2D ED	5	15	30	45	3D ED	5	15	30	45
5	0	7.32	14.16	10.97	5	0	13.22	19.56	17.78
15	7.32	0	6.86	3.70	15	13.22	0	7.30	4.76
30	14.16	6.86	0	3.20	30	19.56	7.30	0	3.24
45	10.97	3.70	3.20	0	45	17.78	4.76	3.24	0

Table 8. Cosine similarity between material feature vectors at different holding times (cells are colored to denote varying degrees of magnitude; T = 2000 °C). A red–yellow–green scheme, where green indicates a maximum value and red indicates a minimum value, is used.

2D CS	5	15	30	45	3D CS	5	15	30	45
5	1	0.9975	0.9914	0.9917	5	1	0.6466	0.5024	0.4827
15	0.9975	1	0.9982	0.9983	15	0.6466	1	0.9832	0.9791
30	0.9914	0.9982	1	1.0000	30	0.5024	0.9832	1	0.9997
45	0.9917	0.9983	1.0000	1	45	0.4827	0.9791	0.9997	1

Table 9. Euclidean distance between material feature vectors at different holding times and 1900 °C. A red–yellow–green scheme, where green indicates a maximum value and red indicates a minimum value, is used.

2D ED	15	30	45	3D CS	15	30	45
15	0	3.72	3.28	15	0	5.84	6.81
30	3.72	0	6.91	30	5.84	0	7.07
45	3.28	6.91	0	45	6.84	7.07	0

Table 10. Cosine similarity between material feature vectors at different holding times and 1900 °C. A red–yellow–green scheme, where green indicates a maximum value and red indicates a minimum value, is used.

2D CS	15	30	45	3D CS	15	30	45
15	1.0000	0.9989	0.9995	15	1.0000	0.9459	0.9152
30	0.9989	1.0000	0.9999	30	0.9459	1.0000	0.9961
45	0.9995	0.9999	1.0000	45	0.9152	0.9961	1.0000

A comparison of the ED at two different temperatures (Tables 6 and 8) facilitates the determination of the optimal temperature, that is, 2000 °C. The disparity between the ED values of 10.44 and 7.69, as well as 13.72 and 10.94, for the two-dimensional (2D) and three-dimensional (3D) vector spaces, respectively, is substantial.

4. Conclusions

The objective of the present research was to study the influence of the titanium carbide proportion in the SiC–TiC system, sintered under a uniaxial pressure of 45 MPa, on its mechanical properties. Additionally, the investigation delved into the optimization of the aforementioned process, with a view to enhancing its efficiency and efficacy.

It was found that the addition of TiC to silicon carbide reduced porosity, almost reaching full density when the titanium carbide proportion was 40 wt.%. Simultaneously, an increase in the TiC content caused an enhancement in fracture toughness from 2.9 MPa·$m^{0.5}$ (for SiC with no TiC addition) to 6.1 MPa·$m^{0.5}$ (for 60 wt.% SiC–40 wt.% TiC composite). The respective values of hardness increased significantly from 2.9 GPa up to 21.5 GPa.

Regarding the temperature effect, an increase in the sintering temperature from 1900 °C to 2000 °C caused an increase in fracture toughness and hardness by approximately 30%. Nevertheless, elevated temperatures of 2100 °C resulted in deleterious effects, manifesting as diminished grain growth and the subsequent deterioration of mechanical properties. A similar effect was observed when the holding time was extended up to 45 min. Following a period of 30 min, it was established that full density had almost been attained and that the characteristics of the sintered composites had reached their maximum levels.

The use of two- and three-dimensional vector spaces of material features was proposed as a novel methodology for the description of manufacturing process optimization. The analysis of the 2D and 3D individual material feature vectors and Euclidean distances between them confirmed the choice of the optimal TiC content, holding time, and temperature achieved through a manual comparison of features. (The cosine similarity disregards the magnitude of the vectors and is thus of limited use.) This new approach facilitates more rapid process optimization. The high heating rates, short holding times, and relatively low sintering temperatures required by electroconsolidation ensure a significant reduction in resource usage. It was demonstrated that the process of densification is significantly accelerated using electroconsolidation.

Furthermore, the changes in both the orientation and magnitude of 2D and 3D material feature vectors for SiC–TiC composites with different TiC percentages are monotonic. This finding indicates that the properties of SiC–TiC composites obtained by electroconsolidation show a smooth change in response to the modification of the percentage of TiC content, which distinguishes this method from classical ones such as SPS or HP. Thus, the material feature vector approach revealed the smooth characteristics offered by the manufacturing process, an advantage that is not offered by the classical methods of SPS or HP. Therefore, it can be deduced that electroconsolidation is associated with more predictable outcomes. It can be hypothesized that material feature vector space may offer enhanced process control. However, this remains to be investigated.

Author Contributions: Conceptualization, J.N.L., T.K., and T.S.; methodology, V.I., E.H., and A.K.; software, M.R. and A.K.; validation, V.I., J.N.L., M.R., N.S., and T.S.; formal analysis, V.I., E.H., M.R., and T.S.; investigation, E.H., N.S., and V.C.; resources, T.K.; data curation, J.N.L.; writing—original draft preparation, V.I.; writing—review and editing, all authors; visualization, J.N.L., E.H., T.K., and V.C.; supervision, V.I.; project administration, N.S.; funding acquisition, N.S. and V.C. All authors have read and agreed to the published version of the manuscript.

Funding: The work was financed from the project #6541030 of the Academy of Sciences of Ukraine.

Institutional Review Board Statement: Not applicable.

Informed Consent Statement: Not applicable.

Data Availability Statement: All data are incorporated into the manuscript.

Acknowledgments: Material specimens were examined at High Energy Technologies LLC.

Conflicts of Interest: Author Natalia Shamsutdinova is employed by the company Plasma Tech LLC. The remaining authors declare that the research was conducted in the absence of any commercial or financial relationships that could be construed as potential conflicts of interest.

References

1. Zhu, B.P.; Go, W.K.; Shen, G.Z.; Zhou, Q.; Sung, K.K. Structure and electrical properties of (111)-oriented $Pb(Mg_{1/3}Nb_{2/3})O_3$-$PbZrO_3$-$PbTiO_3$ thin film for ultra-high-frequency transducer applications. *IEEE Trans. Ultrason. Ferroelectr. Freq. Control* **2011**, *58*, 1962–1967. [CrossRef] [PubMed]
2. Zhang, W.; Shima, H.; Takano, F.; Takenaka, M.; Akinaga, H.; Nimori, S. Magnetron sputtering deposition of $Pr_{12.5}Fe_{77.5}B_{10}$ thin films and substrate temperature dependence of the magnetic properties. *J. Phys. D* **2009**, *42*, 025004. [CrossRef]
3. Zhu, B.P.; Fei, C.; Wang, C.; Zhu, Y.; Yang, X.; Zheng, H.; Zhou, Q.; Shung, K.K. Self-Focused AlScN Film Ultrasound Transducer for Individual Cell Manipulation. *ACS Sens.* **2017**, *1*, 172–177. [CrossRef] [PubMed]
4. Zhu, B.P.; Zhang, Z.; Ma, T.; Yang, X.; Li, Y.; Shung, K.K.; Zhou, Q. (100)-Textured KNN-based thick film with enhanced piezoelectric property for intravascular ultrasound imaging. *Appl. Phys. Lett.* **2015**, *106*, 173504. [CrossRef]
5. Zhu, B.P.; Wang, Z.; Zhang, Y.; Yu, Z.S.; Shi, J.; Xiong, R. Low temperature fabrication of the giant dielectric material $CaCu_3Ti_4O_{12}$ by oxalate coprecipitation method. *Mater. Chem. Phys.* **2009**, *113*, 746–748. [CrossRef]
6. Sui, X.; He, W.; Zuo, C.; Chen, Q.; Gu, G. 10.6 μm Infrared light photoinduced insulator-to-metal transition in vanadium dioxide. *Infrared Phys. Technol.* **2014**, *64*, 103–107. [CrossRef]
7. Choyke, W.J.; Matsunami, H.; Pensl, G. *Silicon Carbide: Recent Major Advances*; Springer: Berlin/Heidelberg, Germany, 2004. [CrossRef]
8. La Via, F.; Alquier, D.; Giannazzo, F.; Kimoto, T.; Neudeck, P.; Ou, H.; Roncaglia, A.; Saddow, S.E.; Tudisco, S. Emerging SiC Applications beyond Power Electronic Devices. *Micromachines* **2023**, *14*, 1200. [CrossRef]
9. Nava, F.; Vanni, P.; Bruzzi, M.; Lagomarsino, S.; Sciortino, S.; Wagner, G.; Lanzieri, C. Minimum ionizing and alpha particles detectors based on epitaxial semiconductor silicon carbide. *IEEE Trans. Nucl. Sci.* **2004**, *51*, 238. [CrossRef]
10. Bernat, R.; Capan, I.; Bakrač, L.; Brodar, T.; Makino, T.; Ohshima, T.; Pastuović, Ž.; Sarbutt, A. Response of 4H-SiC Detectors to Ionizing Particles. *Crystals* **2021**, *11*, 10. [CrossRef]
11. Lei, X.; Kane, S.; Cogan, S.; Lorach, H.; Galambos, L.; Huie, P.; Mathieson, K.; Kamins, T.; Harris, J.; Palanker, D. SiC protective coating for photovoltaic retinal prosthesis. *J. Neural Eng.* **2016**, *13*, 13046016. [CrossRef]
12. Zorman, C.A.; Parro, R.J. Micro- and nanomechanical structures for silicon carbide MEMS and NEMS. *Phys. Stat. Sol.* **2008**, *245*, 1404. [CrossRef]
13. Lohrmann, A.; Iwamoto, N.; Bodrog, Z.; Castelletto, S.; Ohshima, T.; Karle, T.J.; Gali, A.; Prawer, S.; McCallum, J.C.; Johnson, B.C. Single-photon emitting diode in silicon carbide. *Nat. Commun.* **2015**, *6*, 7783. [CrossRef] [PubMed]
14. Castelletto, S. Silicon carbide single-photon sources: Challenges and prospects. *Mater. Quantum Technol.* **2021**, *1*, 023001. [CrossRef]
15. Aldalbahi, A.; Li, E.; Rivera, M.; Velazquez, R.; Altalhi, T.; Peng, X.; Feng, P.X. A new approach for fabrications of SiC based photodetectors. *Sci. Rep.* **2016**, *6*, 23457. [CrossRef]
16. Cheung, R. *Silicon Carbide Microelectromechanical Systems for Harsh Environments*; Imperial College Press: London, UK, 2006. [CrossRef]
17. Heuer, A.H.; Fryburg, G.A.; Ogbuji, L.U.; Mitchell, T.E.; Shinozaki, S. Beta-alpha transformation in polycrystalline SiC: 1, microstructural aspects. *J. Am. Ceram. Soc.* **1978**, *61*, 406–412. [CrossRef]
18. Dresch, A.B.; Venturini, J.; Arcaro, S.; Montedo, O.R.K.; Bergmann, C.P. Ballistic ceramics and analysis of their mechanical properties for armour applications: A review. *Ceram. Int.* **2021**, *47*, 8743. [CrossRef]
19. Chen, B.; Niu, L.; Chai, J.; Bai, Z.; Zhang, J.; Liu, D.; Lu, X.; Zhu, Y. Experimental investigation of microstructure and mechanical properties of β-SiC with various sintering additives supplemented by first-principles calculations. *Ceram. Int.* **2025**, in press. [CrossRef]

20. Shin, S.; Kim, M.; Kim, M.; Kim, U.; Kim, S.-G.; Kwak, Y.; Cho, J. Ultrafast high-temperature sintering of reaction-bonded SiC with Y_2O_3-Al_2O_3 sintering additives. *Mater. Lett.* **2025**, *382*, 137956. [CrossRef]
21. Gevorkyan, E.S.; Nerubatskyi, V.P.; Vovk, R.V.; Chyshkala, V.O.; Kislitsa, M.V. Structure Formation in Silicon Carbide–Alumina Composites during Electroconsolidation. *J. Superhard Mater.* **2022**, *44*, 339. [CrossRef]
22. Selvam, J.D.; Dinaharan, I.; Rai, R.S. Matrix and Reinforcement Materials for Metal Matrix Composites. *Encycl. Mater. Compos.* **2021**, *2*, 615–639. [CrossRef]
23. Jiang, D.L.; Wang, J.H.; Li, Y.L.; Ma, L.T. Studies on the strengthening of silicon carbide-based multiphase ceramics I: The SiC TiC system. *Mater. Sci. Eng. A* **1989**, *109*, 401. [CrossRef]
24. Ivzhenko, V.V.; Gevorkyan, E.S.; Kosenchuk, T.O. Sintering and Properties of Materials Synthesized on the Basis of Silicon, Boron, and Titanium Carbides by Electrospark Sintering. *J. Superhard Mater.* **2021**, *43*, 405. [CrossRef]
25. Liu, J.; Li, Y.; Cheng, C.; Li, W.; Wu, W.; Jin, Y. Study on the toughening mechanism of in-situ synthesis $(Ti_xZr_{1-x})B_2$ in solid-state sintered SiC composite ceramics. *J. Eur. Ceram. Sci.* **2023**, *43*, 760–767. [CrossRef]
26. Wan, C.; Xia, P.; Hua, J.; Chen, X.; Lang, W.; Xue, W.; Geng, Y.; Fang, T. Fabrication of SiC composites by selective laser sintering and reactive melt infiltration. *Ceram. Int.* **2025**, in press. [CrossRef]
27. Hao, B.; Fan, L.; Li, K.; Fu, Y.; Guo, X.; Guan, L.; Yang, S.; Zhang, M.; Sun, D.; An, L. Hot oscillatory pressing of SiC ceramics with B_4C and C as sintering additives. *Ceram. Int.* **2024**, *50*, 53628–53634. [CrossRef]
28. Yang, C.; Li, X.; Wei, Y.; Gao, Y.; Liu, M. Effect of He ions irradiation at 650 °C on microstructural evolution, chemical bonding changes and hardening of pressureless solid-state sintered SiC. *Ceram. Int.* **2025**, in press. [CrossRef]
29. Ivzhenko, V.; Vovk, R.; Hevorkian, E.; Kosenchuk, T.; Chyshkala, V.; Nerubatskyi, V.; Cherniavskyi, V.; Shamsutdinova, N. Investigation of Electrospark Sintering of Composites of SiC–TiC, SiC–VC Systems. *Materials* **2025**, *18*, 508. [CrossRef]
30. Liu, W.; Peng, J.; Liu, J.; Li, J.; Liu, B.; Fang, Q. Investigation of the sintering behavior of nanoparticulate SiC by molecular dynamics simulation. *Mater. Today Commun.* **2024**, *41*, 110654. [CrossRef]
31. Lei, P.; Yu, M.; Gucci, F.; Huang, Z.; Fu, R.; Zhang, D. Numerical simulation of heat transfer during spark plasma sintering of porous SiC. *Ceram. Int.* **2024**, *50*, 19620–19630. [CrossRef]
32. Hevorkian, E.S.; Nerubatskyi, V.P.; Rucki, M.; Kilikevicius, A.; Mamalis, A.G.; Samociuk, W.; Morozow, D. Electroconsolidation Method for Fabrication of Fine-Dispersed High-Density Ceramics. *Nanotechnol. Percept.* **2024**, *20*, 100–113.
33. Piotrkiewicz, P.; Zygmuntowicz, J.; Wachowski, M.; Cymerman, K.; Kaszuwara, W.; Midor, W.A. Al_2O_3-Cu-Ni Composites Manufactured via Uniaxial Pressing: Microstructure, Magnetic, and Mechanical Properties. *Materials* **2022**, *15*, 1848. [CrossRef] [PubMed]
34. Latosińska, M.; Latosińska, J.N. Favipiravir Analogues as Inhibitors of SARS-CoV-2 RNA-Dependent RNA Polymerase, Combined Quantum Chemical Modeling, Quantitative Structure–Property Relationship, and Molecular Docking Study. *Molecules* **2024**, *29*, 441. [CrossRef] [PubMed]
35. Latosińska, J.N.; Latosińska, M.; Seliger, J.; Žagar, V.; Apih, T. Butterfly Effect in Cytarabine: Combined NMR-NQR Experiment, Solid-State Computational Modeling, Quantitative Structure-Property Relationships and Molecular Docking Study. *Pharmaceuticals* **2024**, *17*, 445. [CrossRef] [PubMed]
36. Chiu, K.A.; Lin, J.F.; Lin, K.Y.; Wu, P.H.; Chen, H.I.; Ko, C.J.; Chen, C.H.; Chang, L. Epitaxial growth of TiC on (0001) 4H-SiC substrate by reactive sputtering. *Thin Solid Films* **2023**, *775*, 139874. [CrossRef]
37. Kamei, K.; Kusunoki, K.; Yashiro, N.; Okada, N.; Tanaka, T.; Yauchi, A. Solution growth of single crystalline 6H, 4H-SiC using Si–Ti–C melt. *J. Cryst. Growth* **2009**, *311*, 855–858. [CrossRef]
38. Lapin, J.; Klimová, A.; Pelachová, T.; Štamborská, M.; Bajana, O. Synergistic effect of Ti, B, Si, and C on microstructure and mechanical properties of as-cast $Al_{0.4}Co_{0.9}Cr_{1.2}Fe_{0.9}Ni_{1.2}(Si, Ti, C, B)_{0.375}$ complex concentrated alloy. *J. Alloys Compd.* **2023**, *934*, 168050. [CrossRef]
39. Magnus, C.; Sharp, J.; Ma, L.; Rainforth, W.M. Ramification of thermal expansion mismatch and phase transformation in TiC-particulate/SiC-matrix ceramic composite. *Ceram. Int.* **2020**, *46*, 20488–20495. [CrossRef]
40. Ahmoye, G.; Bucevac, D.; Krstic, V.D. Mechanical properties of reaction sintered SiC-TiC composite. *Ceram. Int.* **2018**, *44*, 14401–14407. [CrossRef]
41. Sun, R.; Zhang, X.; Hao, X.; Hu, W.; Wei, X.; Song, X.; Zhang, Z.; Ying, P.; Zhao, S.; Wang, Y.; et al. Simultaneously enhanced toughness and hardness of nanocrystalline SiC sintered under high pressure. *J. Eur. Ceram. Sci.* **2025**, *45*, 116829. [CrossRef]
42. Lou, Z.; Li, Y.; Zou, Q.; Luo, W.; Gu, H.; Li, Z.; Luo, Y. In-situ fabrication and characterization of TiC matrix composite reinforced by SiC and Ti_3SiC_2. *Ceram. Int.* **2023**, *49*, 20849–20859. [CrossRef]
43. Ueki, M.; Endo, H. Mechanical Properties of Particulate Reinforced SiC-based Ceramic Composites. *ISIJ Int.* **1992**, *32*, 943–952. [CrossRef]

44. Cheng, L.X.; Xie, Z.P.; Liu, G.W. Spark plasma sintering of TiC-based composites toughened by submicron SiC particles. *Ceram. Int.* **2013**, *39*, 5077–5082. [CrossRef]
45. Cabrero, J.; Audubert, F.; Pailler, R. Fabrication and characterization of sintered TiC-SiC composites. *Eur. Ceram. Soc.* **2011**, *32*, 313–320. [CrossRef]

materials

MDPI

Article

As-Cast Magnesium Alloys with Ca Addition as a Replacement for Magnesium Alloys with Rare Earth Metals

Tomasz Rzychoń [1] and Agnieszka Fornalczyk [2,*]

[1] Department of Materials Technologies, Faculty of Materials Engineering, Silesian University of Technology, 40-019 Katowice, Poland; tomasz.rzychon@polsl.pl

[2] Department of Production Engineering, Faculty of Materials Engineering, Silesian University of Technology, 40-019 Katowice, Poland

* Correspondence: agnieszka.fornalczyk@polsl.pl

Abstract: This article evaluates the possibility of replacing creep-resistant magnesium Mg-Zn-RE-Zr alloys (EZ33) with Mg-Al-Ca-Sr alloys. (1) Background: Mg alloys with RE metals show excellent properties. Due to their high cost, new, more economical Mg alloys are being developed. Replacing RE metals with cheaper elements such as Al and Ca allows us to obtain high mechanical properties at elevated temperatures due to the tendency to form stable intermetallic phases. (2) Methods: Microstructure analysis (LM, SEM, TEM, and XRD) was performed and mechanical properties were tested at ambient and elevated temperatures. (3) Results: Increasing the Ca content and decreasing the Al content leads to the formation of a continuous skeleton of high-melting and brittle Ca-rich Laves phases and Sr-rich intermetallic phases and the formation of plate-like precipitates of the C15 phase inside the α-Mg solid solution. The crystallographic orientation of plate-like precipitates contributes to the blocking of dislocations in slip systems activated at elevated temperatures. Increasing the Ca and Sr content allows for the regulation of the Al concentration in the α-Mg, providing solution strengthening and stability of the α-Mg solid solution. These factors contribute to a significant improvement in creep resistance of Mg-Al-Ca-Sr alloys. (4) Conclusions: The strength properties and elongation at ambient temperature of the Mg alloys with Ca and Sr addition are comparable to those of the EZ33 alloy, and due to the presence of lighter alloying elements, a better specific strength is achieved. Ca-rich Mg-Al-Ca-Sr alloys exhibit better creep resistance at temperatures of up to 200 °C compared to the EZ33 alloy.

Keywords: magnesium alloys; gravity casting; creep properties; microstructure

1. Introduction

Both the aviation and automotive industries show great interest in as-cast magnesium alloys, which are characterized by low density (about 1.8 g/cm^3), high specific strength, good castability, machinability, weldability in controlled atmospheres, and high vibration damping capacity. The range of applications of magnesium alloys is limited by their low modulus of elasticity, flammability, low plasticity, high susceptibility to electrochemical corrosion, and low creep resistance [1].

As-cast magnesium alloys with aluminum, zinc, and manganese (AZ91, AM50, and AM60) have found the greatest number of industrial applications, especially in the automotive industry [2–6]. A significant disadvantage limiting the wider use of magnesium–aluminum alloys is the low maximum operating temperature under long-term loads, which should not exceed 120 °C. It is due to the presence of the $Mg_{17}Al_{12}$ phase, characterized by unfavorable morphology and low melting point. For this reason, in the last two decades, intensive research has been carried out on cast magnesium alloys intended for operation at temperatures higher than 120 °C, which could be used for automatic transmission housings, automotive engine blocks, or engine intake housings in aviation. Creep resistance is improved by incorporating strontium, silicon, calcium, and rare earth metals into Mg-Al alloys or by developing aluminum-free alloys, e.g., Mg-RE-Zr, Mg-Gd-Nd-Zr, Mg-Y-RE-Zr, Mg-Sc-Mn, and Mg-Ag-Nd-Zr. These alloying elements cause the elimination of the low-melting $Mg_{17}Al_{12}$ phase and the formation of intermetallic phases characterized by high stability of the crystal structure and morphology at temperatures above 120 °C.

According to the current state of knowledge, assuming the criterion of maximum operating temperature, cast magnesium alloys can be divided into three groups [7–15]:

- Common Mg-Al alloys with the addition of zinc or manganese (AZ91, AM50, and AM60), which can be high-pressure- or gravity-cast (operating temperature up to 120 °C);
- Mg-Al-RE (AE44), Mg-Al-Ca (AXM53), Mg-Al-Sr (AJ62), and Mg-Al-Si (AS21) die-cast alloys and Mg-Sr-Mn and Mg-RE-Zn-Zr (EZ33) gravity-cast alloys (operating temperature up to 200 °C);
- Mg-Gd-Nd-Zr (Elektron 21), Mg-Y-RE-Zr (WE43), Mg-Ag-Nd-Zr (QE22), and Mg-Sc-Mn gravity-cast alloys (operating temperature up to 250 °C; alloys containing scandium up to 300 °C).

Among the alloying elements used in magnesium alloys, rare earth metals have the most beneficial effect on creep resistance. However, due to the high cost of rare earth elements, efforts are being made to develop new magnesium alloys that do not contain these elements [16]. Therefore, research is conducted on magnesium alloys with cheap alloying elements, such as Al, Ca, Sr, Si, Mn, Zn, and Sn, in order to obtain Mg alloys with similar mechanical properties at ambient and elevated temperature comparable to those of Mg alloys with rare earth metals [6,17].

An example of a material with good creep resistance up to 200 °C is the EZ33 alloy, which is designed for gravity casting. It has been successfully used in the aviation industry. Meanwhile, it is possible to replace the EZ33 alloy with much cheaper Mg alloys [18–20]. It has already been shown that the addition of Ca to Mg-Al alloys significantly contributes to the improvement in creep resistance of die-cast alloys [1]; however, the influence of this element on the mechanical properties of sand castings is poorly understood.

This article compares the microstructure and mechanical properties of gravity-cast Mg-Al-Ca-Sr alloys with commercial sand-cast EZ33 (Mg-Zn-RE-Zr) magnesium alloys and assesses the possibility of replacing the EZ33 alloy with cheaper magnesium alloys consisting of Al and Ca.

2. Materials and Methods

The magnesium alloy containing Al, Ca, Sr, and Mn was prepared (Table 1), and its composition was analyzed using a SPECTROMAX optical emission spectrometer (AMETEK, Berwyn, IL, USA). Commercially pure Mg, Al, and Mn were used, and Sr and Ca were added in the form of Al-10 wt% Sr and Al-85 wt% Ca master alloys, respectively. Melting of the alloys was performed by induction melting in an alumina crucible under the protection of an argon atmosphere. The melt was maintained at 730 °C for 3 min and then poured into sand molds. Induction melting was also used to melt EZ33 ingots, but casting was performed at a temperature of 780 °C, in accordance with the manufacturer's recommendations.

Table 1. Chemical composition of Mg-Al-Ca-Sr and EZ33 alloys (wt%). The standard deviation is shown in brackets.

Alloy	Al	Ca	Sr	Mn	Si	Fe	Mg
AXJM51	4.98 (0.38)	1.24 (0.15)	0.75 (0.12)	0.19 (0.05)	0.22 (0.06)	0.0013 (0.002)	Balance
AXJM53	5.11 (0.19)	3.01 (0.23)	0.68 (0.22)	0.16 (0.08)	0.18 (0.07)	0.0013 (0.001)	Balance
AXJM71	6.80 (0.32)	1.22 (0.13)	0.70 (0.16)	0.13 (0.09)	0.12 (0.1)	0.0012 (0.0003)	Balance
AXJM73	7.09 (0.34)	3.05 (0.24)	0.61 (0.14)	0.10 (0.07)	0.10 (0.08)	0.0007 (0.001)	Balance
AXJM91	9.01 (0.51)	1.13 (0.24)	0.81 (0.22)	0.19 (0.08)	0.26 (0.07)	0.0011 (0.0011)	Balance
AXJM93	8.67 (0.50)	2.89 (0.28)	0.60 (0.1)	0.16 (0.07)	0.16 (0.05)	0.001 (0.001)	Balance
Alloy	**Zn**	**RE**	**Zr**	**Mn**	**Fe**	**Ni, Cu**	**Mg**
EZ33	2.70	3.18	0.53	0.02	0.002	<0.001	Balance

RE—mischmetal mixture 50Ce-25La-20Nd-5Pr (wt%).

The microstructure of the AXJM series and the EZ33 alloys was analyzed through light microscopy (LM) using an Olympus GX-71 microscope (Olympus, Tokyo, Japan)

Microstructural observations at higher magnification were carried out using an FE SEM Hitachi S-3400N (Hitachi, Tokyo, Japan) scanning electron microscope and Titan 80-300 (FEI, Hilsboro, OR, USA) and Hitachi HD-2300A (Hitachi, Tokyo, Japan) scanning transmission electron microscopes. Chemical composition microanalysis of the intermetallic compounds was performed through energy-dispersive X-ray spectroscopy (EDS) using an EDS Thermo Noran spectrometer (Thermo Fisher Scientific, Waltham, MA, USA). In order to analyze the chemical composition of the α-Mg solid solution and plate-like precipitates of the C15 phase, electron energy-loss spectroscopy (EELS) was performed using a Titan 80-300 (FEI, Hilsboro, OR, USA) microscope.

Thin foils for microscopic observation were prepared using two methods:

- Electrolytic polishing (double-stream) using a Tenupol-5 polisher: The electrolyte used was composed of 5.3 g of lithium chloride, 11.16 g of magnesium perchlorate, 500 mL of methanol, and 100 mL of 2-butoxy-ethanol. Polishing was performed at −45 °C at a voltage of 20 V.
- FIB method using a Quanta 3D 200i device: The following operating parameters were used: resolution in FIB mode at 7 nm at an accelerating voltage of 30 kV, beam current of 1 pA, and magnification range from 50× to 200,000×. SEM imaging was performed with a resolution down to 3 nm at an accelerating voltage of 30 kV in high vacuum (HiVac), 12 nm at an accelerating voltage of 3 kV in low vacuum (LowVac), and 3 nm at an accelerating voltage of 30 kV in environmental mode.

The volume fraction of the phases and grain size was measured through quantitative metallography using METILO 14.1 software (Silesian University of Technology, Katowice, Poland).

XRD patterns were obtained using a Philips X'Pert 3 Powder (Philips, Eindhoven, The Netherlands) diffractometer with a copper anode. Registration was performed by 0.02° stepwise regression for 2θ ranging from 10 to 140°. Phase identification was performed using the ICDD PDF-4+ database (ICDD, Philadelphia, PA, USA). The Rietveld method was used to determine the lattice parameters of the α-Mg phase. The structure models for the α-Mg, $Mg_{17}Sr_2$, Al_4Sr, $Mg_{17}Al_{12}$, Al_2Ca, Mg_2Ca, and $Al_{1.34}CaMg_{0.66}$ phases were taken from the ICDD PDF-4+ database.

Tensile and creep tests were performed on a Zwick/Roell Kappa 50DS electromechanical testing machine (ZwickRoell, Ulm, Germany). The creep tests were carried out under constant load in a temperature range between 180 °C and 250 °C and at stresses between 30 and 90 MPa. The creep test conditions were selected based on the temperature and stress levels experienced during the service life of engine hulls used in helicopters. Samples were heated up in a resistance furnace, and the temperature was measured using thermocouples attached to the sample at three positions of the gauge part to maintain the temperature within 3 °C. The creep strain was measured by extensometers, which were attached directly to the gauge section of the specimens. The initial distance between the extensometer arms was 50 mm. The length of the specimen was 100 mm, the gauge length was 60 mm, and the diameter of the reduced section was 6 mm. Short creep tests (100 h or until rupture) were used to evaluate the creep properties of Mg-Al-Ca-Sr alloys. Samples with the same geometry were also used in tensile tests. The strain rate was 4 mm/min, and a contact extensometer was used to measure the extension of a specimen under load. Force and the displacement of extensometer arms was continuously recorded during the test using testXpert II testing software (ZwickRoell, Ulm, Germany).

3. Results and Discussion

3.1. Microstructure of Mg-Al-Ca-Sr Alloys

The microstructure of gravity-cast Mg-Al-Ca-Sr alloys consists of the α-Mg matrix and intermetallic compounds at the interdendritic regions (Figure 1). Additionally, in alloys consisting of 3% Ca, the fine precipitates inside α-Mg were observed. An increase in the content of alloying elements leads to an increase in the content of intermetallic phases and the formation of continuous interdendritic networks.

The thin lamellar phase $(Mg,Al)_2Ca$ (marked with the letter A in Figure 2) and the thick lamellar phase $(Al,Mg)_2Ca$ (letter B in Figure 2) exhibit a similar chemical composition (Table 2) and are characterized by a hexagonal crystal structure of the C14 and C36 type (Figure 3), respectively. The lattice parameters a_0 of these compounds are comparable, while the lattice parameter c_0 of the C14 phase is almost twice that of the $(Mg,Al)_2Ca$-C36 phase. Therefore, during identification with the use of selected-area electron diffraction on a transmission electron microscope, it is necessary to obtain the appropriate orientation of the thin foil in relation to the incident electron beam, which enables the determination of interplanar distances d_{0001}. From a practical point of view, the identification of the C14 and C36 phases can be carried out for an incident electron beam B = hki0. The results of the electron diffraction analysis for both phases are shown in Figure 3.

Table 2. Results of the chemical composition microanalysis for precipitates in the tested alloys (at.%); EDS-SEM. Standard deviation is given in brackets.

Phase	Mg	Al	Ca	Sr	Mn
$(Al,Mg)_2Ca$ (C36)	10.0 (1.7)	57.6 (5.6)	30.4 (2.8)	2.0 (0.24)	-
$(Mg,Al)_2Ca$ (C14)	17.5 (2.1)	54.2 (3.7)	26.5 (2.3)	1.8 (0.19)	
$(Mg,Al)_{17}Sr_2$	71.8 (3.9)	19.5 (1.8)	2.5 (1.1)	6.2 (0.13)	-
$(Al,Mg)_4Sr$	47.8 (3.9)	41.2 (3.9)	3.2 (1.2)	7.8 (0.25)	-
Al_2Ca (C15)	75.2 (7.8)	16.0 (2.4)	8.8 (1.8)	-	-
$Mg_{17}Al_{12}$	70.7 (3.1)	28.5 (1.9)	0.7 (0.2)	0.1 (0.11)	-

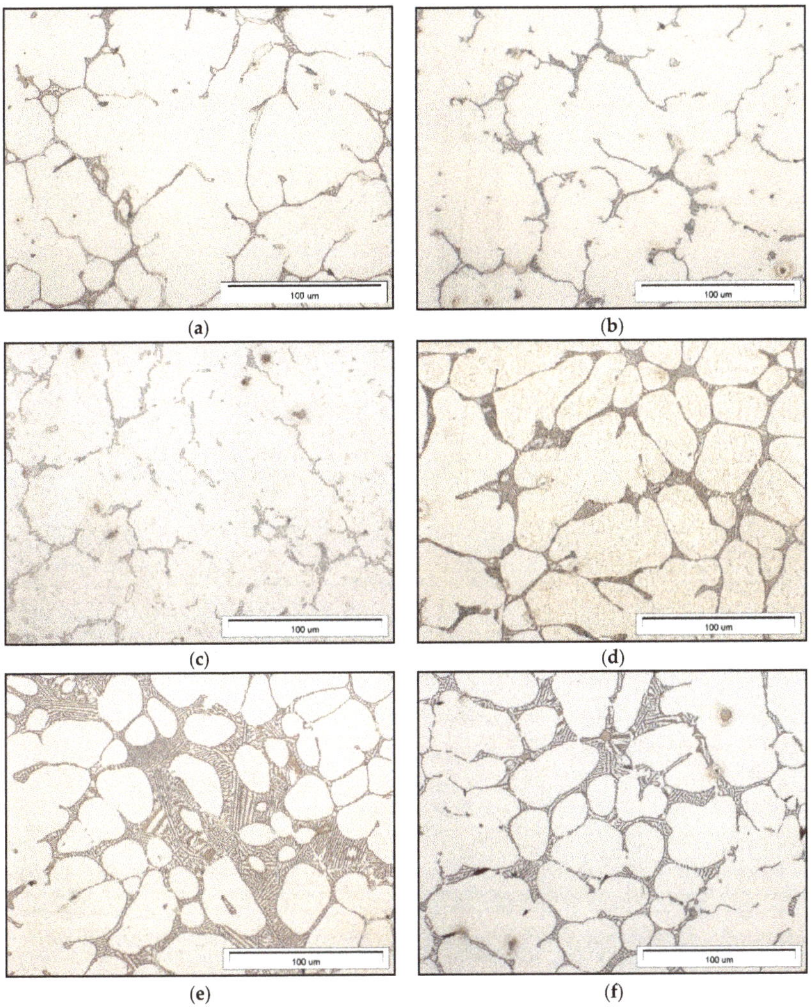

Figure 1. Microstructure of AXJM51 (**a**), AXJM71 (**b**), AXJM91 (**c**), AXJM53 (**d**), AXJM73 (**e**), and AXJM93 (**f**) magnesium alloys.

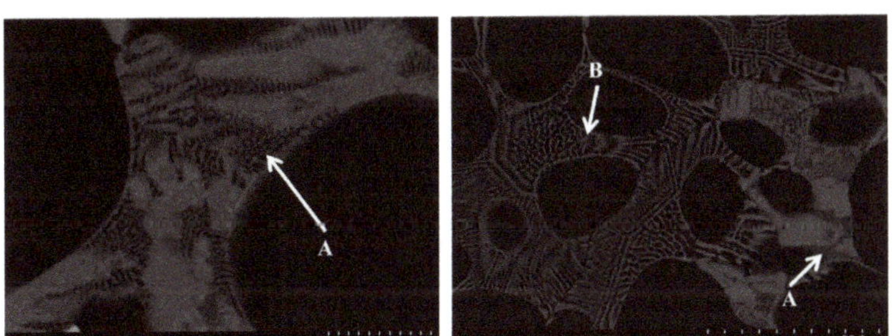

Figure 2. Thin lamellar C14 (A) phase and thick lamellar C36 (B) phase in AXJM53 alloy; SEM.

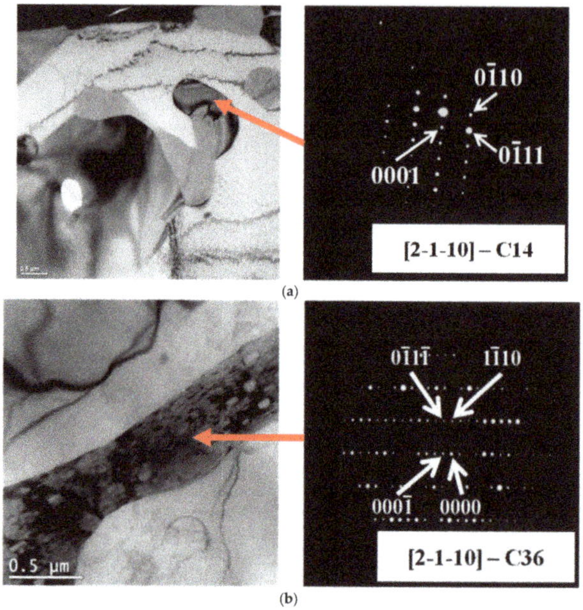

Figure 3. Microscopic images and SAED patterns of C14 phase (**a**) and C15 phase (**b**) obtained by beam parallel to [hki0] axis; B II $[2\bar{1}\bar{1}0]$, TEM, AXJM53 alloy.

In Mg-Al-Ca-Sr alloys containing 1 wt% Ca, the chemical composition of the $(Mg,Al)_2Ca$ phase with a C36 structure is homogeneous, while in the alloys containing 3 wt% Ca, spherical or rectangular Mg-rich clusters were observed inside the C36-$(Mg,Al)_2Ca$ phase (Figure 4). These clusters were also observed in Ref. [21], where the islands were classified as α-Mg solid solution, while in [22], it was shown that Mg-rich clusters have the C36 crystal structure due to the large solubility range of Al and Mg in the C36-$(Mg,Al)_2Ca$ phase. The role of clusters does not seem to have a major impact on the mechanical properties; however, at high temperatures (above 500 °C), they contribute to the formation of cracks inside the C36 phase.

Tiny plate-like precipitates inside the α-Mg solid solution were identified as the Al_2Ca phase with a cubic C15-type crystal structure (Figure 5), which are characterized by a crystallographic relationship with the $\{111\}_{C15}||(0001)_{\alpha\text{-Mg}}$ $[01\bar{1}]_{C15}||[01\bar{1}0]_{\alpha\text{-Mg}}$. It indicates that these precipitates lie in the basic plane of the hexagonal crystal system of the α-Mg solid solution [23,24].

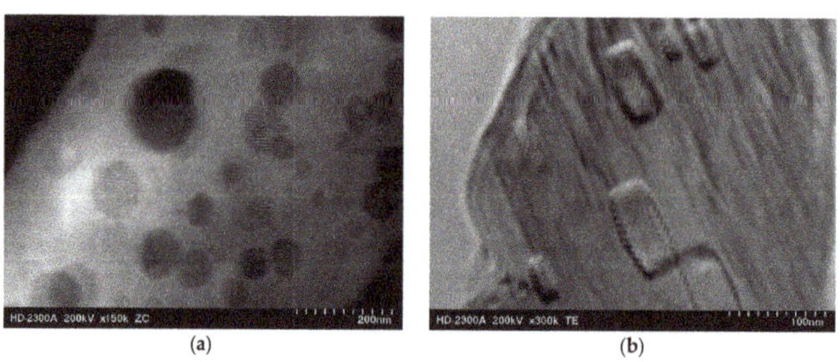

Figure 4. Spherical clusters with moiré fringes observed inside the C36 phase (**a**) and rectangular clusters with moiré fringes observed inside the C36 phase (**b**). STEM, AXJM53 alloy.

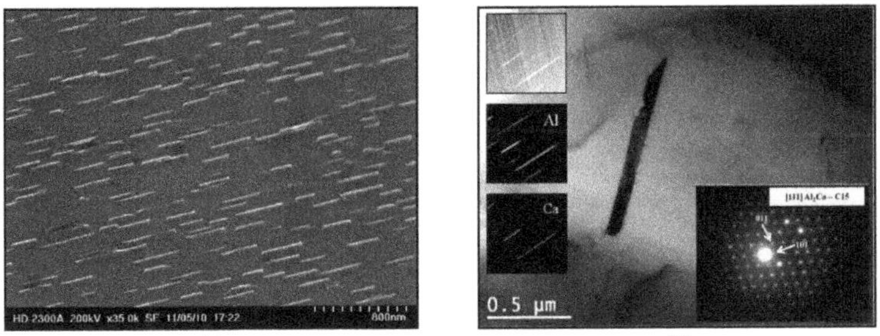

(**a**) (**b**)

Figure 5. Plate-like precipitates of the Al_2Ca (C15) phase inside the α-Mg solid solution; AXJM53 alloy. STEM image (**a**); TEM image with the SAED pattern and surface distribution of elements (**b**).

Minor amounts of Sr dissolve in Ca-rich phases (Table 2) and form two types of intermetallic compounds. Detailed phase identification and microanalysis of the chemical composition showed the presence of the $(Al,Mg)_4Sr$ phase forming a lamellar eutectic with a solid solution of α-Mg (Figure 6), in which Ca can dissolve up to about 3 at.% (Table 2). In addition to the $(Al,Mg)_4Sr$ phase, in Mg-Al-Ca-Sr alloys there is another phase with a high Sr content, which is characterized by an irregular morphology (hereinafter, this phase will be described as the "massive phase") and is identified as the $Mg_{17}Sr_2$ compound (Figure 7).

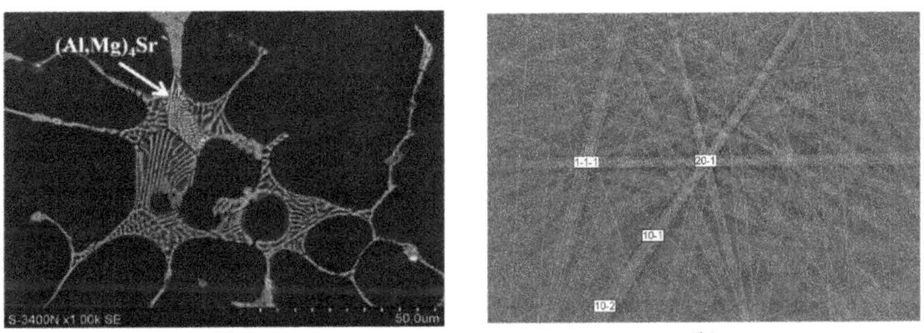

(**a**) (**b**)

Figure 6. $(Al,Mg)_4Sr$ phase in AXJM51 alloy (**a**); indexing of pattern as Al_4Sr phase, MAD = 0.74°, orientation = (158.5; 93.9; 29.4), crystallographic plane $(\bar{5}\bar{3}0)$ (**b**).

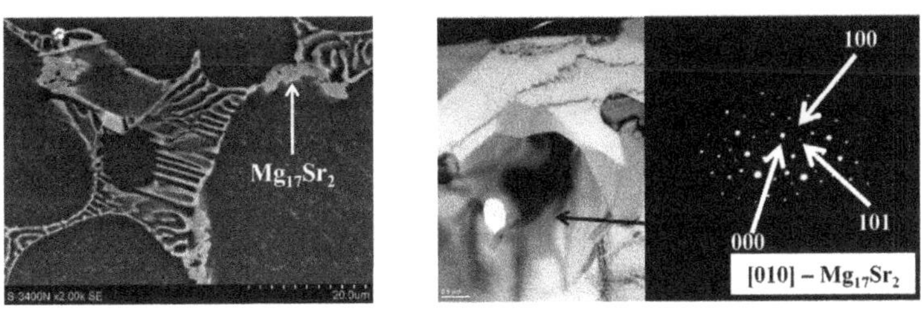

Figure 7. The massive $Mg_{17}Sr_2$ phase in AXJM53 alloy and SAED pattern of this compound (TEM), B II [-12-10].

The $Mg_{17}Al_{12}$ phase, which forms a partially divorced eutectic with the α-Mg solid solution, was also identified (Figure 8). Low amounts of Ca (up to 1 at.%) also dissolve in this intermetallic compound. In terms of creep resistance, the presence of Ca in the crystal structure of the $Mg_{17}Al_{12}$ phase may be beneficial due to the increase in its stability at elevated temperatures [25].

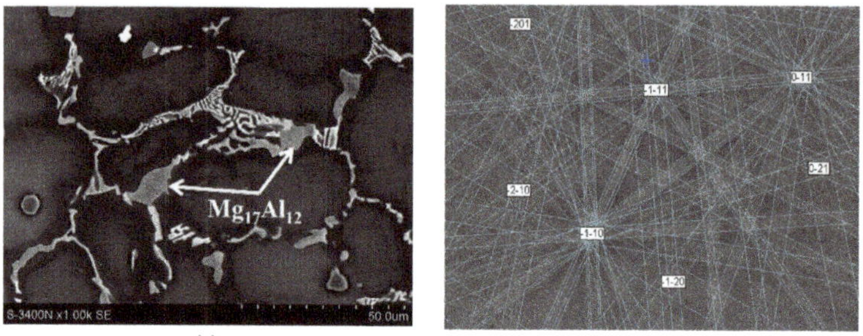

(a) (b)

Figure 8. Microstructure of AXJM91 alloy with $Mg_{17}Al_{12}$ phase and supersaturated areas of α-Mg solid solution (darker areas) (**a**); indexing of pattern as $Mg_{17}Al_{12}$ phase, MAD = 0.41°, orientation = (351.2; 193; 25)°, crystallographic plane $(\bar{6}\,\bar{2}\,1)$ (**b**).

Figure 9 and Table 3 show the influence of the Ca/Al ratio on the volume fraction of intermetallic phases and the content of Al dissolved in the α-Mg solid solution. For a Ca/Al value lower than 0.16, the $Mg_{17}Al_{12}$ phase, the Al-rich α-Mg solid solution, the C36-type $(Al,Mg)_2Ca$ phase, and the $(Al,Mg)_4Sr$ phase are observed in the microstructure. As the Ca/Al ratio increases, the volume fraction of the $Mg_{17}Al_{12}$ phase and the Al content in α-Mg decrease, and the volume fraction of the C36-type phase and the massive phase $(Mg,Al)_{17}Sr_2$ increase. Obviously, this is related to the presence of Ca and Sr, which have a greater chemical affinity to Al than to Mg. Therefore, phases from the Al-Ca and Al-Sr systems are formed first during the solidification of the alloy, then the concentration of aluminum in liquid metal is reduced and the $Mg_{17}Al_{12}$ phase cannot be formed at a lower temperature.

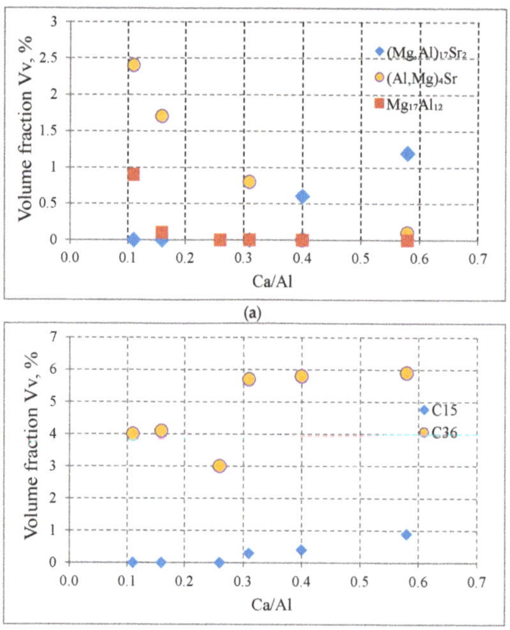

Figure 9. The influence of the Ca/Al ratio on the content of Sr-rich and $Mg_{17}Al_{12}$ phases (**a**) and content of C15 and C36 phases (**b**) in Mg-Al-Ca-Sr alloys.

In all the tested alloys, the dominant intermetallic phase is the C36-$(Al,Mg)_2Ca$ phase (Table 3). It is also worth noting that the increase in the Ca/Al ratio affects the content of Sr-rich phases, causing a decrease in the content of the $(Al,Mg)_4Sr$ phase and an increase in the content of the $(Mg,Al)_{17}(Sr,Ca)_2$ phase due to the lower formation temperature of the massive phase.

In addition to the phase composition, an important role in the strengthening of the tested alloys is also played by the presence of alloying elements in the α-Mg solid solution. Reducing the Ca/Al ratio favors an increase in the content of Al dissolved in α-Mg (Figure 10) due to the lower content of Al, which is consumed by the Ca-rich intermetallic phases formed during crystallization. Although the gravity casting process is characterized by relatively slow cooling rates, the content of Al dissolved in α-Mg exceeds the maximum solubility at ambient temperature (1.1 at.% [26]). In alloys with a low Ca/Al ratio (Mg-9Al-1Ca-0.8Sr), strong Al microsegregation was found in the dendrites of the α-Mg solid solution (Figure 11). The high content of Al (even 9 at.%) in α-Mg favors the precipitation processes of the $Mg_{17}Al_{12}$ phase at elevated temperatures.

Table 3. Volume fraction of intermetallic compounds and grain size of α-Mg solid solution determined by quantitative metallography and Al content in α-Mg solid solution. The standard deviation is given in brackets and, in the case of grain size, the coefficient of variation in %.

Alloy	Ca/Al Ratio	Al Content in α-Mg, at.%	α-Mg Grain Size, μm	Phase Composition	Volume Fraction, %
AXJM51	0.26	1.5 (0.18)	58.1 (19.6)	$(Al,Mg)_2Ca$—C36	3.0 (0.24)
				$(Mg,Al)_{17}Sr_2$	2.0 (0.18)
				$(Al,Mg)_4Sr$	0.1 (0.07)
				Al_8Mn_5	0.1 (0.08)
AXJM53	0.58	1.5 (0.16)	75.2 (34.7)	$(Al,Mg)_2Ca$—C36	5.9 (0.47)
				Al_2Ca—C15	0.9 (0.4)
				$(Mg,Al)_{17}Sr_2$	1.2 (0.11)
				$(Al,Mg)_4Sr$	0.1 (0.03)
				$(Mg,Al)_2Ca$—C14	0.9 (0.05)
				Al_8Mn_5	0.1 (0.02)
AXJM71	0.16	2.9 (0.25)	59.0 (25.0)	$(Al,Mg)_2Ca$—C36	4.1 (0.33)
				$(Al,Mg)_4Sr$	1.7 (0.1)
				Al_8Mn_5	0.1 (0.06)
				$Mg_{17}Al_{12}$	0.1 (0.04)
AXJM73	0.40	2.0 (0.32)	113.7 (68.9)	$(Al,Mg)_2Ca$—C36	5.8 (0.41)
				Al_2Ca—C15	0.4 (0.3)
				$(Mg,Al)_{17}Sr_2$	0.6 (0.04)
				$(Mg,Al)_2Ca$—C14	0.3 (0.04)
				Al_8Mn_5	0.1 (0.01)
AXJM91	0.11	3.7 (0.55)	66.2 (25.9)	$(Al,Mg)_2Ca$—C36	4.0 (0.28)
				$(Al,Mg)_4Sr$	1.9 (0.12)
				Al_8Mn_5	0.1 (0.02)
				$Mg_{17}Al_{12}$	0.9 (0.16)
AXJM93	0.31	2.4 (0.17)	77.6 (28.9)	$(Al,Mg)_2Ca$—C36	5.7 (0.45)
				Al_2Ca—C15	0.3 (0.1)
				$(Al,Mg)_4Sr$	0.8 (0.1)
				Al_8Mn_5	0.1 (0.06)

Both Ca and Sr exhibit very low maximum solubility in magnesium, 0.3 at.% and 0.5 at.%, respectively [27–30], which additionally decreases with lowering temperature. The results of the microanalysis of the chemical composition using an electron beam with a diameter of several nanometers (EELS and TEM) show that in the AXJM53, AXJM73, and AXJM93 alloys, Ca not only participates in the formation of intermetallic phases but also dissolves in the α-Mg solid solution. Confirmation of this can also be found in the changing lattice parameters of the α-Mg solid solution (Figure 12). In alloys containing 1 wt% Ca, the lattice parameter a_0 correlates quite well with Vegard's rule (blue line) for binary Mg-Al alloys. In the case of alloys with 3 wt% Ca, the lattice parameter a_0 is larger, which is related to the expansion of the crystal lattice caused by the presence of Ca (as well as Sr) atoms with a larger atomic radius (Ca—197.4 pm; Sr—215.1 pm) than Mg atoms (159.9 pm).

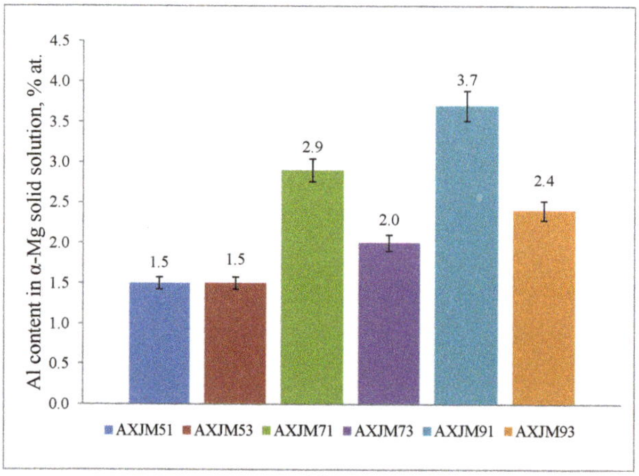

(**a**)

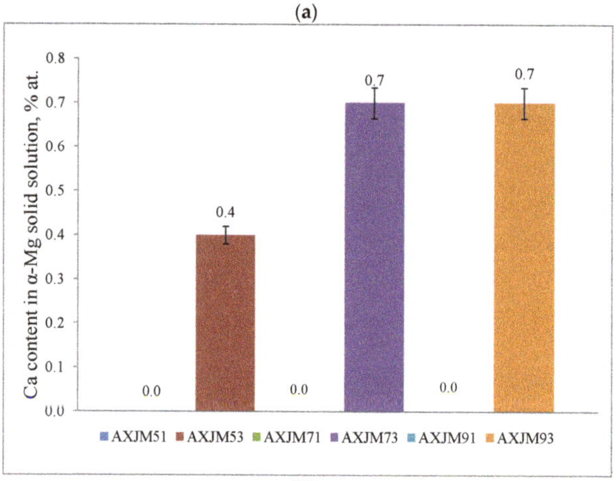

(**b**)

Figure 10. Al content (**a**) and Ca content (**b**) in the α-Mg solid solution of the tested alloys. Results regarding chemical composition in the center of α-Mg grains.

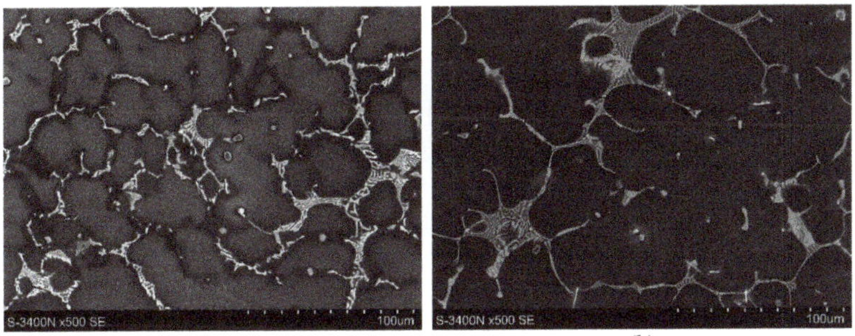

(**a**) (**b**)

Figure 11. Microstructure of AXJM91 alloy with areas (darker areas) of α-Mg solid solution consisting of up to 9 at.%. Inside dendrites, the average Al content is 3.7 at.% (**a**). AXJM51 alloy, in which the strong microsegregation of Al in α-Mg was not observed (**b**).

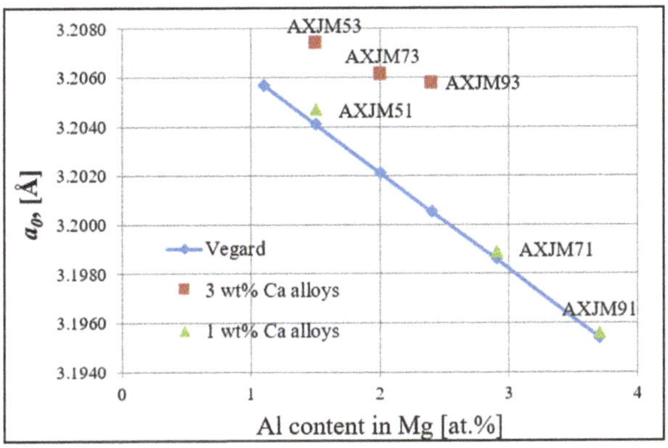

Figure 12. Results of the measurements of the crystal lattice parameters of the α-Mg matrix as a function of the aluminum content dissolved in the α-Mg solid solution (the graph shows only the results for the a_0 parameter; the nature of changes for the c_0 parameter is similar).

It seems that increasing the Ca in Mg-Al-Ca-Sr alloy content contributes to the obtainment of a coarse-grained structure (Figure 13, Table 3), although the results are not conclusive, and for some batches of castings, such a trend was not observed. There is also no clear opinion in publications regarding the influence of Ca on α-Mg grain size [31–38]. The discrepancies concern both the effect on grain size and the critical Ca content causing the formation of a fine-grained structure. In [33–35,39], it was found that Ca strongly refines the grain, and in [36,37,40], significant effects were not reported. Grain refinement caused by the addition of Ca is the result of heterogeneous nucleation on particles of the Al_2Ca (C15) phase [31], as well as the result of microsegregation of alloying elements inside the α-Mg solid solution grains [38]. The critical content of Ca needed to obtain the maximum effect of fine α-Mg grains may vary within the range of 0.4–1 wt%. It is found that the most favorable results occur at 0.4 wt% [38–41] or 1 wt% [42], and a further increase in the Ca content does not affect the grain size of the α-Mg solid solution. On the other hand, exceeding the critical concentration of Ca results in the formation of a coarse-grained structure; however, the literature data differ among studies. It has been noted that significant grain growth will occur when the Ca content in the Mg-Al alloys exceeds 0.8% wt% [31], 2.0 wt% [43], or 4.0 wt% [44]. The reasons for the differing results are primarily the metallurgical purity of the alloys, as well as methodological errors consisting in the incorrect interpretation of the grain diameter and the diameter of dendritic cells. Therefore, the method of preparing metallographic specimens in order to reveal grain boundaries is important, which requires proper etching and observation using polarized light in light microscopy or the EBSD method. It seems that, in the tested alloys, the content of Al dissolved in the α-Mg solid solution will have a large impact. In [41], it was shown that the increase in the content of Al dissolved in a solid solution strongly affects the refinement of the microstructure, which results from the high tendency of this element to undergo microsegregation. Another factor conducive to the formation of a coarse-grained microstructure may be the formation of the $(Al,Mg)_2Ca$ phase of the C36 type instead of the Al_2Ca phase of the C15 type, which in alloys with a low Ca content (below 0.5% Ca) is a heterogeneous crystallization nucleus.

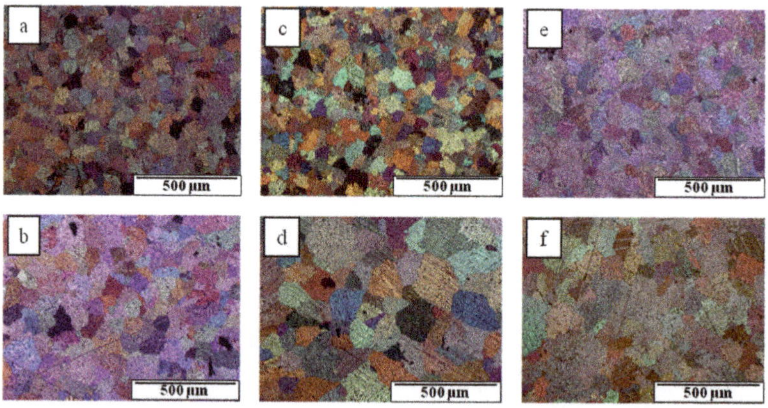

Figure 13. Microstructures with revealed grains of α-Mg solid solution in AXJM51 (**a**), AXJM53 (**b**), AXJM71 (**c**), AXJM73 (**d**), AXJM91 (**e**), and AXJM93 (**f**) alloys.

3.2. Mechanical Properties of Mg-Al-Ca-Sr Alloys

The mechanical properties of the Mg-Al-Ca-Sr alloys, as well as those of the reference EZ33 alloy casted into sand molds, are presented in Table 4. AXJM71 and AXJM91 alloys show the best strength properties at ambient temperature. Alloys containing 3 wt% Ca have a higher hardness compared to the EZ33 alloy, as well as higher tensile strength and comparable ductility.

Table 4. Test results of mechanical properties of Mg-Al-Ca-Sr and EZ33 alloys at ambient temperature and 180 °C. Standard deviation given in parentheses.

Alloy	HV2	Hardness of α-Mg HV0.025	21 °C			180 °C		
			R_m, MPa	$R_{p0.2}$, MPa	A_5, %	R_m, MPa	$R_{p0.2}$, MPa	A_5, %
AXJM51	49(6)	45(1)	138(6)	101(5)	1.9(0.2)	107(4)	71(4)	9.8(1.2)
AXJM53	58(1)	50(2)	135(4)	104(3)	2.1(0.3)	123(8)	84(7)	3.2(0.6)
AXJM71	50(1)	53(2)	174(6)	113(5)	3.1(0.3)	119(7)	77(5)	5.5(0.7)
AXJM73	60(6)	52(2)	137(5)	105(4)	2.1(0.2)	119(3)	88(5)	2.9(0.4)
AXJM91	56(2)	56(1)	165(6)	109(6)	2.9(0.4)	102(2)	78(6)	12.5(1.8)
AXJM93	60(7)	54(1)	149(7)	102(4)	2.5(0.3)	115(7)	81(3)	3.9(0.9)

In Mg-Al-Ca-Sr alloys with an Al content above 7 wt%, an increase in the Ca content from 1 to 3 wt% reduces the tensile strength due to grain growth, an increase in the volume fraction of brittle intermetallic phases, and a decrease in the aluminum content in the α-Mg solid solution. In alloys containing 5 wt% Al, increasing the calcium content does not affect the tensile strength. The increase in the volume fraction of the brittle, eutectic phase and grain growth is compensated by the presence of plate-like Al_2Ca precipitates inside the grains of the α-Mg solid solution.

Based on the microhardness measurements of the α-Mg solid solution (Table 4, HV0.025), the effectiveness of solution and precipitation hardening with the Al_2Ca (C15) lamellar phase was determined. In Mg alloys containing 5 wt% Al (AXJM51 and AXJM53), the Al content in α-Mg is about 1.5 at.%. Increasing the calcium content from 1 wt% to 3 wt% increases the microhardness of the α-Mg solid solution by about 11%. This is due to the appearance of plate-like precipitates of the C15 phase. In Mg alloys containing 1 wt% Ca, plate-like precipitates in the α-Mg solid solution were not found, while the Al content in α-Mg increased from 1.5 at.%. (Mg-5Al-1Ca-0.8Sr) up to 3.7% at. (Mg-9Al-1Ca-0.8Sr). This increases the hardness of the solid solution by more than 20%. It can therefore be concluded that, in the Mg-Al-Ca-Sr alloys, Al dissolved in α-Mg strengthens the alloy more effectively than the lamellar precipitates of the Al_2Ca (C15) phase. The slight influence of the plate-like precipitates of the Al_2Ca phase on the strengthening of the Mg-Al-Ca-Sr alloys is related to its low volume fraction and the crystallographic relationship of the $\{111\}_{C15}||(0001)_{\alpha\text{-Mg}}$ $[01\bar{1}]_{C15}||[0\bar{1}10]_{\alpha\text{-Mg}}$. Such a relationship causes the plate-like precipitates lying in the basic plane of the hexagonal lattice of the α-Mg solid solution not to be effective in blocking the movement of dislocations in the primary slip system of the A3 crystal lattice [45].

3.3. Creep Properties of Mg-Al-Ca-Sr Alloys

The potential application of Mg-Al-Ca-Sr alloys includes loaded elements of aircraft engines that can operate up to a temperature of about 180 °C. For this reason, creep resistance testing is of great importance. The creep curves of the AXJM alloys obtained at stresses of 60 MPa and 75 MPa at 180 °C are shown in Figure 14 (all results obtained in the creep tests are presented in Table 5). Two stages of creep can be distinguished for all curves: primary creep, in which the creep rate decreases, and steady-state creep, in which the strengthening and recovery processes are balanced. At a stress of 90 MPa, the third stage of creep is also observed, characterized by the formation of microcracks at the grain boundaries and at the interface phase and their rapid growth, which leads to the cracking of samples.

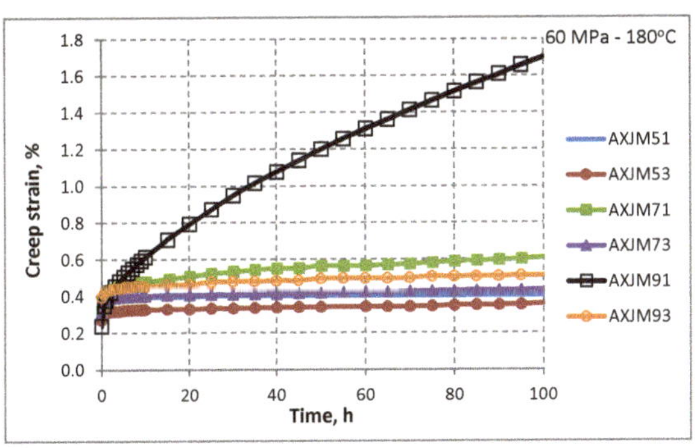

Figure 14. Creep curves of the Mg-Al-Ca-Sr alloys obtained at a temperature of 180 °C and at a stress of 60 MPa.

To assess the creep resistance of magnesium alloys determined in short-term creep tests (100–120 h), it is assumed that satisfactory creep resistance is obtained when the creep rate is in the order of 10^{-9} 1/s and the creep strain does not exceed 1% [5]. It should be emphasized that the steady-state creep rate is the most important feature characterizing the material's resistance to creep and enables the design of elements with the desired durability. Obviously, in industrial conditions, these values may vary depending on the use of these alloys.

Ca has a beneficial effect on creep resistance of sand-cast Mg-Al alloys, similar to that in die-cast alloys [5]. Alloys consisting of 3 wt% Ca show very good creep resistance (low creep rates and creep strain) up to 200 °C and at a stress not exceeding 75 MPa. The improvement in creep resistance is also visible in alloys containing 1 wt% Ca; however, it is not sufficient for industrial applications.

Table 5. Creep test results of Mg-Al-Ca-Sr alloys.

Alloy	Stress, MPa	Temp., °C	Time Test, h	Instant Strain ε_n, %	Creep Strain ε, %	Creep Rate $\dot{\varepsilon}$, 1/s
AXJM53	45	180	100	0.15	0.28	1.1×10^{-10}
AXJM73	45	180	100	0.16	0.27	9.0×10^{-11}
AXJM91	45	180	100	0.11	0.92	8.5×10^{-9}
AXJM51	60	180	100	0.57	0.94	2.7×10^{-9}
AXJM53	60	180	100	0.31	0.36	1.0×10^{-9}
AXJM71	60	180	100	0.32	0.61	3.4×10^{-9}
AXJM73	60	180	100	0.31	0.43	1.3×10^{-9}
AXJM91	60	180	100	0.24	1.7	3.1×10^{-8}
AXJM93	60	180	100	0.39	0.51	1.5×10^{-9}
AXJM51	75	180	100	1.03	3.56	4.0×10^{-8}
AXJM53	75	180	100	0.50	1.38	9.9×10^{-9}
AXJM71	75	180	100	0.61	5.59	1.0×10^{-7}
AXJM73	75	180	100	0.54	1.15	7.9×10^{-9}
AXJM91	75	180	100	0.44	6.1	1.3×10^{-7}
AXJM93	75	180	100	0.31	1.93	3.3×10^{-8}
AXJM53	90	180	16	1.37	5.17	4.9×10^{-7}
AXJM71	90	180	23	1.51	12.46	9.1×10^{-7}
AXJM73	90	180	31	1.47	6.1	8.4×10^{-8}
AXJM91	90	180	9	1.2	7.2	1.3×10^{-5}
AXJM51	60	200	100	0.63	1.27	1.03×10^{-8}
AXJM53	60	200	100	0.33	0.44	2.78×10^{-9}
AXJM71	60	200	100	0.34	0.98	1.43×10^{-8}
AXJM73	60	200	100	0.32	0.54	3.61×10^{-9}
AXJM91	60	200	100	0.28	6.22	1.61×10^{-7}
AXJM93	60	200	100	0.43	0.64	4.43×10^{-9}

For most metallic materials, at a creep temperature above approximately $0.3T_M$ (T_M—melting point), thermally activated processes (climbing and sliding dislocations and diffusion) are initiated. In the case of the tested alloys, the temperature of the creep tests was close to $0.3T_M$ (the solidus temperature of the tested alloys is in the range of 429–513 °C, and the liquidus temperature is in the range of 596–613 °C). Therefore, it was expected that Mg-Al-Ca-Sr alloys would strain-harden due to the increase in dislocation density and that the creep rate would decrease. However, the results of the creep tests (Figure 14) indicate the occurrence of steady-state creep, in which the dislocation density growth processes are balanced by the recovery processes. Therefore, the power law of creep and the Arrhenius Equation (1) were used to describe the research results in order to determine the dominant creep mechanisms [5]:

$$\dot{\varepsilon} = A\sigma^{n} \exp\left(\frac{-Q_c}{RT}\right) \quad (1)$$

where A is the constant, R is the gas constant, σ is the creep stress, n is the stress exponent, Q_c is the activation energy, and T is the temperature.

Based on the knowledge of the n and Q_c parameters, the dominant mechanisms of deformation during the creep process can be identified. For the AXJM91 alloy (Figure 15) in the stress range of 45–75 MPa at 180 °C, the power exponent n is 5.3, indicating the dislocation climbing as the main strain mechanism. Due to the large grain size of the α-Mg solid solution in the AXJM91 alloy (d ≈ 47 μm), the contribution of grain boundary slip to the total deformation should be insignificant. For alloys with a higher Ca content (AXJM53 and AXJM73), the power exponent n values are 8.1 and 8.7, which may be related to the presence of fine precipitates of the Al_2Ca phase inside the grains of the α-Mg solution. These n values also indicate the dominant role of dislocation climbing in the creep deformation in the range of 45–75 MPa and at a temperature of 180 °C. It has often been demonstrated that, for magnesium alloys containing dispersed precipitates within grains, the values of the exponent n are in the range of 7–9 [5,46]. The presence of stable, dispersed phase particles of high hardness, constituting obstacles limiting the movement of the dislocations, and the dislocation can overcome these obstacles by climbing over long distances.

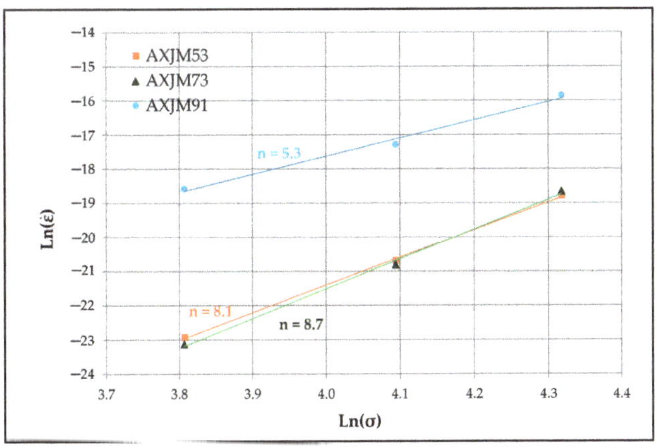

Figure 15. Relationship between creep rate and stress for selected Mg-Al-Ca-Sr alloys at 180 °C and in the stress range of 45–90 MPa.

In alloys in which a higher value of the n exponent was observed, a greater increase in dislocation density Δρ after 100 h of creep at 180 °C was also found (Table 6). It may be due to the presence of plate-like Al_2Ca precipitates, which, by blocking the movement of dislocations, generate new dislocations according to the Frank–Reed mechanism.

The activation energy Q_c is a useful indicator that allows us to determine the creep rate controlling mechanisms. Depending on temperature and stress in metallic materials, dislocation creep (0.3Tt < T < 0.6Tt) and diffusion creep (T > 0.6Tt) can occur [47]. The results of many studies on the creep of magnesium and its alloys [47–54] have allowed the determination of values of creep activation energy for diffusion and dislocation processes. For diffusion processes, the values of Q_c are as follows: Al diffusion in Mg ($Q_c \approx$ 142–147 kJ/mol), lattice diffusion ($Q_c \approx$ 135 kJ/mol), grain boundary diffusion ($Q_c \approx$ 60–80 kJ/mol), pipe diffusion ($Q_c \approx$ 92 kJ/mol), and discontinuous precipitation of the $Mg_{17}Al_{12}$ phase ($Q_c \approx$ 22–30 kJ/mol). For dislocation processes, the values of Q_c are as follows: dislocation climbing ($Q_c \approx$ 135 kJ/mol) and transverse slip ($Q_c \approx$ 200 kJ/mol). For grain boundary slip, Q_c values are in the range of 40–60 kJ/mol.

Table 6. Results of dislocation density measurements using the modified Warren–Averbach method (XRD) of the tested alloys in the as-cast condition and after creep at a temperature of 180 °C and a stress of 60 MPa. The standard deviation is given in brackets.

Alloy	Alloy State	Dislocation Density ρ [1/m²]	Δρ [%]
AXJM53	Sand-cast	2.9×10^{13} (1.2×10^{12})	21
	60 MPa/180 °C	3.5×10^{13} (1.9×10^{12})	
AXJM73	Sand-cast	3.4×10^{13} (2.1×10^{12})	18
	60 MPa/180 °C	4.0×10^{13} (1.7×10^{12})	
AXJM91	Sand-cast	9.7×10^{13} (2.6×10^{12})	3
	60 MPa/180 °C	1.0×10^{14} (3.7×10^{12})	

In alloys containing 3% Ca, the activation energy for steady-state creep is in the range of 91–96 kJ/mol (Figure 16), which suggests that plastic deformation during creep is controlled by pipe diffusion. Higher values of the activation energy Q_c were obtained for alloys containing AXJM51 and AXJM71 (119–128 kJ/mol), indicating that the creep deformation is mainly controlled by the dislocation climbing. In the case of the AXJM91 alloy (Q_c = 147 kJ/mol), the Al diffusion in the α-Mg solid solution or the transverse slip of the edge dislocations makes a greater contribution to the creep deformation.

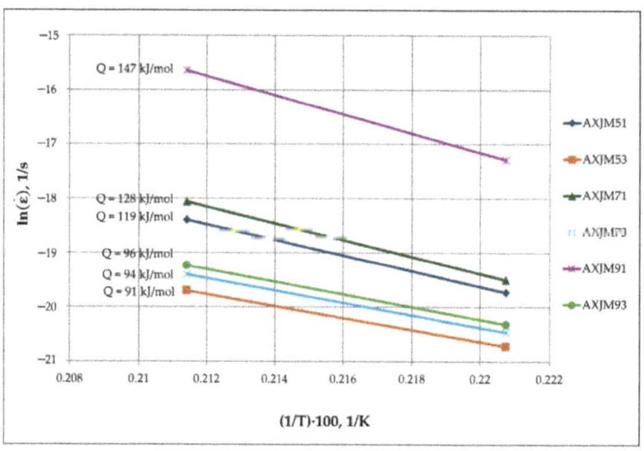

Figure 16. The relationship between the creep rate and the temperature of the tested alloys of Mg-Al-Ca-Sr at a stress of 60 MPa in the temperature range of 180–200 °C.

In magnesium alloys, especially die-cast alloys, grain boundary slip also plays an important role in the creep conditions. In studies on die-cast Mg-Al-Ca-Sr alloys [55], i.e., with fine-grained structure (grain diameter d ≈ 10 μm), it was shown that, at a temperature of 175 °C in the stress range of 40–70 MPa, creep is controlled by grain boundary slipping. In the case of Mg-Al-Ca-Sr alloys after sand casting, the obtained results indicate that grain boundary slip plays a rather marginal role. This may be related to a much larger grain size (d ≈ 40–60 μm) of sand-cast alloys and thus a smaller surface of grain boundaries. Moreover, a continuous skeleton of hard and stable intermetallic compounds will contribute to reduction in the grain boundary creep.

3.4. Influence of Microstructure on Creep Properties of Mg-Al-Ca-Sr Alloys

Observations conducted using a transmission electron microscope were used to understand the deformation behavior of the alloys at elevated temperatures. The tested alloys in the as-cast state are characterized by high dislocation density ($\rho \approx 10^{-13}$ [1/m^2] (Table 6) and the presence of stacking faults (Figure 17). The tendency to form a cellular dislocation structure is slight in magnesium alloys due to low stacking fault energy (SFE); therefore, after creep, the split dislocations are mainly visible in the microstructure, indicating the cross-slip mechanism that occurs when a dislocation changes its slip plane to a different crystallographic plane (Figure 18). This mechanism of dislocation movement is observed in metals when the creep temperature is lower than 0.5Tt and the value of the creep activation energy Q_c is lower than the value of the self-diffusion activation energy (for Mg, Q_{SD} = 135 kJ/mol) [47].

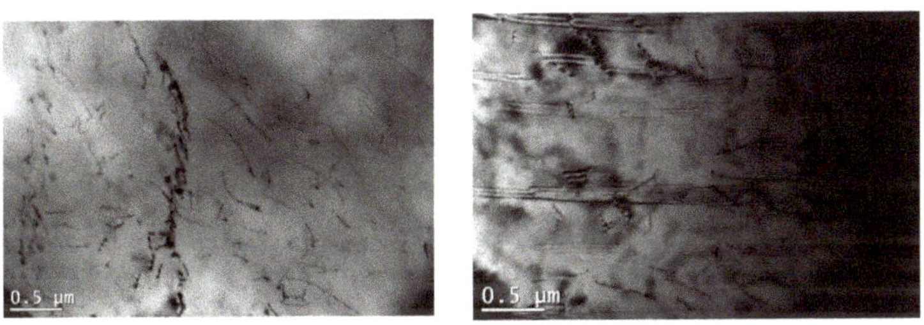

(**a**) (**b**)

Figure 17. Dislocations (**a**) and stacking faults (**b**) in sand-cast AXJM53 alloy.

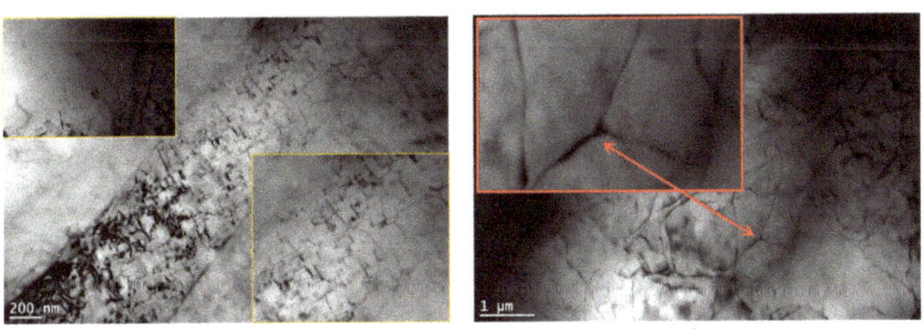

(**a**) (**b**)

Figure 18. Dislocation substructure in AXJM91 alloy (**a**) and split dislocations in AXJM53 alloy (**b**) suggest the occurrence of cross-slip after creep at 180 °C and at 60 MPa for 100 h.

Dislocations moving both in the basic slip system $\langle 11\bar{2}0\rangle$ (0001) and in the pyramidal slip system $\langle 11\bar{2}0\rangle$ {10-11} can be observed in samples crept at 180 °C (Figure 19), indicating the significant contribution of edge dislocation climbing to the creep deformation of alloys containing 1% Ca. In alloys with a higher calcium content (3% Ca), dislocation climbing is also observed, but the elongation of these alloys in creep tests is clearly lower compared to Ca-poor alloys (Table 5). According to the activation energy calculations, the pipe diffusion dominates in the overall deformation, contributing to dislocation climbing.

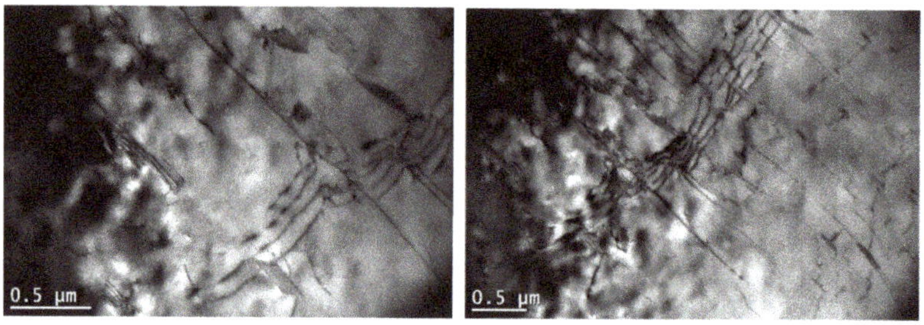

Figure 19. AXJM71 alloy with visible <a> and <a + c> dislocations that indicate climb and cross slip of dislocations in AXJM71 alloy after creep at 180 °C and at 60 MPa for 100 h.

A significant influence on the creep resistance of Mg-Al-Ca-Sr alloys containing 3% Ca is exerted by plate-like Al_2Ca-C15 precipitates located in the basic crystallographic planes (0001) of the α-Mg matrix. Their effect on blocking the movement of dislocations in the slip plane is not significant at ambient temperature, but at elevated temperatures, where additional slip systems are activated and the movement of dislocations is possible in prismatic and pyramidal planes, effects (avoiding precipitates and cross slip) indicating blocking of dislocation movement by C15 precipitates are clearly observed (Figure 20).

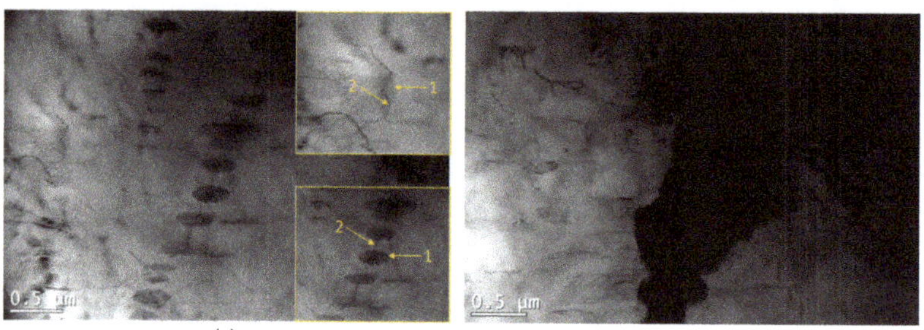

(a) (b)

Figure 20. Blocking of the movement of dislocation motion plate-shaped precipitates of the Al_2Ca (C15) phase during creep at 180 °C at 60 MPa in AXJM53 (**a**) and accumulation of dislocations in the vicinity of $(Al,Mg)_2Ca$ (C36) in the AXJM53 alloy after creep at 180 °C and at 60 MPa for 100 h (**b**). 1—precipitate, 2—bent dislocation.

In the vicinity of intermetallic phases forming eutectics (C36, C14, $(Al,Mg)_4Sr$, and $(Mg,Al)_{17}Sr_2$) characterized by high melting point and high hardness, dislocation accumulations were found (Figure 20b). At ambient temperature, a large accumulation of dislocations favors the nucleation of cracks at the interface, which results in low tensile strength. At temperatures above 150 °C, dynamic recrystallization may occur in these areas, and crack formation may be delayed.

Observations of the microstructure of the tested alloys after creep tests at a temperature of 180 °C and a stress range of 60–75 MPa showed no significant changes in the morphology of the phases $(Al,Mg)_2Ca$-type C36, $(Mg,Al)_{17}Sr_2$, and $(Al,Mg)_4Sr$ forming eutectics with α-Mg solid solution (Figure 21). Analysis of the phase composition using methods with the use of electron diffraction (TEM) and X-ray diffraction (XRD) exhibited no changes in the phase composition of the crept alloys. This indicates the high stability of the intermetallic phases $(Al,Mg)_2Ca$-type C36, $(Mg,Al)_{17}Sr_2$, and $(Al,Mg)_4Sr$ at a temperature of 180 °C. The high thermal stability of the intermetallic phases forming eutectics was also confirmed after long-term annealing (1000 h) at 200 °C.

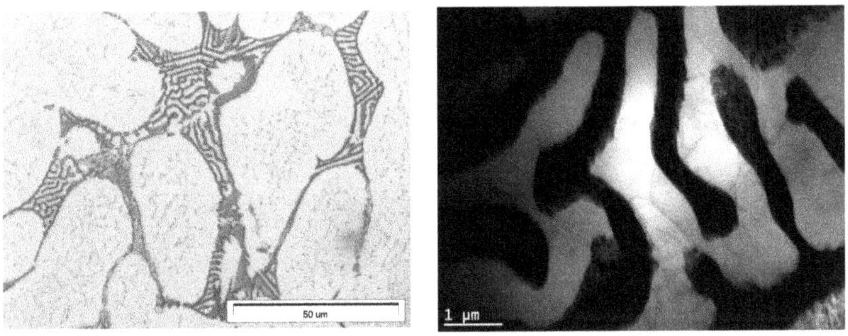

(**a**) (**b**)

Figure 21. Microstructure of AXJM73 alloy after creep at 180 °C and at 75 MPa for 100 h (**a**); microstructure of AXJM53 alloy after creep at 180 °C and at 60 MPa for 100 h (**b**).

While eutectic phases are stable, microstructural changes occur inside the grains of the α-Mg solid solution. After creep tests were performed on the AXJM53 alloy at a temperature of 180 °C for 100 h, a higher volume fraction of lamellar Al_2Ca phase precipitations (V_V = 1.4%) was found inside the grains of the α-Mg solid solution compared to the as-cast alloy (V_V = 0.9%) (Table 7) due to the precipitation process of the plate-like Al_2Ca (C15) phase from the α-Mg solid solution. Moreover, the initial stages of spheroidization of the C15 precipitates are visible (Figure 22), which will contribute to faster destruction of alloys under creep conditions. Precipitation of this phase was also observed after creep at 180 °C in other Mg-Al-Ca-Sr alloys containing 3 wt% Ca.

Table 7. Volume fraction of phases in Mg-Al-Ca-Sr alloys after creep at 180 °C and 60 MPa stress. Standard deviation given in brackets.

Alloy	State	$(Al,Mg)_2Ca$—C36	Al_2Ca—C15	$Mg_{17}Al_{12}$	$(Mg,Al)_{17}Sr_2$	$(Al,Mg)_4Sr$
AXJM51	As-cast	3.0 (0.24)	-	-	2.0 (0.18)	0.1 (0.07)
	60 MPa/180 °C	3.2 (0.36)	-	-	2.1 (0.45)	0.1 (0.10)
AXJM53	As-cast	5.9 (0.47)	0.9 (0.5)	-	1.2 (0.11)	0.1 (0.03)
	60 MPa/180 °C	5.7 (0.89)	1.4 (0.5)	-	1.4 (0.11)	0.12 (0.1)
AXJM71	As-cast	4.1 (0.33)	-	0.1 (0.04)	-	1.7 (0.1)
	60 MPa/180 °C	4.2 (0.94)	-	0.4 (0.08)	-	1.7 (0.3)
AXJM73	As-cast	5.8 (0.41)	0.4 (0.3)	-	0.6 (0.04)	-
	60 MPa/180 °C	6.0 (0.92)	0.9 (0.3)		0.8 (0.15)	
AXJM91	As-cast	4.0 (0.28)	-	0.9 (0.16)	-	1.9 (0.12)
	60 MPa/180 °C	3.8 (0.38)	-	1.9 (0.17)	-	1.8 (0.25)
AXJM93	As-cast	5.7 (0.45)	0.3 (0.1)	-	-	0.8 (0.1)
	60 MPa/180 °C	5.9 (0.28)	0.7 (0.3)	-	-	0.9 (0.4)

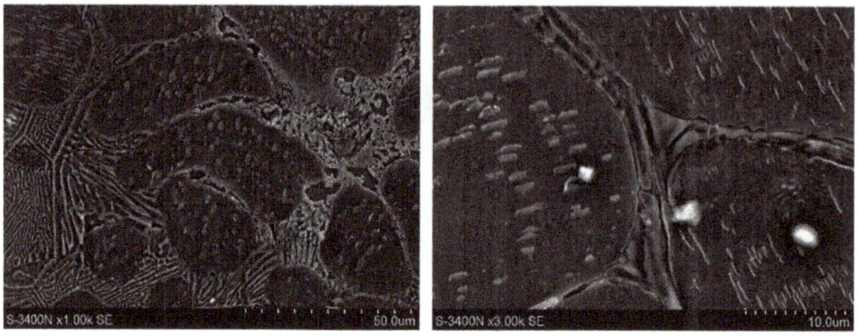

Figure 22. Microstructure of AXJM53 alloy after creep at 180 °C and at 75 MPa for 100 h after strong etching in 3% HNO_3 solution in alcohol.

In the AXJM91 and AXJM71 alloys crept at a temperature of 180 °C, plate-like precipitates of the $Mg_{17}Al_{12}$ phase were observed in the supersaturated areas of the α-Mg solid solution surrounding the eutectics (Figure 23). This leads to the depletion of the α-Mg solid solution in aluminum. Obviously, the content of the precipitated $Mg_{17}Al_{12}$ phase is lower in the AXJM71 alloy in comparison to the AXJM91 alloy due to the lower Al amount dissolved in the α-Mg matrix.

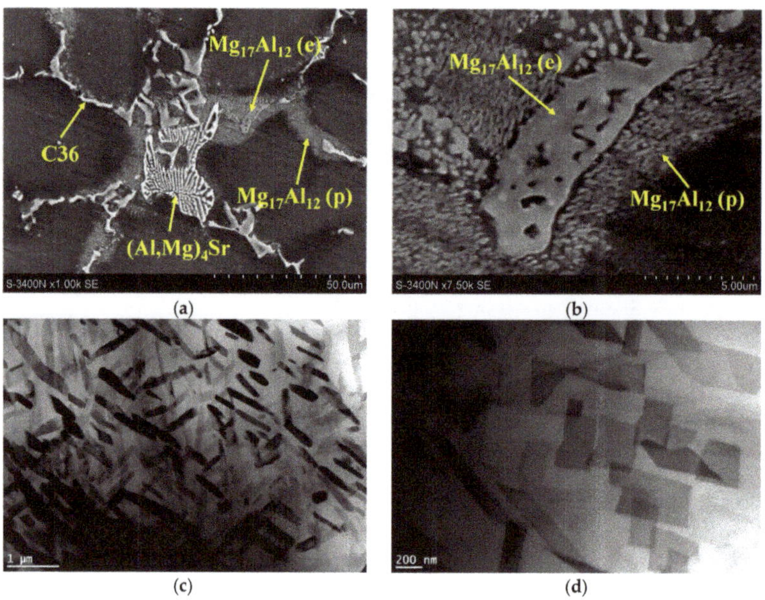

Figure 23. Microstructure of AXJM91 alloy after creep at 180 °C and at 75 MPa for 100 h with eutectic $Mg_{17}Al_{12}$ (e) phase and plate-shaped precipitates of $Mg_{17}Al_{12}$ (p) phase arising in supersaturated areas of α-Mg solid solution surrounding eutectics, SEM (**a**,**b**); plate-shaped precipitates of $Mg_{17}Al_{12}$ phase (**c**,**d**).

The destruction of the tested alloys during creep is caused primarily by the initiation of cracks at the interfaces and the propagation of these cracks (Figure 24). As the creep strain increases, the stress concentration on the precipitates increases, which leads to their cracking and loss of connection with the alloy's matrix. In Mg-Al-Ca-Sr alloys containing 3 wt% Ca, the first cracks are observed mainly at the α-Mg/C36 interface. In the AXJM91 alloy, creep cracks are initiated at the interface between the α-Mg solid solution and the $Mg_{17}Al_{12}$ phase (Figure 25b).

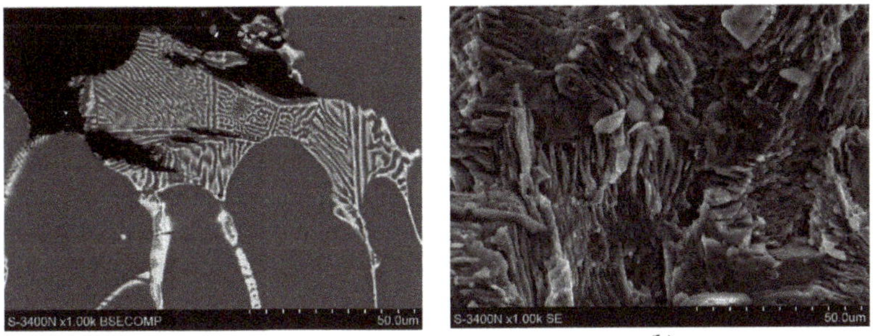

(a) (b)

Figure 24. Cracks occurring during creep of the AXJM53 alloy at 180 °C at 90 MPa (**a**). The propagation of the cracks occurs mainly along the precipitations of the $(Al,Mg)_2Ca$ (C36) phase (**b**).

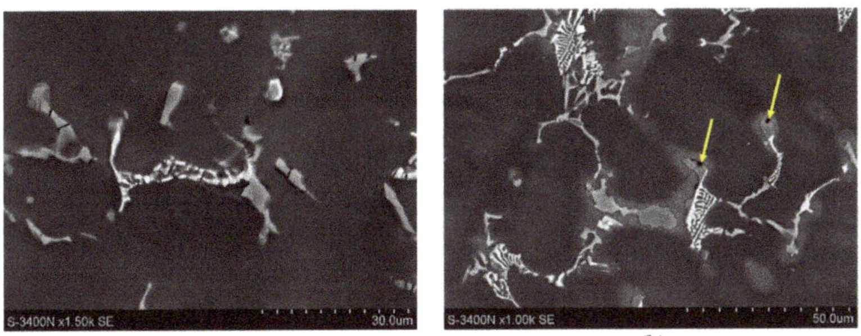

(a) (b)

Figure 25. Cracks occurring during creep of the AXJM71 alloy at 180 °C at 75 MPa (**a**); voids appearing at the interface α-Mg/$Mg_{17}Al_{12}$ in the AXJM91 alloy after creep at 180 °C at a stress of 75 MPa for 100 h (**b**).

It is well known that the $Mg_{17}Al_{12}$ phase adversely affects creep resistance, which is confirmed by the obtained test results. The precipitation processes of this phase during creep cause a decrease in the Al concentration in the solid solution, facilitating lattice diffusion and further contributing to the reduction in creep resistance. Therefore, the creep resistance of the AXJM91 alloy is much lower than that of the other tested alloys.

The formation of cracks at the α-Mg/C6 interface is related to the concentration of dislocations, the high hardness of the C36 phase, and the low ability to accommodate plastic deformation. The lamellar structure of this phase will contribute to slower crack development, and it seems that the presence of this phase is more advantageous than the presence of massive phases such as $(Mg,Al)_{17}(Sr,Ca)_2$.

3.5. Comparison of Mg-Al-Ca-Sr Alloys with EZ33 Magnesium Alloy

The microstructure and mechanical properties of the as-cast EZ33 magnesium alloy have been extensively described [9,18–20]. The typical microstructure of this alloy after gravity casting into sand molds consists of α-Mg solid solution dendrites and a $(Mg,Zn)_{12}RE$ ($Mg_{12}Nd$ type) compound at interdendritic areas (Figure 26). This intermetallic compound exhibits good stability of morphology and chemical composition up to 250 °C. Solution strengthening caused by the presence of Zn atoms in the α-Mg solid solution and the presence of the thermally stable $(Mg,Zn)_{12}RE$ phase strengthens the grain boundaries and contributes to the obtainment of quite satisfactory mechanical properties at ambient temperature (Table 8) and good creep resistance at temperatures up to 200 °C and stresses up to 60 MPa (Table 9). It is worth noting that the mechanical properties of sand castings are about 10% lower compared to alloy casts into permanent molds [9]. Therefore, the EZ33 alloy can be used for elements operating at elevated temperatures (up to 200 °C) in the aviation and automotive industries. Mechanical properties can be slightly improved by applying T5 heat treatment consisting of annealing at 170–200 °C for 10–16 h with cooled air due to the formation of Zn_2Zr_3 precipitates inside the α-Mg solid solution.

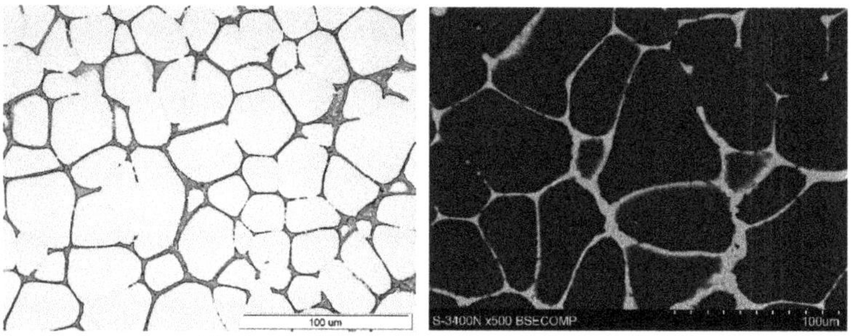

(a) (b)

Figure 26. Microstructure of as-cast EZ33 magnesium alloy: LM (**a**) and SEM (**b**).

Table 8. Mechanical properties of sand-cast EZ33 alloy (T5—heat treatment 10 h at 170 °C and cooled air).

Casting Method	Heat Treatment	R_m MPa	$R_{p0.2}$ MPa	Elongation A_{10}, %
Sand casting	T5	119–140	80–95	3

Table 9. Creep test results of sand-cast EZ33 magnesium alloy (T5—heat treatment 10 h at 170 °C and cooled air).

Alloy	Heat Treatment	Stress, MPa	Temperature, °C	Test Time, h	Creep Strain ε, %	Creep Rate $\dot{\varepsilon}$, 1/s
EZ33	-	60	180	100	0.89	5.0×10^{-9}
EZ33	-	60	200	100	1.14	7.7×10^{-9}
EZ33	-	60	225	87	5.9	2.4×10^{-8}
EZ33	T5	60	180	100	0.64	3.6×10^{-9}
EZ33	T5	60	200	100	0.90	6.5×10^{-9}
EZ33	T5	75	200	100	1.91	2.4×10^{-8}
EZ33	T5	30	250	100	0.82	1.4×10^{-8}
EZ33	T5	40	250	100	3.3	6.9×10^{-8}

The most advantageous mechanical properties among the tested alloys are those of the AXJM93 alloy, while the AXJM53 and AXJM73 alloys also demonstrate the required creep resistance up to 200 °C and acceptable strength properties at ambient temperature. Mg-Al-Ca-Sr alloys with a 3 wt% Ca content have another advantage: they contain lighter alloying elements than the EZ33 alloy; hence, the specific strength is more advantageous

(Table 10). In the case of AXJM53 and AXJM73 alloys, the disadvantage is the relatively low elongation. Increasing the Al content to 9% contributes to the obtainment of plastic properties similar to those of the EZ33 alloy. It should be emphasized that the use of Mg-Al-Ca-Sr alloys in aviation requires additional tests, including those for fatigue strength, and possibly requires applying protective coatings against corrosion.

Table 10. Specific strength, yield strength, elongation, and creep properties of Mg-Al-Ca-Sr and EZ33 alloys.

Alloy	Density ρ, g/cm^3	Specific Strength R_m/ρ, MPa·cm^3/g	Specific Yield Strength $R_{p0.2}/\rho$, MPa·cm^3/g	Elongation, %	Creep Strain at 200 °C and 60 MPa After 100 h, %	Creep Rate $\dot{\varepsilon}$, at 200 °C and 60 MPa After 100 h, 1/s
AXJM51	1.81	76.4	55.9	1.9	1.27	1.03×10^{-8}
AXJM71	1.82	95.5	62.0	3.1	0.98	1.43×10^{-8}
AXJM91	1.85	89.2	58.9	2.9	6.22	1.61×10^{-7}
AXJM53	1.80	75.0	57.8	2.1	0.44	2.78×10^{-9}
AXJM73	1.82	75.4	57.8	2.1	0.54	3.61×10^{-9}
AXJM93	1.84	81.2	55.6	2.5	0.64	4.43×10^{-9}
EZ33 *	2.08	57.2	38.5	2.9	0.90	6.5×10^{-9}
EZ33 **	2.08	67.3	43.3	3.0	-	-

* Own research, ** manufacturer data [9].

The advantage of using Mg-Al-Ca-Sr alloys, apart from the lower price of alloying elements, is the casting process at a lower temperature. In the case of the EZ33 alloy, melting is carried out at a temperature of 780–815 °C, while for Mg-Al-Ca-Sr alloys, it is possible to carry out the process at a temperature of 730 °C. The time of preparation of the liquid alloy in both cases is similar. Other advantages related to melting Mg-Al-Ca-Sr alloys include the non-flammability of these alloys during melting (Mg has a very low ignition temperature), which makes the casting process safer. However, it is necessary to use protective atmospheres in order to obtain the desired chemical composition of the alloy and to avoid an excessive number of non-metallic inclusions.

Despite the many advantages of Mg-Al-Ca-Sr alloys, mold sticking and hot cracking can be a serious problem during the casting process. This problem was mainly observed during die casting of Mg-Al-Ca alloys or gravity casting into metal molds and most frequently concern Mg-Al alloys with a low calcium content (up to approx. 1–2% Ca). At higher Ca contents, susceptibility to hot cracking and the formation of sticking is not greater than in typical magnesium alloys (AM50 and AZ91) [56–59]. No problems were found in the case of Mg-Al-Ca-Sr alloys gravity-cast into sand molds. After casting in an induction furnace into a mold based on quartz bentonite mass, the surface of the castings was free of sticking and cracks, and the surface quality was of a satisfactory level (Figure 27a). It is worth noting that the tests of Mg-Al-Ca-Sr alloys conducted in industrial conditions with melting in an electric furnace and casting into a mold based on furan mass showed the absence of hot cracks and a high susceptibility to the formation of sticking, regardless of the chemical composition of the Mg-Al-Ca-Sr alloys. The sticking was easily removed after sand blastin; however, staining remained visible (Figure 27b). The surface quality of the Mg-Al-Ca-Sr alloy after casting in industrial conditions was clearly worse compared to that of the EZ33 alloy, but it can be improved by an appropriate selection of molding materials.

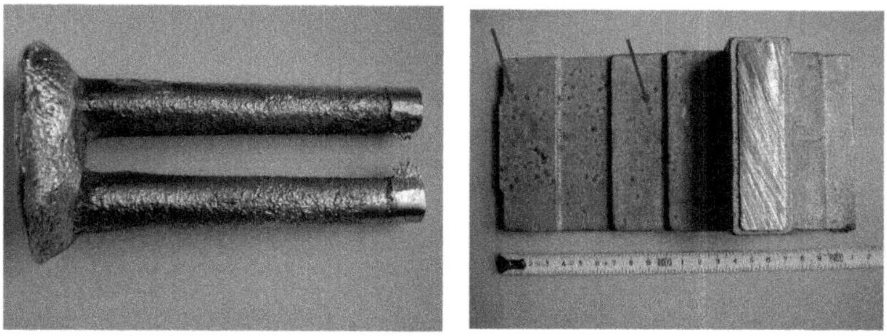

(**a**) (**b**)

Figure 27. AXJM53 alloy after casting in an induction furnace using a mold made of bentonite mass (**a**); AXJM53 alloy after casting in industrial conditions using an electric furnace and a mold made of furan mass (**b**).

4. Conclusions

The aim of the research was to develop a magnesium alloy consisting of cheap alloying elements, which after gravity casting into sand molds will be characterized by mechanical properties similar to those of the EZ33 alloy at temperatures up to 200 °C. The aim was achieved by appropriate selection of alloying elements such as Al, Ca, and Sr. This allowed us to obtain alloys that exhibit the presence of a skeleton of intermetallic phase precipitates stable at elevated temperature, strengthened α-Mg solid solution by alloying elements (especially Al), and plate-like precipitates of the C15 Laves phase inside the α-Mg solid solution.

Based on the research conducted and analysis of the results, the following conclusions were formulated:

1. Increasing the Ca concentration in Mg-Al-Sr alloys causes a decrease in the volume fraction of the $Mg_{17}Al_{12}$ phase, the degree of supersaturation of the α-Mg solid solution, and the formation of Laves Ca-rich $(Mg,Al)_2(Ca,Sr)$ (C36) and $(Mg,Al)_2Ca$ (C14) phases and Sr-rich $(Al,Mg)_4Sr$ and $(Mg,Al)_{17}Sr_2$ phases. For large Ca/Al concentration ratios, additional plate-like precipitates of the Al_2Ca (C15) Laves phase appear inside the dendrites of the α-Mg solid solution.
2. Alloys containing 1% Ca are characterized by good mechanical properties at ambient temperature, but their creep resistance at 200 °C is insufficient. Increasing the Ca content to 3% causes a decrease in mechanical properties at ambient temperature while achieving satisfactory creep resistance at 200 °C. The most favorable mechanical properties at ambient and elevated temperatures are achieved by the AXJM93 alloy (R_m = 149 MPa, $R_{p0.2}$ = 102 MPa, A = 2.5%, creep properties at a temperature of 200 °C and at a stress of 60 MPa—$\dot{\varepsilon}$ ~10^{-9} 1/s, ε = 0.64%), which can be used as a substitute for the EZ33 alloy due to its comparable mechanical properties at ambient temperature, slightly better creep resistance at temperatures up to 200 °C, and lower density.
3. Factors contributing to the improvement in the creep resistance of Mg-Al-Ca-Sr alloys at 200 °C include the following: solution strengthening caused by the dissolution of Al and Ca in α-Mg, precipitation strengthening resulting from the presence of plate-like precipitates of the Al_2Ca (C15) phase with crystallographic orientation of the type $\{111\}_{C15}||(0001)_{\alpha\text{-Mg}}$ $[01\bar{1}]_{C15}||[0\bar{1}10]_{\alpha\text{-Mg}}$, and the presence of high-melting Laves phases in interdendritic areas. Conversely, the $Mg_{17}Al_{12}$ phase and the presence of supersaturated α-Mg solid solution, in which the Al content significantly exceeds the solubility limit of Al in Mg at 200 °C, exhibit an unfavorable effect on creep resistance.

4. In the case of 1% Ca alloys, dislocation climbing was found to be the dominant creep mechanism at 180 °C. In the 3% Ca alloys, the steady-state creep activation energy (Q_c) values indicate that creep deformation is controlled by pipe diffusion. Grain boundary slip, which is a significant contributor to creep deformation in die-cast Mg-Al-Ca-Sr alloys, plays a marginal role in sand-cast alloys due to the larger grain size.

Author Contributions: Conceptualization, T.R. and A.F.; methodology, T.R. and A.F.; software, T.R.; validation, T.R.; formal analysis, A.F.; investigation, T.R.; resources, T.R.; data curation, T.R.; writing—original draft preparation, T.R. and A.F.; writing—review and editing, T.R. and A.F.; visualization, T.R. All authors have read and agreed to the published version of the manuscript.

Funding: The present work was supported by the Polish Ministry of Science and Higher Education under project No. 11/030/BK_24/1177 and strategic project No. POIG.01.01.02-00-015/09 (FSB-71/RM3/2010).

Institutional Review Board Statement: Not applicable.

Informed Consent Statement: Not applicable.

Data Availability Statement: The original contributions presented in this study are included in the article. Further inquiries can be directed to the corresponding author.

Acknowledgments: The authors are grateful to B. Chmiela (Silesian University of Technology, Katowice) and P. Skupień for their assistance and useful discussion.

Conflicts of Interest: The authors declare no conflict of interest.

References

1. Friedrich, H.E.; Mordike, B.L. *Magnesium Technology. Metallurgy, Design Data, Applications*; Springer-Verlag: Berlin/Heidelberg, Germany, 2006.
2. Avedesian, M.M.; Baker, H. *ASM Specialty Handbook, Magnesium and Magnesium Alloys*; ASM International: Novelty, OH, USA, 1999.
3. Mordike, B.L.; Ebert, T. Magnesium, properties–applications–potential. *Mater. Sci. Eng. A* **2001**, *302*, 37–45. [CrossRef]
4. Friedrich, H.; Schumann, S. Research for a "new age of magnesium" in the automotive industry. *J. Mater. Process. Technol.* **2001**, *117*, 276–281. [CrossRef]
5. Luo, A.A. Recent magnesium alloy development for elevated temperature applications. *Int. Mater. Rev.* **2004**, *49*, 13–30. [CrossRef]
6. Luo, A.A. Applications: Aerospace, automotive and other structural applications of magnesium. In *Advances in Wrought Magnesium Alloys*; Woodhead Publishing Limited: Cambridge, UK, 2013; pp. 266–316.
7. Aghion, E.; Bronfin, B.; von Buch, F.; Schumann, S.; Friedrich, H. Newly Developed Magnesium Alloys for Powertrain Applications. *JOM* **2003**, *55*, 30–33. [CrossRef]
8. Aghion, E.; Bronfin, B.; Eliezer, D.; Schumann, S. The Art of Developing New Magnesium Alloys for High Temperature Applications. In Proceedings of the Second Osaka International Conference on Platform Science and Technology, Osaka, Japan, 26–30 January 2003; pp. 40–44.
9. Elektron ZRE1 Datasheet 450. Available online: https://magnesium.elektron.com (accessed on 18 March 2025).
10. Rzychoń, T.; Kiełbus, A. Optimisation of Hot-Chamber Die-Casting Process of AM60 Alloy Using Taguchi Method. *Materials* **2024**, *17*, 6256. [CrossRef]
11. Tariq, H.M.R.; Ishtiaq, M.; Kang, H.-H.; Chaudry, U.M.; Jun, T.-S. A Critical Review on the Comparative Assessment of Rare-Earth and Non-Rare-Earth Alloying in Magnesium Alloys. *Metals* **2025**, *15*, 128. [CrossRef]
12. Tuz, L.; Novák, V.; Tatíček, F. Evaluation of the Microstructure and Properties of As-Cast Magnesium Alloys with 9% Al and 9% Zn Additions. *Materials* **2025**, *18*, 10. [CrossRef]
13. Al-Samman, T.; Letzig, D.; Yi, S. Microstructure–Mechanical Properties and Application of Magnesium Alloys. *Metals* **2021**, *11*, 1958. [CrossRef]
14. Tan, J.; Ramakrishna, S. Applications of Magnesium and Its Alloys: A Review. *Appl. Sci.* **2021**, *11*, 6861. [CrossRef]

15. Yang, Q.; Wu, X.; Qiu, X. Microstructural Characteristics of High-Pressure Die Casting with High Strength–Ductility Synergy Properties: A Review. *Materials* **2023**, *16*, 1954. [CrossRef]
16. Klupa, A. Cenniejsze niż złoto? Metale ziem rzadkich w światowej strategii gospodarczej. *Przegląd Strateg.* **2012**, *1*, 239–251. [CrossRef]
17. Koltygin, A.V.; Bazhenov, V.E.; Belova, E.A.; Nikitina, A.A. Development of a magnesium alloy with good casting characteristics on the basis of Mg-Al-Ca-Mn system, having Mg-Al2Ca structure. *J. Magnes. Alloys* **2013**, *1*, 224–229. [CrossRef]
18. Rzychoń, T.; Kiełbus, A.; Szala, J. Microstructure, Castability, Microstructural Stability and Mechanical Properties of ZRE1 Magnesium Alloy. *Arch. Metall. Mater.* **2012**, *57*, 246–252. [CrossRef]
19. Ahmad, R.; Sheggaf, Z.M.; Asmael, M.B.A.; Hamzah, M.Z. Effect of yttrium addition on microstructure and hardness of cast ZRE1 magnesium alloy. *J. Eng. Appl. Sci.* **2016**, *11*, 7689–7693.
20. Tukiat, I.S.T.; Yusuf, N.K.; Khaireez, H.; Al-Alimi, S.; Lajis, M.A.; Shamsudin, S.; Ruhaizat, N.E. Microstructure and Mechanical Properties of Magnesium ZRE1 (Mg-Zn-Zr) Alloy with Rare Earth Element (Samarium) Addition. *Int. J. Technol.* **2024**, *15*, 49–62. [CrossRef]
21. Li, J.; Zhou, X.; Breen, A.; Peng, Z.; Su, J.; Kürnsteiner, P.; Correa, M.J.D.; Lipińska-Chwałek, M.; Wang, H.; Holec, D.; et al. Elucidation of formation and transformation mechanisms of Ca-rich Laves phase in Mg-Al-Ca-Mn alloys. *J. Alloys Compd.* **2022**, *928*, 167–177. [CrossRef]
22. Rzychoń, T. Characterization of Mg-rich clusters in the C36 phase of the Mg-5Al-3Ca-0.7Sr-0.2Mn alloy. *J. Alloys Compd.* **2014**, *598*, 95–105. [CrossRef]
23. Suzuki, A.; Saddock, N.D.; Riester, L.; Lara-Curzio, E.; Jones, J.W.; Pollock, T.M. Effect of Sr Additions on the Microstructure and Strength of a Mg-Al-Ca Ternary Alloy. *Metall. Mater. Trans. A* **2007**, *38*, 420–427. [CrossRef]
24. Bai, J.; Sun, Y.; Xue, F.; Xue, S.; Qiang, J.; Zhu, T. Effect of Al contents on microstructures, tensile and creep properties of Mg–Al–Sr–Ca alloy. *J. Alloys Compd.* **2007**, *437*, 247–253. [CrossRef]
25. Kondori, B.; Mahmudi, R. Effect of Ca additions on the microstructure and creep properties of a cast Mg–Al–Mn magnesium alloy. *Mater. Sci. Eng. A* **2017**, *700*, 438–447. [CrossRef]
26. Murray, J.L. *Binary Alloy Phase Diagrams*, 2nd ed.; Massalski, T.B., Ed.; ASM International: Materials Park, OH, USA, 1990; Volume 1, pp. 169–171.
27. Ninomiya, R.; Yukawa, H.; Morinaga, M.; Kubota, K. An electronic approach to the prediction of the mechanical properties of magnesium alloys. *J. Alloys Compd.* **1994**, *215*, 315–323. [CrossRef]
28. Burke, E.C. Solid solubility of calcium in magnesium. *J. Met. Trans. AIME* **1955**, *203*, 285–286. [CrossRef]
29. Chartrand, P.; Pelton, A.D. Critical evaluation and optimization of the thermodynamic properties and phase diagrams of the Al-Mg, Al-Sr, Mg-Sr, and Al-Mg-Sr systems. *J. Phase Equilib.* **1994**, *15*, 591–605. [CrossRef]
30. Mezbahul-Islam, M.; Mostafa, A.O.; Medraj, M. Essential Magnesium Alloys Binary Phase Diagrams and Their Thermochemical Data. *J. Mater.* **2014**, *2014*, 1–33. [CrossRef]
31. Jiang, B. Grain refinement of Ca addition in a twin-roll-cast Mg–3Al–1Zn alloy. *Mater. Chem. Phys.* **2012**, *133*, 611–616. [CrossRef]
32. Nagasivamuni, B.; Ravi, K.R. An analytical approach to elucidate the mechanism of grain refinement in calcium-added Mg–Al alloys. *J. Alloys Compd.* **2015**, *622*, 789–795. [CrossRef]
33. Liu, S.F.; Li, B.; Wang, X.H.; Su, W.; Han, H. Refinement effect of cerium, calcium and strontium in AZ91 magnesium alloy. *J. Mater. Process. Technol.* **2009**, *209*, 3999–4004. [CrossRef]
34. Park, J.P.; Kim, M.G.; Yoon, U.S. Microstructures and mechanical properties of Mg-Al-Zn-Ca alloys fabricated by high-frequency electromagnetic casting method. *J. Mater. Sci.* **2009**, *44*, 47–54. [CrossRef]
35. Kim, J.H.; Kang, N.E.; Yim, C.D.; Kim, B.K. Effect of calcium content on the microstructural evolution and mechanical properties of wrought Mg–3Al–1Zn alloy. *Mater. Sci. Eng. A* **2009**, *525*, 18–29. [CrossRef]
36. Masoudpanah, S.M.; Mahmudi, R. Effects of rare-earth elements and Ca additions on the microstructure and mechanical properties of AZ31 magnesium alloy processed by ECAP. *Mater. Sci. Eng. A* **2009**, *526*, 22–30. [CrossRef]
37. Cao, P.; Qian, M.; StJohn, D.H. Native grain refinement of magnesium alloys. *Scr. Mater.* **2005**, *53*, 841–844. [CrossRef]
38. Lee, Y.C.; Dahle, A.K.; StJohn, D.H. The role of solute in grain refinement of magnesium. *Metall. Mater. Trans. A* **2000**, *31A*, 2895–2906. [CrossRef]
39. Hirai, K.; Somekawa, H.; Takigawa, Y.; Higashi, K. Effects of Ca and Sr addition on mechanical properties of a cast AZ91 magnesium alloy at room and elevated temperature. *Mater. Sci. Eng. A* **2005**, *403*, 276–280. [CrossRef]
40. Li, P.; Tang, B.; Kandalova, E.G. Microstructure and properties of AZ91D alloy with Ca additions. *Mater. Lett.* **2005**, *59*, 671–675. [CrossRef]
41. StJohn, D.H.; Easton, M.A.; Qian, M.; Taylor, J.A. Grain Refinement of Magnesium Alloys: A Review of Recent Research, Theoretical Developments, and Their Application. *Metall. Mater. Trans. A* **2013**, *44A*, 2935–2949. [CrossRef]

42. Pekguleryuz, M.; Kaya, A.A. Creep Resistant Magnesium Alloys for Powertrain Applications. *Adv. Eng. Mater.* **2003**, *5*, 866–878. [CrossRef]
43. Wang, Q.D.; Chen, W.D.; Zeng, X.Q.; Lu, Y.Z.; Ding, W.J.; Zhu, Y.P.; Xu, X.P.; Mabuchi, M. Effects of Ca addition on the microstructure and mechanical properties of AZ91 magnesium alloy. *J. Mater. Sci.* **2001**, *36*, 3035–3040.
44. Harandi, S.E.; Mirshahi, M.; Koleini, S.; Idris, M.H.; Jafari, H.; Kadir, M.R.A. Effect of calcium content on the microstructure, hardness and in-vitro corrosion behavior of biodegradable Mg–Ca binary alloy. *Mater. Res.* **2013**, *16*, 11–18. [CrossRef]
45. Nie, J.F. Precipitation and Hardening in Magnesium Alloys. *Metall. Mater. Trans. A* **2012**, *43*, 3891–3939. [CrossRef]
46. Peng, Z.; Qudong, W.; Chunquan, Z.; Yanping, Z. Effects of strontium and titanium on the microstructure, tensile properties and creep behavior of AM50 alloys. *Mater. Sci. Eng. A* **2007**, *444*, 318–326.
47. Pekguleryuz, M.; Celikin, M. Creep Resistance in Magnesium Alloys. *Int. Mater. Rev.* **2010**, *55*, 197–217. [CrossRef]
48. Shi, L.; Northwood, D.O. Strain-hardening and recovery during the creep of pure polycrystalline magnesium. *Acta Metall. Mater.* **1994**, *42*, 871–877. [CrossRef]
49. Miller, W.K. Creep of die cast AZ91 magnesium at room temperature and low stress. *Metall. Mater. Trans. A* **1991**, *22A*, 873–877. [CrossRef]
50. Regev, M.; Aghion, E.; Rosen, A.; Bamberger, M. Creep studies of coarse-grained AZ91D magnesium castings. *Mater. Sci. Eng. A* **1998**, *252A*, 6–16. [CrossRef]
51. Regev, M.; Aghion, E.; Rosen, A. Creep studies of AZ91D pressure die casting. *Mater. Sci. Eng. A* **1997**, *234*, 123–126. [CrossRef]
52. Terada, Y.; Itoh, D.; Sato, T. Creep rupture properties of die-cast Mg–Al–Ca alloys. *Mater. Chem. Phys.* **2009**, *113*, 503–506. [CrossRef]
53. Celikin, M.; Kaya, A.A.; Pekguleryuz, M. Effect of manganese on the creep behavior of magnesium and the role of α-Mn precipitation during creep. *Mater. Sci. Eng. A* **2012**, *534*, 129–141. [CrossRef]
54. Maruyama, K.; Suzuki, M.; Sato, H. Creep strength of magnesium-based alloys. *Metall. Mater. Trans. A* **2002**, *33*, 875–882. [CrossRef]
55. Luo, A.A.; Balogh, M.P.; Powell, B.R. Creep and microstructure of magnesium–aluminum–calcium based alloys. *Metall. Mater. Trans. A* **2002**, *33A*, 567–574. [CrossRef]
56. Kondori, B.; Mahmudi, R. Effect of Ca additions on the microstructure, thermal stability and mechanical properties of a cast AM60 magnesium alloy. *Mater. Sci. Eng. A* **2010**, *527*, 2014–2021. [CrossRef]
57. Liu, Y.; Wang, Z.; Li, S.; Ding, N.; Liu, K.; Du, W. Hot Cracking Behaviors of Mg-Zn-Er Alloys with Different Er Contents. *Materials* **2023**, *16*, 3546. [CrossRef] [PubMed]
58. Powell, B.R.; Luo, A.A.; Tiwari, B.L.; Rezhets, V. The Die Castability of Calcium Containing Magnesium Alloys: Thin-Wall Computer Case. In *Magnesium Technology 2002;* TMS: Pittsburgh, PA, USA, 2002; pp. 123–129.
59. Cao, G.; Kou, S. Hot Tearing of Ternary Mg-Al-Ca Alloy Castings. *Metall. Mater. Trans. A* **2006**, *37A*, 3647–3663. [CrossRef]

materials

MDPI

Article

Free Radical Copolymerization of *N*-Isopropylacrylamide and 2,3-Dihydroxypropyl Methacrylate: Reaction Kinetics and Characterizations

Zhishu Chen [1,2] and Chao Zhang [1,2,*]

1 School of Biomedical Engineering, Sun Yat-sen University, Shenzhen Campus, Shenzhen 518107, China; chenzhsh6@mail2.sysu.edu.cn

2 Guangdong Provincial Key Laboratory of Sensor Technology and Biomedical Instrument, Sun Yat-sen University, Shenzhen 518107, China

* Correspondence: zhchao9@mail.sysu.edu.cn

Abstract: Poly(*N*-isopropylacrylamide) (PNIPAm) undergoes a sharp phase transition in aqueous solutions at around 32 °C, which is called the lower critical solution temperature; the tuning of the LCST of PNIPAm could be achieved by the copolymerization of *N*-isopropylacrylamide (NIPAm) with other hydrophilic/hydrophobic monomers to regulate the solvation state of PNIPAm and meet the requirements of possible applications. Herein, a hydrophilic monomer, 2,3-dihydroxypropyl methacrylate (DHPMA), w introduced to regulate the phase transition behavior of PNIPAm via free radical copolymerization. A series of poly(*N*-isopropylacrylamide-*co*-2,3-dihydroxypropyl methacrylate) (P(NIPAm-*co*-DHPMA)) was synthesized and characterized. The reaction kinetics were investigated in detail. In this copolymerization, the reactivity ratios of DHPMA and NIPAm were found to be 3.09 and 0.11, suggesting that DHPMA had greater preference for homopolymerization than for copolymerization, while NIPAm had greater preference for copolymerization than for homopolymerization. The phase transition temperature of P(NIPAm-*co*-DHPMA) copolymers varied from 31 to 42 °C by controlling the content of DHPMA in the copolymers from 0 to 58 mol%. Finally, the good cytocompatibility of P(NIPAm-*co*-DHPMA) was confirmed. These results provide insights into designing thermo-responsive polymers with suitable responsive behaviors that meet the requirements of different applications.

Keywords: poly(*N*-isopropylacrylamide); 2,3-dihydroxypropyl methacrylate; reaction kinetics; reactivity ratio

1. Introduction

Thermo-responsive polymers drastically change their physicochemical properties when the temperature reaches their phase transition temperatures, which has gained attention for applications in various fields [1]. Among these polymers, poly(*N*-isopropylacrylamide) (PNIPAm) is one of the most popular and classic thermo-responsive polymers, with 57 years of research history since the first systematic study of the phase diagram of PNIPAm was reported by Heskins and Guillet in 1968 [2]. Its lower critical solution temperature (LCST) had been identified to be approximately 32 °C in aqueous solutions [1–3], which is close to the physiological temperature, attracting interest in biomedical fields. The molecular chain of PNIPAm undergoes a coil-to-globule transition [4] driven by entropy changes in aqueous solutions in response to the increase in temperature, which transitions through the LCST [5,6]. These microscopic changes lead to the aggregation of PNIPAm-based materials and finally result in clear changes in the macroscopic properties of PNIPAm-based materials such as volume shrinkage [7], a decrease in light transmittance [8], conductivity changes [9], and a decrease in hydrophilicity [10]. Based on these fascinating properties, PNIPAm-based materials have been developed in biosensing [11,12], drug delivery [13,14], and tissue engineering [15,16].

The different requirements for thermo-responsive polymers in various applications pose a challenge to develop novel thermo-responsive polymers with a rational design. Copolymerization is a powerful strategy to build novel thermo-responsive polymers based on known thermo-responsive polymers with on-demand phase transition behavior for various applications [17–20]. Typically, hydrophilic comonomers increase PNIPAm solvation, raising the phase transition temperature. In contrast, hydrophobic comonomers reduce solvation, lowering the transition temperature [21]. However, the molecular interactions between comonomers also play an essential role in the phase transition behavior of the copolymer [5]. For example, 2-hydroxyethyl methacrylate (HEMA) is hydrophilic. However, it decreases the phase transition temperature of PNIPAm due to the hydrogen bond between the hydroxyl groups and carbonyl groups, which makes the copolymer more hydrophobic [22]. Zhou et al. investigated the micro-dynamic mechanism of the phase transition behavior of poly(*N*-isopropylacrylamide-*co*-2-hydroxyethyl methacrylate) hydrogels and found that the breakage of the hydrogen bond between the carbonyl groups of HEMA and water molecules as well as the breakage of hydrogen bonds between the carbonyl groups of HEMA and the amide groups of NIPAm at low temperatures created a hydrophobic environment in the hydrogel. This led to the breakage of the hydrogen bonds between the amide group of NIPAm and water molecules at lower temperatures and finally led to a lower phase transition temperature than PNIPAm [22]. The amide groups of NIPAm and the hydroxyl groups of HEMA then formed hydrogen bonds with each other [22]. Therefore, it is necessary to understand the regulation effect of comonomers on the phase transition behavior of PNIPAm to develop novel PNIPAm-based polymers on demand.

The reaction kinetics of copolymerization play an important role in optimizing the polymerization process to obtain the copolymer required [23]. Kinetic parameters such as the reactivity ratios of copolymerization are important to control the final structures of the copolymer [24,25]. Mathematical fitting methods such as the Fineman–Ross method [26], Tidwell–Mortimer method [27], and Kelen–Tüdös method [28], in combination with experimental kinetic data, are used to determine the reactivity ratio.

2,3-Dihydroxypropyl methacrylate (DHPMA), otherwise known as glycerol monomethacrylate [29], is a promising monomer in biomedical fields that has been used in commercialized contact lenses [30,31] and has many studies in biosensing [32], cell culture [33], and tissue engineering [34]. Its homopolymers have been recognized as a potential candidate of poly(ethylene glycol) in biomedical fields because of its hydrophilic, biocompatible, non-toxic, and potential stealth effect [25,33,35]. In contrast to HEMA, DHPMA has one more hydroxyl group and has been found to increase the phase transition temperature higher than PNIPAm after the reversible addition-fragmentation chain transfer (RAFT) copolymerization of DHPMA and NIPAm [25]. Saldívar-Guerra et al. investigated the reactivity ratios of DHPMA and other comonomers, including NIPAm, and found that the reactivity ratio of DHPMA was 2.55, while the reactivity ratio of NIPAm was 0.11 [25]. However, the reactivity ratios of DHPMA and NIPAm remain to be confirmed due to the existence of 22 mol% of a DHPMA isomer, 1,3-dihydroxypropyl methacrylate, in the previous report. The reaction kinetics of the free radical copolymerization of NIPAm and DHPMA remained unexplored.

In this study, a series of poly(*N*-isopropylacrylamide-*co*-2,3-dihydroxypropyl methacrylate) (P(NIPAm-*co*-DHPMA)) was synthesized via free radical copolymerization. The chemical structure of P(NIPAm-*co*-DHPMA) was carefully confirmed by nuclear magnetic resonance (NMR) spectroscopy. The reaction kinetics of the copolymerization were investigated using ^{1}H nuclear magnetic resonance (^{1}H NMR) spectra and elemental analyses. The phase transition behavior of P(NIPAm-*co*-DHPMA) in an aqueous solution was evaluated. As the results show, a series of poly(*N*-isopropylacrylamide-*co*-2,3-dihydroxypropyl methacrylate) with different contents of NIPAm and DHPMA units were successfully synthesized. The conversion and the reactivity ratios of the comonomers were produced. These findings provide a design for P(NIPAm-*co*-DHPMA) copolymers with precisely controlled phase transition behavior matching different criteria.

2. Materials and Methods

2.1. Materials

2,3-Dihydroxypropyl methacrylate (DHPMA) (95%; Beijing HwrkChemical Technology Co., Ltd., Beijing, China) was further purified using vacuum distillation to ensure a purity above 99% (confirmed by ^{1}H NMR) before use. *N*-isopropylacrylamide (NIPAm) (98%; Macklin, Shanghai, China) was recrystallized from *n*-Hexane (≥99.7%; Guangzhou Chemical Reagent Factory, Guangzhou, China)/methylbenzene (≥99.7%; Guangzhou Chemical Reagent Factory, Guangzhou, China) (50/50, v/v) via dissolution at 40 °C and then cooled under refrigeration. 2,2′-Azobis(2-methylpropionitrile) (AIBN) (98%; Macklin, Shanghai, China) was recrystallized from ethanol (≥99.7%; Guangzhou Chemical Reagent Factory, Guangzhou, Guangdong, China) via dissolution at 50 °C and then cooled under refrigeration. Anhydrous dimethyl sulfoxide (DMSO) (99.7%; Macklin, Shanghai, China), copper(I) chloride (CuCl) (99%; Bidepharm, Shanghai, China), 1,4-dinitrobenzene (98%; Sigma-Aldrich, Darmstadt, Germany), and deuterated dimethyl sulfoxide (DMSO-d_6) (99.9% with 0.03% TMS; Cambridge Isotope Laboratories, Inc., Tewksbury, MA, USA) were used as received. The poly(methacrylic acid methyl ester) (PMMA) standard and *N,N*-dimethylformamide (DMF; 99.8%) were products of Aladdin, Shanghai, China.

NIH 3T3 cells were a gift from the Hospital of Stomatology, Sun Yat-sen University. High-glucose Dulbecco's Modified Eagle Medium (DMEM; Gibco Life Technologies, Inc., Grand Island, NY, USA) supplemented with 10% fetal bovine serum (FBS; Gibco Life Technologies, Inc., Grand Island, NY, USA) and 1% penicillin–streptomycin (Gibco Life Technologies, Inc., Grand Island, NY, USA) was used as the culture medium for NIH 3T3 cells. A Cell-Counting Kit-8 (CCK-8) was obtained from Beyotime Biotechnology, Shanghai, China. A Live/Dead Viability/Cytotoxicity Assay Kit for Animal Cells was obtained from KeyGEN BioTECH, Nanjing, Jiangsu, China.

2.2. Synthesis of P(NIPAm-co-DHPMA)

P(NIPAm-*co*-DHPMA) was synthesized via free radical copolymerization. In a typical reaction, 33.0 mmol of comonomers in total and 0.5 mol% AIBN (Table S1) were dissolved in 30.0 mL DMSO in a Schlenk flask. The solution was first purged with nitrogen (99.99%) for 30 min, followed by three vacuum/nitrogen purge cycles. Then, the flask was immersed in a 70 °C water bath with magnetic stirring. At predetermined time points (set as 0, 10, 20, 30, 60, and 120 min), 200 µL of the reaction mixture was sampled and then mixed with 400.0 µL DMSO-d_6 containing 72.0 µg CuCl and 3.0 mg 1,4-dinitrobenzene. The mixtures were subjected to NMR measurements to determine the conversion of monomers at certain time points. After 2 h, the reaction was quenched in ice water and the reaction mixture was dialyzed against deionized (DI) water in a dialysis tube (molecular weight cut-off (MWCO) of 1000 Da) for 72 h with frequent changes of DI water. Finally, the dialysate was lyophilized before characterization.

2.3. Structural Characterization

The chemical structure of the copolymerization product was investigated using NMR spectroscopy. ^{1}H NMR spectra, ^{13}C nuclear magnetic resonance (^{13}C NMR) spectra, distortionless enhancement by polarization transfer (DEPT; θ = 135°) spectra, ^{1}H-^{1}H correlation (^{1}H-^{1}H COSY) spectra, and ^{1}H-^{13}C heteronuclear single quantum coherence (HSQC) spectra were recorded using a Bruker Advance 400 spectrometer (Bruker, Billerica, MA, USA). All NMR measurements were conducted using DMSO-d_6 as the solvent and TMS as the internal standard. The experimental settings of ^{1}H NMR, ^{13}C NMR, DEPT, ^{1}H-^{1}H COSY, and HSQC were as follows:

For ^{1}H NMR, the pulse sequence, relaxation delay, number of scans, and spectral size were set as zg30, 1.00 s, 16, and 64 K, respectively. For ^{13}C NMR, the pulse sequence, relaxation delay, number of scans, and spectral size were set as zgpg30, 1.55 s, 3600, and 64 K, respectively. For DEPT, the pulse sequence, relaxation delay, number of scans, and spectral size were set as dept135, 1.00 s, 2048, and 64 K, respectively. For ^{1}H-^{1}H COSY, the pulse sequence, relaxation delay, number of scans, and spectral size were set as cosygpdf, 1.00 s, 16, and (1024, 1024), respectively. For HSQC, the pulse sequence, relaxation delay, number of scans, and spectral size were set as hsqcedetgp, 0.9050 s, 16, and (1024, 1024), respectively.

2.4. Molecular Molar Mass

The molecular molar mass and polydispersity index (PDI) of the copolymerization products were determined using gel permeation chromatography (GPC) on an Agilent 1260 high-performance liquid chromatography system (Agilent Technologies, Inc., Santa Clara, CA, USA) with a Plgel 5 µm MIXED-C column (Agilent Technologies, Inc., Santa Clara, CA, USA). DMF was selected as the eluent with a flow of 1.0 mL/min at 35 °C. A light-scattering detector, a differential refractive index detector, and a viscometer were used to estimate the molar mass of the product and the results were calibrated using PMMA standards. The synthesized products were dissolved in DMF at a concentration of 1.0 mg/mL and filtered through a 0.22 µm membrane for each GPC characterization.

2.5. Conversion of Monomers

The conversion of the monomers was determined using ^{1}H NMR spectroscopy. Before testing, the longitudinal relaxation times (T_1) of NIPAm (δ = 5.54 ppm) and DHPMA (δ = 5.67 ppm) were measured following the inversion recovery method using nonlinear fitting with the following formula:

$$I(\tau) = I(0) + \mathrm{P}\exp(-\frac{\tau}{\mathrm{T}_1}) \quad (1)$$

where τ is the relaxation time, $I(t)$ is the integral area of the peaks after a relaxation time of τ, P is a constant, and T_1 is the longitudinal relaxation time.

The T_1 of 1,4-dinitrobenzene (δ = 8.46 ppm) from the literature was used in this study [36]. The value of the relaxation delay (D_1) was decided by using the largest T_1 of NIPAm (δ = 5.54 ppm), DHPMA (δ = 5.67 ppm) (Figure S1), and 1,4-dinitrobenzene (δ = 8.46 ppm).

For ^{1}H NMR spectroscopy, the pulse sequence, D_1, number of scans, and spectral size were set as zg, 30 s, 32, and 64 K, respectively.

1,4-dinitrobenzene was set as the internal standard compound and its characteristic peak at 8.46 ppm was normalized as 1.00. The conversions of NIPAm and DHPMA were calculated using the normalized peak areas of NIPAm (δ = 5.54 ppm) and DHPMA (δ = 5.67 ppm) at different time points. The conversion of the monomers was calculated using the following formulae:

$$\alpha_{NIPAm,x} = \frac{A_{NIPAm,0} - A_{NIPAm,x}}{A_{NIPAm,0}} \times 100\% \quad (2)$$

$$\alpha_{\mathrm{DHPMA},x} = \frac{A_{\mathrm{DHPMA},0} - A_{\mathrm{DHPMA},x}}{A_{\mathrm{DHPMA},0}} \times 100\% \quad (3)$$

$$\alpha_{\mathrm{Total},x} = \frac{A_{\mathrm{DHPMA},0} + A_{\mathrm{NIPAm},0} - A_{\mathrm{NIPAm},x} - A_{\mathrm{DHPMA},x}}{A_{DHPMA,0} + A_{NIPAm,0}} \times 100\% \quad (4)$$

Here, $\alpha_{\mathrm{NIPAm},x}$, $\alpha_{\mathrm{DHPMA},x}$, and $\alpha_{\mathrm{Total},x}$ are the conversion of NIPAm, DHPMA, and total monomers at x min, respectively. $A_{\mathrm{NIPAm},0}$ and $A_{\mathrm{DHPMA},0}$ are the normalized peak areas of NIPAm (δ = 5.54 ppm) and DHPMA (δ = 5.67 ppm) at 0 min, while $A_{\mathrm{NIPAm},x}$ and $A_{\mathrm{DHPMA},x}$ are the normalized peak areas of NIPAm (δ = 5.54 ppm) and DHPMA (δ = 5.67 ppm) at x min.

2.6. Composition of the Copolymer

The C/H/N and O contents in the copolymer were determined using a Vario EL Cube (Elementar, Frankfurt, Germany) elemental analyzer and a Vario EL Cube CHNS/O (Elementar, Frankfurt, Germany) elemental analyzer, respectively; each type of copolymer was tested twice and the results were averaged (Table S2).

The molar ratio (F) of the DHPMA unit (d[M_1]) and NIPAm unit (d[M_2]) in the copolymer were calculated using the following formula:

$$F = \mathrm{d}[\mathrm{M}_1]/\mathrm{d}[\mathrm{M}_2] = [([\mathrm{O}]/16.00)/([\mathrm{N}]/14.01) - 1]/4 \quad (5)$$

Here, [O] and [N] are the mass fractions of elements O and N in the copolymer, respectively.

2.7. Reactivity Ratios

After an 8 min reaction, samples were collected and quenched immediately to terminate the polymerization and calculate the reactivity ratios. The conversions of monomers were quantified to be less than 10% using the ^{1}H NMR method. The composition of the copolymer was calculated using the elemental analysis results (Table S2). The reactivity ratios of NIPAm and DHPMA were calculated using the Fineman–Ross method [26]. The concentration ratio (f) of DHPMA ([M_1]) and NIPAm ([M_2]) was ascertained as follows:

$$f = \frac{A_{DHPMA,x}}{A_{NIPAm,x}} \quad (6)$$

The molar ratio (F) of the DHPMA unit (d[M_1]) and NIPAm unit (d[M_2]) in the copolymer was calculated using the content of element N and element O obtained from the elemental analysis.

The reactivity ratios of DHPMA (r_1) and NIPAm (r_2) were obtained by a linear fitting with the following equation:

$$f(F-1)/F = f^2/F \times r_1 - r_2 \quad (7)$$

Sequence length distributions of NIPAm and DHPMA in the copolymer were calculated using the following formulae:

$$\underline{N}(DHPMA)_x = (\frac{r_1}{r_1 + 1/A})^{x-1}(\frac{1}{r_1 A + 1}) \quad (8)$$

$$\underline{N}(NIPAm)_x = (\frac{r_2}{r_2 + A})^{x-1}(\frac{1}{r_2 / A + 1}) \quad (9)$$

where x represents the sequence length of the xDHPMA sequence (NIPAm-DHPMA$_x$-NIPAm) or xNIPAm sequence (DHPMA-NIPAm$_x$-DHPMA) in the copolymer and $\underline{N}$(DHPMA)$_x$ and $\underline{N}$(NIPAm)$_x$ represent the percentage of each sequence length in the DHPMA sequence and NIPAm sequence, respectively. A represents the theoretical feeding ratios of DHPMA and NIPAm ([DHPMA]/[NIPAm]).

2.8. Cloud Point

To determine the phase transition temperature of the copolymer, the transmittance of 0.2 wt% P(NIPAm-*co*-DHPMA) in deionized water was recorded at 500 nm in the temperature range of 25 °C to 50 °C with a step of 0.5 °C using a 721 Visible Spectrometer (Tianjin Taisite Analytical Instrument Co., Ltd., Tianjin, China). The cloud point (T_{cp}) was defined as the temperature corresponding with the midpoint of the transmittance. The phase transition range was defined as the temperature region where the transmittance decreased. Triplicate samples were measured for each step.

2.9. In Vitro Cytotoxicity

To assess the cytotoxicity of P(NIPAm-*co*-DHPMA), 5×10^3 NIH 3T3 cells in 100 µL of the culture medium were seeded in each well of a 96-well plate and incubated at 37 °C in a 5% CO_2 atmosphere for 24 h; then, the culture medium was replaced by 100 µL of a fresh medium containing P(NIPAm-*co*-DHPMA) at various concentrations (10 mg/mL and 0 mg/mL (Control group)) and the cells were incubated for another 24 h. Subsequently, the culture medium was replaced by 110 µL of a fresh medium containing 10 µL of the CCK-8 solution and incubated for an additional 1 h, then the absorbance of each well at 450 nm was measured using an AMR-100 microplate reader (Allsheng Instruments Co., Ltd., Hangzhou, China); five duplicates were measured for each group. The cell viability was calculated according to the following formula:

$$Cell\ viability\% = (A_{sample} - A_{background})/(A_{control} - A_{background}) \times 100 \quad (10)$$

The data were then analyzed via a one-way ANOVA with Tukey's post hoc test using OriginPro 2024. The difference between each group was considered to be statistically significant at $p < 0.05$.

Calcein-AM/PI staining was used to observe the live and dead cells after co-culture with P(NIPAm-*co*-DHPMA). Briefly, NIH 3T3 cells were seeded in a 24-well plate at a density of 5×10^4 cells per well. The cells were incubated in 1.0 mL of the culture medium in a 5% CO_2 atmosphere at 37 °C for 24 h. Then, the culture medium was replaced by 1.0 mL of fresh medium containing P(NIPAm-*co*-DHPMA) (10 mg/mL and 0 mg/mL (Control group)). The cells were incubated for another 24 h. Subsequently, the cells were stained and observed under an Olympus IX71 fluorescence microscope (Olympus, Tokyo, Japan).

3. Results and Discussion

3.1. Chemical Structure of the Copolymers

Before copolymerization, DHPMA was vacuum-distilled and the structure was confirmed by ^{1}H NMR (Figure S2). Generally, DHPMA contains minor isomers, 1,3-dihydroxypropyl methacrylate, because the isomerization reaction of DHPMA continues to occur during storage, which involves the migration of the ester group [37,38]. The equilibrium levels of DHPMA and 1,3-dihydroxypropyl methacrylate have been reported to be 90% and 10%, even as monomer units in the polymer [37,38].

A series of P(NIPAm-*co*-DHPMA) copolymers with different feeding ratios were synthesized using free radical copolymerization and were named PND 8-2, PND 7-3, and PND 6-4, respectively. Their chemical structures were confirmed by ^{1}H NMR and ^{13}C NMR spectra (Figures 1a,b and S3). The attribution of the peaks in the ^{1}H NMR and ^{13}C NMR spectra was evidenced by DEPT, ^{1}H-^{1}H COSY, and HSQC spectra (Figure 1c–e; Tables S3 and S4) and are summarized in Table 1.

Table 1. Chemical shifts and their attribution from the ^{1}H NMR and ^{13}C NMR spectra of the PND 7-3 copolymer.

^{1}H NMR		^{13}C NMR	
Chemical Shift [ppm]	**Attribution**	**Chemical Shift [ppm]**	**Attribution**
7.38	1H, –NH– [39]	176.44	C=O$_{DHPMA}$
4.99	1H, –CH$_2$–CH–OH	173.88	C=O$_{NIPAm}$
4.74	1H, –CH–CH$_2$–OH	69.56	–CH$_2$–CH–OH
3.90	2H, –O–CH$_2$–	66.17	–O–CH$_2$–
3.83	1H, –CH–(CH$_3$)$_2$	63.11	–CH–CH$_2$–OH
3.70	1H, –CH$_2$–CH–OH	60.04	–CH(CH$_2$OH)$_2$
3.54	–CH(CH$_2$OH)$_2$	45.33	–C(CH$_3$)–CH$_2$–
3.40	2H, –CH–CH$_2$–OH	41.81	–CH–CH$_2$–
1.91	1H, –CH–CH$_2$–	40.31	–CH–(CH$_3$)$_2$
1.40	2H, –CH–CH$_2$–	22.72	–CH–(CH$_3$)$_2$
1.04	3H, –CH–(CH$_3$)$_2$	22.45	–CH–CH$_2$–
0.85	3H, –C–CH$_3$	17.52	–C–CH$_3$

The carbon atom of methyne (f′) was located at 76.37 ppm in ^{13}C NMR, while the proton of methyne (f′) was located in the 4.61–4.83 ppm region in ^{1}H NMR and was covered by the signal of other peaks [40]. The peaks of the 1,3-dihydroxypropyl methacrylate units in the copolymers were weak compared with the NIPAm and DHPMA units, indicating that the main chemical structure of P(NIPAm-*co*-DHPMA) comprised NIPAm units and DHPMA units.

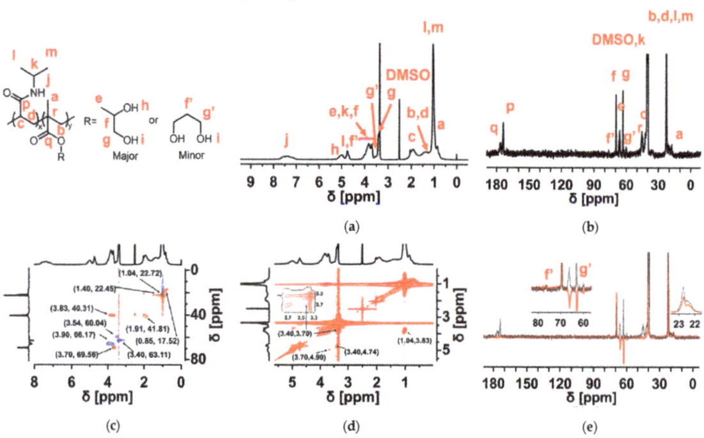

Figure 1. Structure of PND 7-3 copolymer: (**a**) ^{1}H NMR, (**b**) ^{13}C NMR, (**c**) HSQC, (**d**) ^{1}H-^{1}H COSY, and (**e**) DEPT spectrum (135°; red) vs. ^{13}C NMR (gray).

The molecular molar mass and distribution of synthesized copolymers were determined using GPC measurements (Figure S4). The number average molar mass (M_n) and PDI were dependent on the feeding ratio of DHPMA/NIPAm (Table S5). When the feeding ratio increased from 2/8 to 3/7 (DHPMA/NIPAm), M_n increased from 57,820 g/mol to 234,160 g/mol, while PDI decreased from 3.79 to 1.82. Further increasing the feed ratio to 4/6 led to a decrease in both M_n (217,880 g/mol) and PDI (1.66).

3.2. *Kinetics of Copolymerization*

The kinetics of copolymerization were monitored using ^{1}H NMR spectroscopy and elemental analyses. In this work, DHPMA contained 10% 1,3-dihydroxypropyl methacrylate after purification. The investigation of the copolymerization of NIPAm and DHPMA at this DHPMA purity was suitable for guiding the industrial production of materials. Based on the T_1 value of DHPMA (δ = 5.67 ppm), NIPAm (δ = 5.54 ppm) (Figure S1), and 1,4-dinitrobenzene (δ = 8.46 ppm), the value of D_1 = 30 s was used to obtain 99% of the equilibrium magnetization [41].

The conversion of monomers was calculated based on the integral of the peaks of the characteristic protons of monomers in the ^{1}H NMR spectra at different copolymerization times (Figures 2a–c and S5–S7). After a 2 h reaction, the conversion of DHPMA reached over 90%, while the conversion of NIPAm was over 60%. The consumption of DHPMA was much faster than NIPAm, suggesting that the DHPMA content in the formed copolymers was higher at the beginning of copolymerization and decreased as the reaction progressed. The total conversion of the comonomers decreased from 94% to 77% when the DHPMA/NIPAm feeding ratio increased from 2/8 to 4/6, suggesting that the reaction rate of copolymerization decreased with an increase in the DHPMA/NIPAm feeding ratio. At a DHPMA/NIPAm feed ratio of 2/8, the conversion of DHPMA reached 100%; that of NIPAm reached approx. 60% after one hour's reaction, suggesting the homopolymerization of NIPAm happened after the complete consumption of DHPMA.

A Fineman–Ross plot was harnessed to derive the reactivity ratios of the comonomers based on the conversion of monomers and the composition of products after an 8 min copolymerization (Figures 2d and S8 and Table S6). All the conversions of the comonomers were confirmed to be no more than 10% after 8 min of copolymerization by ^{1}H NMR. The reactivity ratio of DHPMA was 3.09, suggesting that DHPMA had greater preference for homopolymerization than for copolymerization. The reactivity ratio of NIPAm was 0.11, indicating that NIPAm had greater preference for copolymerization than for homopolymerization. DHPMA has previously been reported to contain minor isomers, 1,3-dihydroxypropyl methacrylate. The reactivity ratio of DHPMA obtained in this work was higher than the value of 2.55 previously reported [25], based on unpurified DHPMA (purity of only 90%). The reactivity ratios of DHPMA and 1,3-dihydroxypropyl methacrylate have been reported to be 0.7 and 1.6, respectively [42]. Therefore, the different values of the reactivity ratio of DHPMA could be attributed to either the lower purity of the starting materials or the actually lower content of 1,3-dihydroxypropyl methacrylate in this work. These results suggest that DHPMA with a lower minor isomer ratio had a higher preference for homopolymerization during the copolymerization of NIPAm and DHPMA.

Based on the reactivity ratios, one could calculate the instantaneous sequences for a copolymer system [25] (Figure 3). The most abundant DHPMA sequence was the NIPAm-DHPMA-NIPAm sequence (Figure 3a). The percentage of the NIPAm-DHPMA-NIPAm sequence decreased from 56.42% to 32.68% with an increase in the DHPMA/NIPAm feeding ratio from 2/8 to 4/6, while the percentage of other DHPMA sequences increased. The most abundant NIPAm sequence was the DHPMA-NIPAm-DHPMA sequence (Figure 3b).

The percentage of the DHPMA-NIPAm-DHPMA sequence increased from 69.44% to 85.84% with an increase in the DHPMA/NIPAm feeding ratio from 2/8 to 4/6, while the percentage of other NIPAm sequences decreased.

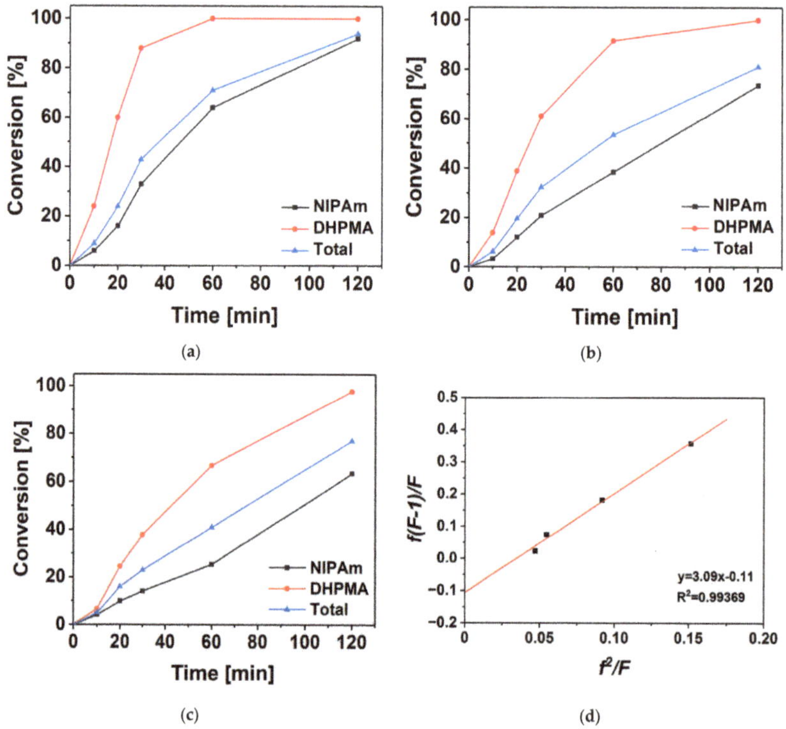

Figure 2. Kinetics of copolymerization. Conversion of (**a**) PND 8-2, (**b**) PND 7-3, and (**c**) PND 6-4 groups. (**d**) Fineman–Ross plot.

3.3. *Phase Transition Behavior*

The phase transition behavior of the P(NIPAm-*co*-DHPMA) aqueous solution was investigated at a macroscopic level using the cloud point method. The P(NIPAm-*co*-DHPMA) aqueous solution turned from clear to turbid upon increasing the temperature to T_{cp}, exhibiting characteristic LCST behavior (Figure 4). The T_{cp} of P(NIPAm-*co*-DHPMA) increased from 33.0 °C to 42.1 °C, with the DHPMA content increasing from 34.21 mol% to 57.98 mol% (Table 2), and was higher than PNIPAm (31.1 °C; this work). The molecular structure of a DHPMA unit comprises two hydroxyl groups and one ester group, which can form more hydrogen bonds compared with NIPAm. Therefore, P(NIPAm-*co*-DHPMA) could form more hydrogen bonds compared with PNIPAm. From the perspective of thermodynamics, the increase in hydrogen bonds between the polymers increased the value of negative mixing enthalpy ($|\Delta H_{mix}|$) and then increased the phase transition temperature, which turned the Gibbs energy of mixing (ΔG_{mix}) from negative to positive [5].

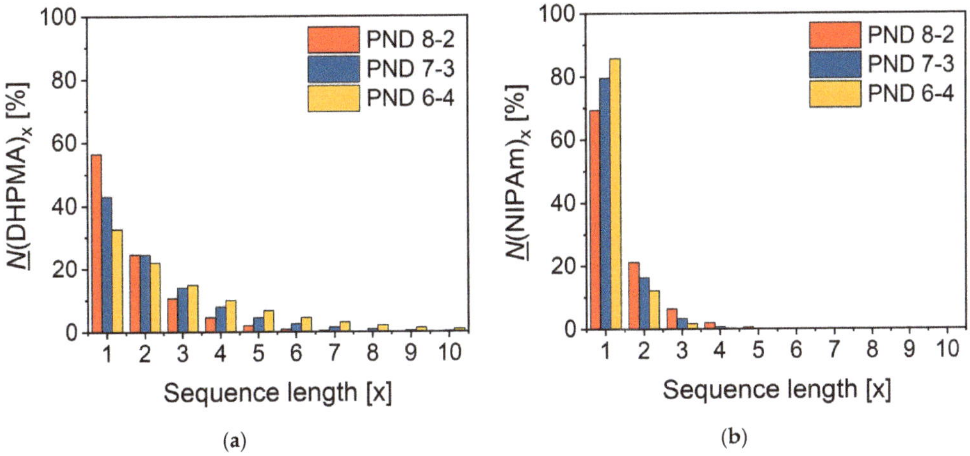

Figure 3. Sequence length distribution for the copolymerization of NIPAm and DHPMA at different feeding ratios. The vertical axis represents the percentage of the sequence length for (**a**) DHPMA and (**b**) NIPAm. A sequence length longer than 10 existed, but the percentage of these sequences was less than 2%.

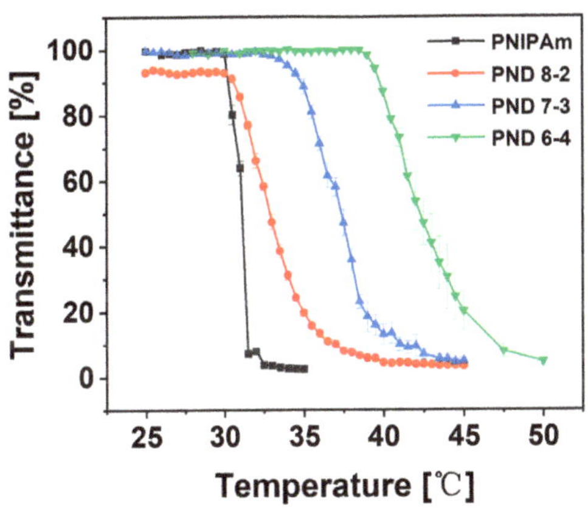

Figure 4. Phase transition behavior of P(NIPAm-*co*-DHPMA).

Table 2. The T_{cp} and phase transition range of the copolymers.

Polymer Name	Feeding Ratio ($[M_1]/[M_2]$)	DHPMA Content [mol%]	Phase Transition Range [°C]	T_{cp} [°C]
PNIPAm	0	0	30.0–32.5	31.1
PND 8-2	0.25	34.21	30.0–40.0	33.0
PND 7-3	0.43	45.65	32.0–43.5	36.5
PND 6-4	0.67	57.98	38.5–50.0	42.1

Compared with reversible addition-fragmentation chain transfer copolymerization [25], the copolymer synthesized using free radical polymerization had a higher PDI because of uncontrollable polymerization. Therefore, the products of free radical copolymerization in this work comprised a series of molecular chains with different DHPMA contents, which finally led to a wider phase transition range [43]. These findings provide a one-step method to fabricate a thermo-responsive polymer for a specific temperature range, which has potential application in biomedical fields. For example, the photothermal treatment temperature can be controlled by the transmittance of a temperature response coverage device [44].

3.4. *In Vitro Cytotoxicity*

To evaluate the cytocompatibility of P(NIPAm-*co*-DHPMA), NIH 3T3 cells were co-cultured with the copolymer for 24 h and stained. Almost no dead cells were detected at concentrations up to 10 mg/mL, and no difference in the cell morphology was observed compared with the Control group (Figure 5a–c). The cell viability in the presence of P(NIPAm-*co*-DHPMA) was above 80% for all studied copolymers and showed no significant difference between the Control group and experimental groups (Figure 5d). These results suggested the good cytocompatibility of the synthesized P(NIPAm-*co*-DHPMA).

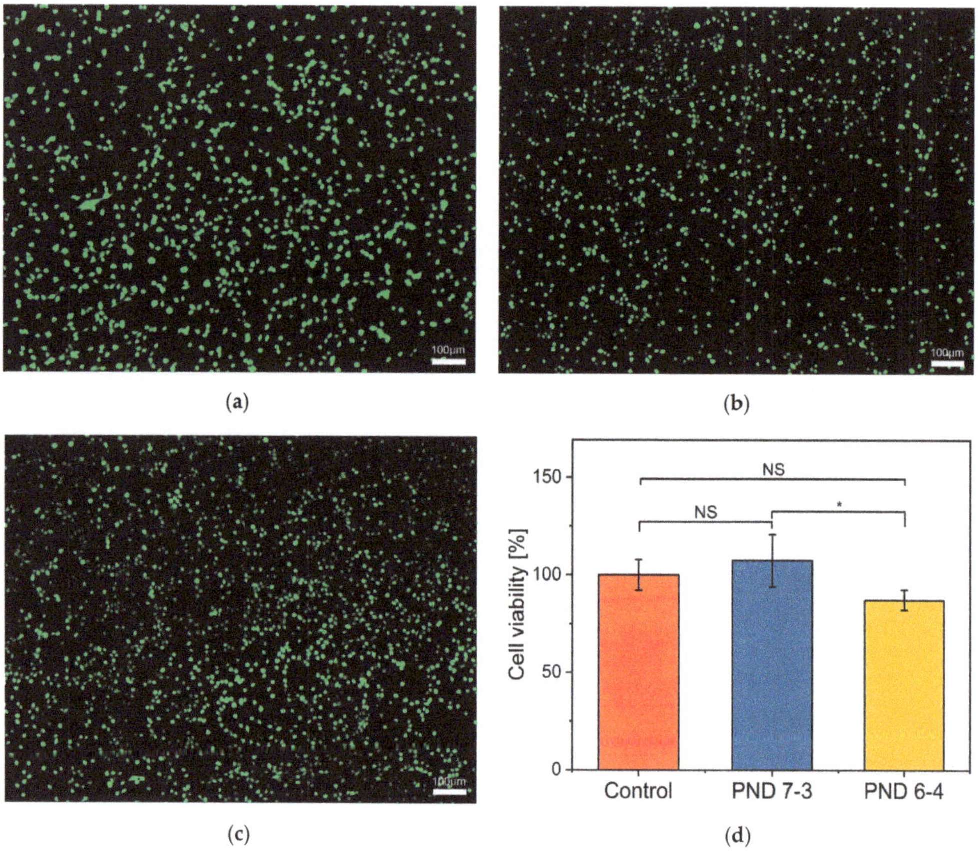

Figure 5. In vitro cytotoxicity of P(NIPAm-*co*-DHPMA). Live/Dead staining of (**a**) Control group, (**b**) PND 7-3, and (**c**) PND 6-4. Scale bar: 100 μm. (**d**) Cell viability. * $p < 0.05$; NS: no significance.

4. Conclusions

In this work, DHPMA with 10% isomers was used to regulate the phase transition behavior of PNIPAm. A series of thermo-responsive copolymers (P(NIPAm-*co*-DHPMA)) were synthesized via free radical copolymerization. Characterizations of the copolymerization process and products were carried out, and the following findings were obtained:

- The reaction kinetics of the free radical copolymerization of NIPAm and DHPMA were investigated. After two hours' copolymerization, the conversion of DHPMA and NIPAm reached at least 90% and 60%, respectively. The reactivity ratios of DHPMA and NIPAm were 3.09 and 0.11, respectively. DHPMA with a lower isomer content had a higher preference for homopolymerization.
- The NIPAm-DHPMA-NIPAm and the DHPMA-NIPAm-DHPMA sequences were the most abundant in the copolymer, suggesting a preferred random distribution of the two structural units in the main chain of the copolymer.
- The cloud point of P(NIPAm-*co*-DHPMA) increased from 31 to 42 °C with an increase in the content of DHPMA from 0 to 58 mol%; these copolymers exhibited a wide phase transition temperature range.
- Finally, the good cytocompatibility of P(NIPAm-*co*-DHPMA) copolymers was confirmed.

Supplementary Materials: The following supporting information can be downloaded at: https://www.mdpi.com/article/10.3390/ma18071614/s1. Figure S1: T_1 and ^{1}H NMR used for fitting; Figure S2: ^{1}H NMR spectrum of DHPMA; Figure S3: ^{1}H NMR and ^{13}C NMR of PND 8-2 and PND 6-4 copolymers; Figure S4: GPC results of P(NIPAm-*co*-DHPMA); Figure S5: ^{1}H NMR of the conversion of the PND 6-4 group at different times; Figure S6: ^{1}H NMR of the conversion of the PND 7-3 group at different times; Figure S7: ^{1}H NMR of the conversion of the PND 8-2 group at different times; Figure S8: ^{1}H NMR of the conversion of the PND 8-2, PND 7.5-2.5, PND 7-3, and PND 6-4 groups after an 8 min reaction; Table S1: Reaction information for the copolymer; Table S2: Elemental analysis; Table S3: Correlation peaks and their attribution in the ^{1}H-^{1}H COSY spectrum; Table S4: Correlation peaks and their attribution in the HSQC spectrum; Table S5: GPC results for P(NIPAm-*co*-DHPMA) at different feeding ratios; Table S6: Information for 8 min copolymerization and the reactivity ratios.

Author Contributions: Conceptualization: C.Z. and Z.C.; methodology: Z.C.; formal analysis: Z.C.; investigation: Z.C.; data curation: Z.C.; writing—original draft preparation: Z.C.; writing—review and editing: C.Z.; visualization: Z.C.; funding acquisition: C.Z. All authors have read and agreed to the published version of the manuscript.

Funding: This research was funded by the National Natural Science Foundation of China (81971760), the Guangzhou Key Research and Development Program (202206010115), and the Foundation of Guangdong Provincial Key Laboratory of Sensor Technology and Biomedical Instrument (No. 2020B121206007).

Institutional Review Board Statement: Not applicable.

Informed Consent Statement: Not applicable.

Data Availability Statement: The original contributions presented in this study are included in the article/Supplementary Materials. Further inquiries can be directed to the corresponding author.

Conflicts of Interest: The authors declare no conflicts of interest.

References

1. Das, A.; Babu, A.; Chakraborty, S.; Van Guyse, J.F.R.; Hoogenboom, R.; Maji, S. Poly(*N*-isopropylacrylamide) and its copolymers: A review on recent advances in the areas of sensing and biosensing. *Adv. Funct. Mater.* **2024**, *34*, 2402432. [CrossRef]
2. Pelton, R. Poly(N-isopropylacrylamide) (PNIPAM) is never hydrophobic. *J. Colloid Interface Sci.* **2010**, *348*, 673–674. [CrossRef]
3. Halperin, A.; Kröger, M.; Winnik, F.M. Poly(*N*-isopropylacrylamide) phase diagrams: Fifty years of research. *Angew. Chem. Int. Ed.* **2015**, *54*, 15342–15367. [CrossRef]
4. Wang, X.; Qiu, X.; Wu, C. Comparison of the coil-to-globule and the globule-to-coil transitions of a single poly(*N*-isopropylacrylamide) homopolymer chain in water. *Macromolecules* **1998**, *31*, 2972–2976. [CrossRef]

5. Zhao, C.; Ma, Z.; Zhu, X.X. Rational design of thermoresponsive polymers in aqueous solutions: A thermodynamics map. *Prog. Polym. Sci.* **2019**, *90*, 269–291. [CrossRef]
6. Tang, L.; Wang, L.; Yang, X.; Feng, Y.; Li, Y.; Feng, W. Poly(*N*-isopropylacrylamide)-based smart hydrogels: Design, properties and applications. *Prog. Mater. Sci.* **2021**, *115*, 100702. [CrossRef]
7. Gregg, A.; De Volder, M.; Baumberg, J.J. Kinetics of light-responsive CNT/PNIPAM hydrogel microactuators. *Small* **2024**, *20*, 2305034. [CrossRef]
8. Wang, K.; Chen, G.; Weng, S.; Hou, L.; Ye, D.; Jiang, X. Thermo-responsive poly(*N*-isopropylacrylamide)/hydroxypropylmethyl cellulose hydrogel with high luminous transmittance and solar modulation for smart windows. *ACS Appl. Mater. Interfaces* **2023**, *15*, 4385–4397. [CrossRef]
9. Cui, X.; Liu, Z.; Zhang, B.; Tang, X.; Fan, F.; Fu, Y.; Zhang, J.; Wang, T.; Meng, F. Sponge-like, semi-interpenetrating self-sensory hydrogel for smart photothermal-responsive soft actuator with biomimetic self-diagnostic intelligence. *Chem. Eng. J.* **2023**, *467*, 143515. [CrossRef]
10. Eklund, A.; Ikkala, O.; Zhang, H. Highly efficient switchable underwater adhesion in channeled hydrogel networks. *Adv. Funct. Mater.* **2024**, *34*, 2214091. [CrossRef]
11. Li, F.; Wang, C.; Guo, W. Multifunctional poly-N-Isopropylacrylamide/DNAzyme microgels as highly efficient and recyclable catalysts for biosensing. *Adv. Funct. Mater.* **2018**, *28*, 1705876. [CrossRef]
12. Kim, H.; Nam, S.; Durukan, M.B.; Unalan, H.E.; Lee, H.J. Self-charging dual-modal sensor for glucose monitoring based on piezoelectric nanowire/microgel hybrid film. *Adv. Funct. Mater.* **2024**, *34*, 2308086. [CrossRef]
13. Li, B.-Y.; Lin, T.-Y.; Lai, Y.-J.; Chiu, T.-H.; Yeh, Y.-C. Engineering multiresponsive alginate/PNIPAM/carbon nanotube nanocomposite hydrogels as on-demand drug delivery platforms. *Small* **2025**, *21*, 2407420. [CrossRef]
14. Lopez-Larrea, N.; Wustoni, S.; Peñas, M.I.; Uribe, J.; Dominguez-Alfaro, A.; Gallastegui, A.; Inal, S.; Mecerreyes, D. PNIPAM/PEDOT:PSS hydrogels for multifunctional organic electrochemical transistors. *Adv. Funct. Mater.* **2024**, *34*, 2403708. [CrossRef]
15. Viola, M.; Valverde, M.G.; Bernal, P.N.; van Trijp, J.P.; Hak, J.; Marco, G.D.; Neumann, M.; Schuurmans, C.C.; van Nostrum, C.F.; Masereeuw, R.; et al. Thermal shrinking of biopolymeric hydrogels for high resolution 3D printing of kidney tubules. *Adv. Funct. Mater.* **2024**, *34*, 2406098. [CrossRef]
16. Ding, H.; Li, B.; Liu, Z.; Liu, G.; Pu, S.; Feng, Y.; Jia, D.; Zhou, Y. Decoupled pH-and thermo-responsive injectable chitosan/PNIPAM hydrogel via thiol-ene click chemistry for potential applications in tissue engineering. *Adv. Healthc. Mater.* **2020**, *9*, 2000454. [CrossRef]
17. Lerch, A.; Käfer, F.; Prévost, S.; Agarwal, S.; Karg, M. Structural insights into polymethacrylamide-based LCST polymers in solution: A small-angle neutron scattering study. *Macromolecules* **2021**, *54*, 7632–7641. [CrossRef]
18. Kawano, S.; Lie, J.; Ohgi, R.; Shizuma, M.; Muraoka, M. Modulating polymeric amphiphiles using thermo- and pH-responsive copolymers with cyclodextrin pendant groups through molecular recognition of the lipophilic dye. *Macromolecules* **2021**, *54*, 5229–5240. [CrossRef]
19. Wang, C.; Li, P.; Zhang, S.; Zhang, G.; Tan, S.; Wu, Y.; Watanabe, M. Azobenzene molecular trigger controlling phase transitions of PNIPAm in ionic liquids and light-controlled adhesiveness. *Macromolecules* **2020**, *53*, 4901–4907. [CrossRef]
20. Vishnevetskaya, N.S.; Hildebrand, V.; Niebuur, B.-J.; Grillo, I.; Filippov, S.K.; Laschewsky, A.; Müller-Buschbaum, P.; Papadakis, C.M. Aggregation behavior of doubly thermoresponsive polysulfobetaine-b-poly(*N*-isopropylacrylamide) diblock copolymers. *Macromolecules* **2016**, *49*, 6655–6668. [CrossRef]
21. Feil, H.; Bae, Y.H.; Feijen, J.; Kim, S.W. Effect of comonomer hydrophilicity and ionization on the lower critical solution temperature of N-isopropylacrylamide copolymers. *Macromolecules* **1993**, *26*, 2496–2500. [CrossRef]
22. Su, G.; Zhou, T.; Liu, X.; Ma, Y. Micro-dynamics mechanism of the phase transition behavior of poly(N-isopropylacrylamide-co-2-hydroxyethyl methacrylate) hydrogels revealed by two-dimensional correlation spectroscopy. *Polym. Chem.* **2017**, *8*, 865–878. [CrossRef]
23. Cao, Z.; Zhang, L.; Zhao, S.; Li, X.; Zhao, B.; Zhou, Y.; Gong, S.; Chen, X.; Feng, X.; Pan, K. Study on polymerization kinetics of copolymerized flame-retardant polyamide 66: Apparent rate constant and reaction activation energy. *Polymer* **2024**, *299*, 126936. [CrossRef]
24. Wu, Z.-C.; Liu, Y.; Wei, W.; Chen, F.-S.; Qiu, G.-X.; Xiong, H.-M. Reaction kinetics in anionic copolymerization: A revisit on Mayo-Lewis equation. *Chin. J. Polym. Sci.* **2016**, *34*, 431–438. [CrossRef]
25. Ventura-Hunter, C.; Lechuga-Islas, V.D.; Ulbrich, J.; Kellner, C.; Schubert, U.S.; Saldívar-Guerra, E.; Rosales-Guzmán, M.; Guerrero-Sánchez, C. Glycerol methacrylate-based copolymers: Reactivity ratios, physicochemical characterization and cytotoxicity. *Eur. Polym. J.* **2022**, *178*, 111478. [CrossRef]

26. Fineman, M.; Ross, S.D. Linear method for determining monomer reactivity ratios in copolymerization. *J. Polym. Sci.* **1950**, *5*, 259–262. [CrossRef]
27. Tidwell, P.W.; Mortimer, G.A. An improved method of calculating copolymerization reactivity ratios. *J. Polym. Sci. Part A Gen. Pap.* **1965**, *3*, 369–387. [CrossRef]
28. Kucharski, M.; Ryttel, A. Copolymerization of alkyl acrylates with triallyl cyanurate: Kelen-Tüdös method applied for determining copolymerization reactivity ratios. *Polymer* **1984**, *25*, 555–558. [CrossRef]
29. Save, M.; Weaver, J.V.M.; Armes, S.P.; McKenna, P. Atom transfer radical polymerization of hydroxy-functional methacrylates at ambient temperature: Comparison of glycerol monomethacrylate with 2-hydroxypropyl methacrylate. *Macromolecules* **2002**, *35*, 1152–1159. [CrossRef]
30. Moreddu, R.; Vigolo, D.; Yetisen, A.K. Contact lens technology: From fundamentals to applications. *Adv. Healthc. Mater.* **2019**, *8*, 1900368. [CrossRef]
31. Fan, X.; Torres-Luna, C.; Azadi, M.; Domszy, R.; Hu, N.; Yang, A.; David, A.E. Evaluation of commercial soft contact lenses for ocular drug delivery: A review. *Acta Biomater.* **2020**, *115*, 60–74. [CrossRef]
32. Wang, C.; Yu, B.; Knudsen, B.; Harmon, J.; Moussy, F.; Moussy, Y. Synthesis and performance of novel hydrogels coatings for implantable glucose sensors. *Biomacromolecules* **2008**, *9*, 561–567. [CrossRef]
33. Patrucco, E.; Ouasti, S.; Vo, C.D.; De Leonardis, P.; Pollicino, A.; Armes, S.P.; Scandola, M.; Tirelli, N. Surface-initiated ATRP modification of tissue culture substrates: Poly(glycerol monomethacrylate) as an antifouling surface. *Biomacromolecules* **2009**, *10*, 3130–3140. [CrossRef]
34. Nalawade, A.C.; Ghorpade, R.V.; Shadbar, S.; Qureshi, M.S.; Chavan, N.N.; Khan, A.A.; Ponrathnam, S. Inverse high internal phase emulsion polymerization (i-HIPE) of GMMA, HEMA and GDMA for the preparation of superporous hydrogels as a tissue engineering scaffold. *J. Mater. Chem. B* **2016**, *4*, 450–460. [CrossRef]
35. Robert-Nicoud, G.; Evans, R.; Vo, C.-D.; Cadman, C.J.; Tirelli, N. Synthesis, self-assembly and (absence of) protein interactions of poly(glycerol methacrylate)–silicone macro-amphiphiles. *Polym. Chem.* **2013**, *4*, 3458–3470. [CrossRef]
36. Rundlöf, T.; Mathiasson, M.; Bekiroglu, S.; Hakkarainen, B.; Bowden, T.; Arvidsson, T. Survey and qualification of internal standards for quantification by ^{1}H NMR spectroscopy. *J. Pharm. Biomed.* **2010**, *52*, 645–651. [CrossRef]
37. Shaw, S.E.; Russo, T.; Solomon, D.H.; Qiao, G.G. An alternative pathway for the hydrolysis of epoxy ester compounds. *Polymer* **2006**, *47*, 8247–8252. [CrossRef]
38. García, N.; Guzmán, J.; Riande, E. Migration phenomena in methacrylic esters and polymers derived from glycerol. *Macromol. Chem. Phys.* **2002**, *203*, 2225–2231. [CrossRef]
39. Huang, Z.-S.; Shiu, J.-W.; Way, T.-F.; Rwei, S.-P. A thermo-responsive random copolymer of poly(NIPAm-co-FMA) for smart textile applications. *Polymer* **2019**, *184*, 121917. [CrossRef]
40. Wolf, F.K.; Hofmann, A.N.; Frey, H. Poly(isoglycerol methacrylate)-*b*-poly(D or L-lactide) copolymers: A novel hydrophilic methacrylate as building block for supramolecular aggregates. *Macromolecules* **2010**, *43*, 3314–3324. [CrossRef]
41. Bharti, S.K.; Roy, R. Quantitative ^{1}H NMR spectroscopy. *TrAC-Trend Anal. Chem.* **2012**, *35*, 5–26. [CrossRef]
42. Madsen, J.; Madden, G.; Themistou, E.; Warren, N.J.; Armes, S.P. pH-responsive diblock copolymers with two different fluorescent labels for simultaneous monitoring of micellar self-assembly and degree of protonation. *Polym. Chem.* **2018**, *9*, 2964–2976. [CrossRef]
43. Kubo, M.; Higuchi, M.; Koshimura, T.; Shoji, E.; Tsukada, T. Control of the temperature responsiveness of poly(*N*-isopropylacrylamide-co-2-hydroxyethyl methacrylate) copolymer using ultrasonic irradiation. *Ultrason. Sonochem.* **2021**, *79*, 105752. [CrossRef]
44. Chu, Y.; Wang, Q.; Lyu, S.; Yuan, B.; Huang, J.; Li, J.; Wang, Y. Thermal-responsive gel-based overheat limiter enabled intelligent photothermal therapy. *Small* **2024**, *20*, 2312140. [CrossRef]

materials

MDPI

Article

Experimental and Numerical Characterization of Electrospun Piezoelectric Polyvinylidene Fluoride Nanocomposites Reinforced with Silver Nanoparticles

Strahinja Milenković [1,2,*], Fatima Živić [2], Nenad Grujović [2], Katarina Virijević [1,2], Aleksandar Bodić [2] and Danilo Petrović [2]

[1] Institute for Information Technologies, University of Kragujevac, 34000 Kragujevac, Serbia; katarina.virijevic@uni.kg.ac.rs

[2] Faculty of Engineering, University of Kragujevac, 34000 Kragujevac, Serbia; zivic@kg.ac.rs (F.Ž.); gruja@kg.ac.rs (N.G.); abodic@uni.kg.ac.rs (A.B.); petrovicdanilo1999@gmail.com (D.P.)

* Correspondence: strahinja.milenkovic@kg.ac.rs

Abstract: This study focuses on preparing piezoelectric polyvinylidene fluoride (PVDF) nanocomposites reinforced with silver nanoparticles (AgNPs) using an electrospinning process. The aim of this study is to assess AgNPs' influence on the piezoelectric properties of PVDF and, therefore, create an optimal piezoelectric composite with enhanced properties, enabling its application in various fields both as sensor and actuator. Because electrical stimuli have proven to have a positive influence in tissue engineering, combined with AgNPs, which have antimicrobial properties, these composites demonstrate a promising opportunity for application as biomedical scaffolds. The proposed scaffolds were characterized by scanning electron microscopy, energy-dispersive X-ray spectroscopy and Fourier transform infrared spectroscopy. In addition, mechanical properties are studied through tensile tests, while piezoelectric response is measured on an in-house built setup coupling mechanical stimuli and electrical response monitoring. An experimental test was combined with numerical simulations through the COMSOL Multiphysics version 6.3 software package, and this paper also presents a short review of the numerical and analytical methods used for the modelling and simulation of piezoelectric composites.

Keywords: PVDF/AgNP composite nanofibers; silver nanoparticles (AgNPs); piezoelectricity; electrospinning; numerical modelling

1. Introduction

Poly(vinylidene fluoride) (PVDF), a piezoelectric polymer, has been extensively studied for usage in sensor and actuation applications [1,2]. Its piezoelectric properties make it particularly suitable for converting mechanical stress and strain into electrical energy, making it valuable for various energy harvesting applications [3,4]. In recent years, research has focused on polar polymers that exhibit piezoelectric, pyroelectric and ferroelectric properties, including PVDF, copolymers of trifluoroethylene and vinylidene fluoride and, as well as aromatic properties, including polyurethanes, vinyl cyanide and acetate, and nylons [5]. PVDF can exist in multiple crystalline phases (α, β, γ, δ and ε), but only the β phase possesses piezoelectric, pyroelectric and ferroelectric characteristics [6–9] suitable for energy harvesting applications.

In order to fabricate PVDF with a high β-phase fraction, certain challenges need to be addressed. The most common method to enhance the β-phase fraction in PVDF is mechanical stretching, which promotes transformation from the α phase [8,10,11]. This transition is also highly affected by the temperature and stretching ratio, which, in turn, impact the degree of crystallinity and microstructure, ultimately influencing the macroscopic properties of said material. PVDF's fabrication, polymerization and process parameters can also affect the formation of certain phase contents [12]. Various fabrication methods have been investigated, including electrospinning, solution casting, spin coating, hot pressing, self-poling, melt blending, soft lithography and additive manufacturing techniques such as 3D printing and solvent evaporation-assisted 3D printing [2,4,13,14].

Electrospinning, as a fabrication method, employs mechanical stretching of a liquid solution droplet under a high electric field to produce long nanofibers [15,16]. Both the stretching process and the applied electric field facilitate dipole alignment in PVDF, leading to a higher β-phase fraction in the resulting nanofiber mats [17–19].

Adding fillers into the PVDF matrix is proven to increase electroactive phase content in PVDF, as well as mechanical properties. This fillers include barium-titanate ($BaTiO_3$) [20], Graphene Oxide (GO) [21], silver nanoparticles (AgNPs) [22–24], zinc oxide (ZnO) 25], etc. In addition, silver nitrate ($AgNO_3$) is a well-known antimicrobial agent that inhibits the growth of microbes; it has proven antimicrobial and anti-cancer effects due to its release of silver ions (Ag+) that disrupt the membranes of cancer cells and microorganisms [2].

Ag's proven cytotoxic effect on cancer cells combined with PVDF piezoelectric properties paves the way for the development of smart, multifunctional materials for biomedical applications such as cancer treatment [2]; tissue engineering for orthopedics [25,26] and wound healing [27]; and biomedical sensors [28]. In our previously published paper [2], we proved that PVDF/AgNPs nanomats show both a prominent effect on the MDA-MB-231 breast cancer cell line and non-toxicity on the healthy MRC-5 cell line, as well as antibacterial activity against *S. aureus* and *P. aeruginosa* at Ag contents of 0.3% Ag, which was one of the reasons why we choose to test PVDF + 0.3% AgNPs in this work.

Fourier transform infrared spectroscopy (FTIR) is an effective method for the characterization of electroactive phases in PVDF [2,12,29]. In addition, Raman spectroscopy [12] and X-ray diffraction [10] are also utilized for the same purposes.

In order to measure piezoelectric response, various methods are reported in the literature since there are no established experimental methods for measuring all piezoelectric properties of a material. Piezoelectric Force Microscopy (PFM) is based on deformation induced by voltage on piezoelectric structures; this method can be compared to Atomic Force Microscopy (AFM) [30]. Generally, piezoelectric properties can be determined using commercially available equipment, but many researchers build their own in-house equipment, such as impedance analyzers [31], or run simulation measurements of mechanical load and electrical voltage [32] or mechanical strain and voltage in quasi-static conditions [33] or under frequent load conditions [21,23,34]. In addition, some authors reported testing piezoelectric nanomats under stochastic test conditions simulating those found in nature [20].

Modelling and simulating piezoelectric composites' piezoelectric response is currently cutting-edge research [20,32,35] focused on establishing methods for easier design and the prediction of material behaviour to tailor it for specific purposes. A short review of numerical and analytical methods will be presented in the next chapter.

In this study, the linear electroelastic behaviour of piezoelectric composites was investigated by applying FEM modelling, where previously experimentally obtained material properties are utilized. For this purpose, COMSOL Multiphysics version 6.3 software was used.

2. Numerical and Analytical Modelling of Piezoelectric Composites

Computational models of piezoelectric composites have been developed and studied to support different aspects of real case applications, as previously listed [36–40]. Each of the specific applications focuses on some of the available properties of the piezoelectric material, such as maximizing the output voltage or power. Hence, it is necessary to optimize the composite structure and performance in order to provide necessary material property for a specific application. Experimental procedures for adjusting the composite properties are tedious and long-lasting. These experiments often lack precise predictability, particularly in the absence of the physical analytical laws for the complex material structures. As a result, numerous repetitions of the experimental samples and corresponding tests are typically required. Computational models and related simulations of the material performance under various conditions have immensely shortened the time for material discoveries. They also provide more comprehensive understanding of the associated phenomena in material behaviour, thus justifying efforts engaged in developing new material models and new methods in simulating their behaviour. This is extremely important especially for conditions that cannot be easily set up in a real lab environment, such as mimicking human tissue environments. However, the complexity of possible piezoelectric composite structures, together with simultaneous coupling of different influential conditions, represents a significant challenge in the development of the computational models.

Theoretical analytical and mathematical models, very often physics-based models, are the foundation for the development of modern computational modelling and simulations, including AI-based predictions for material behaviour. Some of the first theoretical frameworks for the prediction of properties of piezoelectric composites with different reinforcements were reviewed in [36], together with the introduction of connectivity theory by Newnham [41], where different standard types of connections between the reinforcement and matrix were suggested, to distinguish different types of piezoelectric composites where 0-3, 1-3, 3-3 and 2-2 composites are amongst the most utilized ones. These defined connectivity patterns further enable the definition of series and parallel models that represent connections between piezoelectric composite layers or reinforcing fibres and particles, thus enabling well-defined approximations in numerical models. In the case of PVDF-based composites, 0-3, 1-3, 3-3 and 2-2 connectivity patterns are of interest. The coupled behaviour present in piezoelectric composites, originating from mechanical and electrical influences, has been a subject of many works, also by introducing specific boundary conditions.

In general, micromechanics models were first used for the modelling of piezoelectric materials, and they are still the most significant theoretical models [36]. The four most prominent micromechanics models for piezoelectric composites are as follows [36,40]:

1. The dilute approximation;
2. The self-consistent method;
3. The Mori–Tanaka mean field method;
4. The differential scheme.

More detailed categorisation of the existing material models used in the modelling of piezoelectric behaviour are given in [40]. This very recent review presented the latest developments of the numerical methods for piezoelectric composites, including homogenization methods [40].

Modelling of the piezoelectric composites, according to the specific approach, can be categorized as follows [40]:

5. Analytical models:
    a. Micromechanics models based on averaging material properties through the constituent phases and with different types of connectivity, where 0-3 represents the random dispersion of elements within the composite that have isotropic properties; 1-3 represents structures with fibres where one observed phase can be connected through series (meaning constant properties within the composite) or parallel connections (where weights are calculated); and 3-3 connectivity represents the 3D distribution of elements, but, unlike the 0-3 type, it has a somewhat arranged orientation and alignment of the phases. However, these models cannot capture localized behaviour within small domains due to the averaged properties, such as commonly appearing fluctuations in properties (both mechanical and electrical) within the small domains in different phases and along their interfaces.
    b. Models based on Eshelby equations:
        i Dilute approximation models;
        ii Mori–Tanaka-based models;
        iii Extended Mori–Tanaka models, such as Mori–Tanaka–Eshelby models;
        iv Self-consistent models and their extended models;
        v Extended rule of mixture models.
    c. Models based on asymptotic homogenization (or periodic homogenization).
6. Numerical models or Finite Element (FE) models, which use representative volume elements (RVEs) throughout the 3D volumetric space of the composite for the calculation of composite properties and responses to external influences, where the definition of the size and shape of RVEs essentially determines the model validity. FEMs can capture localized behaviour to some degree, depending on the defined RVEs, with several established approaches, also related to the computational resources needed for calculations:
    a. Linear FE models (usually for elastic behaviour or small deformations);
    b. Nonlinear FE models (large deformations, elasto-plastic material or nonlinear contacts);
    c. The representative volume element method;
    d. The theory of periodic boundary conditions;
    e. Calculation methods for the effective coefficients;
    f. Optimization of piezoelectric behaviour with different possible methods, such as the Discrete Material Optimization (DMO) method.

However, the accuracy of each model is closely related to the defined boundary conditions that, on the other hand, strongly depend on the observed specific application case. The developed micromechanics models are also used for the calculations of the effective composite properties, such as the effective elasticity modulus, effective piezoelectric coupling coefficients and dielectric coefficients. Each of the currently existing models have certain limitations for use. For example, the self-consistent method is not suitable for composites where there is a large difference in the moduli of the matrix and reinforcements. The number of reinforcements, or their volume fractions, have a very significant influence on the model accuracy. For example, the Mori–Tanaka and dilute methods are relevant methods for volume fractions lower than 20%.

The Mori–Tanaka method utilizes averaged stress and strain, strain concentrations in the reinforcements that are dependent on the plastic domain within the material microstructure, strain and load distribution in the plain, electrostatic field and the overall shape of the studied sample [42]. The dilute model assumes that there is a large space and no interactions between the reinforcement inclusions. Hence, it can be assumed that concentration tensors in reinforcements can be observed as independent of their volume fractions, which also means that this method is not suitable for high volume fractions. Determination of the concentration tensor fields can be performed in different ways, such as using Green's function [43]. The differential method deals with nonlinear problems, thus requiring large amounts of computational resources for solving ordinary differential equations. When considering computational resources and explicit solutions for the effective material properties, the Mori–Tanaka and dilute methods stand out for their easy use and are very often used in material modelling in general, including piezoelectric composites [38–40].

A summary of the advantages and limitations of different modelling approaches is given in Table 1.

From a theoretical point of view, it is possible to design PVDF-based piezoelectric composites with different connectivity patterns, as defined by [41]. However, in real case applications, reinforcements in the form of continuous long unidirectional fibres, or as nano and micro particles and their hybrid structural forms, are among the most commonly used composites. They can be efficiently represented by 0-3, 1-3 and 2-2 types of connections within the two-phase, three-phase and multi-phase composites, including porous structures.

Table 1. Summary of advantages and limitations of different modelling approaches.

Model Type	Advantages	Limitations
Micromechanics Models	- Estimate effective properties based on constituent phases and connectivity types (0-3, 1-3, 3-3). - Simple and efficient for initial approximations.	- Averaging approach may not capture localized behaviours. - Cannot model property fluctuations at small scales, such as phase interfaces.
Models Based on Eshelby's Equations	- Provide detailed predictions of inclusion effects on overall composite properties. - Suitable for modelling different volume fractions and shape effects.	- Limited by assumptions about phase interactions. - Accuracy depends on inclusion shape and volume fraction approximations.
Dilute Approximation Models	- Simple and computationally efficient. - Works well for low volume fractions of inclusions (assumes weak interactions). - Provides explicit expressions for electroelastic moduli.	- Assumes that inclusions do not interact, making it unreliable for higher volume fractions. - Fails to account for stress and strain interactions between inclusions.
Mori–Tanaka-Based Models	- More accurate than the dilute approximation for moderate inclusion concentrations. - Provides explicit estimates for effective electroelastic moduli. - Works well for aligned inclusions in a dominant matrix phase.	- Assumes uniform stress/strain fields in inclusions, which may not always be realistic. - Assumes that the matrix phase dominates the response, making it less reliable for composites with a high inclusion fraction. - Does not always predict transverse moduli accurately.
Extended Mori–Tanaka Models (e.g., Mori–Tanaka–Eshelby)	- Incorporates Eshelby's solution for a more precise estimation of inclusion effects.	- Limited in capturing strong interactions between phases.

Table 1. *Cont.*

Model Type	Advantages	Limitations
Self-Consistent Models	- Suitable for composites with moderate inclusion volume fractions. - Captures some level of phase interaction. - Incorporates interaction effects among inclusions. - Provides more realistic predictions compared to the dilute approximation.	- More complex and computationally intensive than simpler models. - Results in an implicit nonlinear algebraic matrix equation for electroelastic moduli, requiring numerical solutions. - Becomes inaccurate when there is a large contrast in the properties of the matrix and inclusions.
Extended Rule of Mixture Models	- Offers a more refined approach to account for phase interactions. - Bridges the gap between simple rule of mixtures and micromechanics models.	- Relies on approximations and assumptions about phase behaviour.
Models Based on Asymptotic Homogenization (or Periodic Homogenization)	- Can capture periodic microstructure effects. - More accurate for composites with ordered phase distribution.	- Computationally demanding. - Requires periodicity assumption, limiting applicability to random structures.
Differential Scheme	- Capable of modelling a wide range of inclusion volume fractions. - More accurate than Mori–Tanaka for high inclusion concentrations. - Accounts for gradual changes in material properties as more inclusions are added.	- Computationally demanding due to the need to solve 81 coupled nonlinear differential equations. - Requires iterative numerical integration.

Other micromechanics models have also been developed and studied, including the very recent approach of using artificial intelligence (AI) for material simulations [44,45]. Different analytical solutions have been implemented in the existing micromechanics models, considering different boundary conditions, to achieve the coupling of the mechanical and electrical phenomena in piezoelectric materials and determine effective composite properties [36,46]. Rigorous boundary conditions are still subject to research [40].

Material models can be further used to predict material behaviour, such as the recent mathematical model for the prediction of the harvested energy by piezoelectric composites [47]. This model can enable the study of different compositions for the optimal material design to suit the intended application. The recent approach focused on damage mechanics through phase-field modelling what is valuable for designing flexible piezoelectronics for a range of applications [48]. These models are relevant for large deformations and can capture nonlinear behaviour, which is a profoundly important property of PDVF-based piezoelectric composites that exhibit large failure strains [49]. Phase-field modelling uses thermodynamic principles together with kinetics and required input data about the microstructure to be able to predict microstructural evolution during the functional piezoelectric behaviour. Phase-field models employ Allen–Cahn and Cahn–Hilliard equations to include the interfaces between the different phases with the composite microstructure [50,51].

There are continuous efforts to upgrade FEM methods, focusing on better efficiency with lower computational resources needed for data processing [52]. Two-dimensional discretization within a single layer is a common FEM approach, but it lacks sufficient accuracy in simulation piezoelectric behaviour, especially for thick composites. Hence, 3D models and new FE models have been developed, often using combinations of 2D methods for mechanical displacements with 3D layerwise-like approximation for the electric potential field that can be used for both thin and moderately thick composites [53].

In 2D models, the distribution of the electric potential through the thickness of the composite is usually assumed to be constant, which can result in larger differences between the model and experimental results in the case of electrospun PVDF-based nanocomposites. Very fine discretization of the thickness can capture nonlinear behaviour and, together with carefully considered electric boundary conditions, the model can be designed with a moderate number of field variables to not require excessive computing power [53]. Commonly assumed linear or constant electric potential fields through the composite thickness can result in numerical solutions (for mechanical displacement and electric potential) that do not converge to the same exact solution, even with a finer meshing [53].

Currently developed 3D FEM models of piezoelectric composites use 3D representative volume elements (RVEs), are suitable for modelling thick piezoelectric structures and, often, can provide the exact solutions through solving the constitutive equations in three-dimensional space under certain prescribed boundary conditions (both the mechanical and electrical boundary conditions) [54]. If focusing on the lower computational resources with retained accuracy, layerwise theory has been used for composites that exhibit laminar structures [53].

Some recent models have combined FEM modelling with extended Eshelby–Mori–Tanaka micromechanics models for determining effective composite properties [54,55]. This model suggests that large piezoelectric coefficients, with low values of mechanical elastic compliance and low dielectric permittivity, are the best combinations for achieving high energy conversion efficiency in the case of energy harvesting applications using flexible piezoelectric composites. Another approach in FEM modelling has utilized Hamilton's variational principle in the specific cases of plates and beams [56]. Coupling of the electric potential with mechanical displacements represents a complex task in 3D FEM modelling and is still under investigation, whereas the definition of the appropriate RVEs has the most profound effects on the model accuracy. In the case of a polymer matrix reinforced with nanostructures such as nanotubes, molecular interactions play a detrimental role and traditional continuum mechanics cannot be used for modelling, with new approaches suggested through the introduction of an effective continuum fibre that can be further used in micromechanics models [57].

Recent approaches in the modelling of PVDF-based piezoelectric composites have considered flexoelectricity, or the coupling of the resulting changing strain gradients and associated electric potential fields [58]. However, when flexoelectricity is considered for anisotropic structures, the contribution of the shear components is still not well understood, since it is hard to determine it in experimental conditions and due to the simultaneous influences of the mechanical properties of the composite.

A micromechanical model has been developed for finding effective properties of hybrid orthotropic composites where a piezoelectric matrix made of polymer is reinforced with piezoelectric particles [39]. This model showed good accuracy for the low volume fractions of reinforcing particles. Anisotropy of PVDF fibre-based composites needs to be considered in material models with several approaches, such as considering specific equations for predicting effective dielectric properties using the Knott model [59], piezoelectric coefficients using the Furukawa model [60,61] , or using other theoretical models [39,62]. Some micromechanical models have considered anisotropy in composite structures and used different homogenization models, such as the models by Voigt [63], variational principles and the derivation of bounds by Hashin and Shtrikman [64], the Halpin–Tsai equations based on the self-consistent method [65,66], the upgraded self-consistent method-like incremental self-consistent (ISC) method [55], the Eshelby–Mori–Tanaka model [39,67], including several extended models [36,68–70]; and the asymptotic homogenization method for developing analytical equations [71]. FEM modelling has also been used for anisotropic structures [39,72–74]. In the case of high volume fractions of the reinforcements, the Eshelby–Mori–Tanaka model showed good accuracy [55].

Kuo and Huang [69] developed analytical equations for the electrostatic field based on the 3D anisotropic inclusion method and provided explicit electrostatic tensors in line with Eshelby tensors. They also developed analytical expressions for the composite effective properties that depend on the volume fraction and phase properties (e.g., reinforcements), such as the orientation angle and shape of the phase, since these showed the most prominent influence on the resulting effective properties [69]. Several recently developed analytical models are shown in [75]. Odegard [72] developed a constitutive model suitable for the different range of volume fractions, different polymer matrices and geometries of reinforcements. An analytical model for 1-3 composites with anisotropic properties was developed by Kar-Gupta and Venkatesh [73]. This model, including a method to determine 45 material constants for piezoelectric composites, showed good accuracy in longitudinal directions. Optimization of the composite design can be realized based on these analytical models [75].

It is challenging to establish a 3D FEM model for composite structures that can simulate the simultaneous influence of electric potential fluctuations and mechanical displacements in micro domains throughout the volume, especially considering the high computational costs of this method of processing data. The definition of RVEs that can represent hybrid non-uniform composite structures is also complex and determines the validity of the material model. A recent approach is to use artificial intelligence (AI) and new machine learning (ML) algorithms that are applicable for a range of thicknesses and compositions, including hybrid composites [76,77]. Data-driven ML material models enable so-called meshless modelling, usually based on the data generated through experiments. Further challenges associated with experimental datasets to be used in AI/ML processing is their diversity of formats and common lack of metadata, indicating the urgent need for standards in materials science related to data and metadata for AI/ML-ready datasets. This is usually overcome by using experimental data from trusted sources, where the most common method is to perform comprehensive experimental tests and create AI/ML-ready datasets in line with data science principles.

New ML algorithms have been developed focusing on reducing the time and computational costs and especially to serve in the optimization of material properties to provide desirable composite strength or strain [78], optimal composite patterns [79] or to address coupling influences and predict the properties of complex composite structures [80]. A decision tree classification ML model, complemented with regression models, can accurately predict various mechanical properties of fibre-based composites, such as the shear modulus or modulus in longitudinal and transverse directions [81]. Artificial neural network (ANN) models with 20 input and 11 output parameters have been used to predict the electromechanical properties of piezoelectric composites with 99.998% accuracy [82]. An ANN model was used to predict conductivity and strain in nanocomposites, reducing the 3.5 h time for FEM simulations to 0.25 s by using an ANN, with excellent accuracy [83]. ML methods have also been used in relation to fabrication technologies and subsequent correlations with resulting material properties [84], such as predicting the fibre diameter of PVDF depending on the electrospinning parameters [85], or to optimize fibre diameter according to the desired material responses [86].

A deep neural network can be used to approximate any function regardless of its physical meaning and, as such, it can replace commonly used numerical FEM-based methods or analytical material models, including for modelling material structures of hybrid piezoelectric nanocomposites [76]. Recent approaches in material modelling have used physics-informed neural networks (PINNs) through solving partial differential equations (PDEs) that govern the material physical system and where specific boundary conditions can be defined [87]. For material modelling through PINNs, it is important to properly define the loss function in relation to PDEs, including boundary conditions in relation to setting collocation points. Physics-informed neural networks can be used to create accurate models for known physical systems, including for FE and continuous domains within the material structure, and can deal with both elastic and inelastic boundary conditions. It is common to use some homogenization methods, such as analytical homogenization solutions that are again based on some finite blocks within the material structure.

RVEs used in FEM models are a relatively simple solution to represent the whole material structure, but they lack the ability to represent dynamic changes within micro and nano domains, meaning that local distribution that considers multiphysics is impossible to represent with RVEs. A new approach is the introduction of a repeating unit cell (RUC) that can be flexible over domains, including nano and micro domains within a composite structure that commonly exhibits non-uniform distributions of reinforcements [88]. The difference between RVEs and RUCs is that RVEs are applied for statistically homogenous structures with rigid boundary conditions for the whole structure, where micro domains are hard to interpret, while RUCs are small blocks of material volume within the arrays periodically appearing throughout the whole material volume that always consider microstructural features and their interdependences, and which exhibit flexible boundary conditions related to these periodic arrays. RUCs are used to build periodic arrays rather than the whole material volume, and these arrays can have different microstructural features, depending on the observed material structure. In the case of RUCs' use, localized homogenization is applied to provide a set of properties such as moduli or local stress fields, instead of one global value for each property, where arrays are commonly approximated by square, hexagonal or rectangular shapes for nanocomposites. Recently introduced RUC-based modelling is suitable for AI data processing due to its stability, significantly lower time needed for processing and very quick convergence, which enables comprehensive parametric study and very fast finding of the optimal material model [88]. However, defining the boundary conditions in the case of RUCs is still very challenging.

A multiphysics deep homogenization neural network (MDHN) has been developed for transversely isotropic piezoelectric composites where RUCs were used with periodic boundary conditions [76]. Changes in the micro domains of electric potential and mechanical displacements are simultaneously considered through PDEs related to both the matrix and reinforcements and observed as separate material phases. Two different neural networks were used to solve PDEs to avoid an otherwise commonly applied artificial interphase layer that is usually introduced into ML algorithms to perform smooth transitions between the matrix and reinforcement phases. Sharp transitions of the stress at the interface of these two distinctly different composite phases was addressed through the loss function and trainable weights.

Very often, FEM-based virtual simulations are used to generate datasets for training AI/ML models in the absence of AI/ML-ready experimental data [79,82,89]. The combination of FEM with AI/ML technology has shown excellent results [82], including a new approach to incorporate an AI surrogate material model into the FE solver for automatic analysis and predictions [90–92]. However, in the case of fibre-reinforced piezoelectric composites, the use of ANNs is still under study. Discoveries in materials science propelled by the use of advanced AI and ML algorithms have gained significant attention, belonging to the novel research area of Materials Informatics that has started to emerge [93,94].

3. Materials and Methods

Poly(vinylidene fluoride) (PVDF, Mw~180.00 by GPC) was purchased from Sigma Aldrich (St. Louis, MO, USA). Acetone (Ac, ≥99.5%) was purchased from Honeywell (Charlotte, NC, USA), and dimethylformamide (DMF, ≥99.5%) and silver nitrate ($AgNO_3$, ≥99.9%) were purchased from Fisher Chemical (Waltham, MA, USA). In this study, as well as in a previous one [2], all chemicals were applied without further purification.

3.1. Fabrication of PVDF Nanofibers with Incorporated AgNPs

For PVDF electrospinning, the earlier-determined [2], most conductive solution was prepared. Parameters are given in Table 2.

Table 2. The solution and electrospinning parameters [2].

Solution		Electrospinning Parameters				
PVDF Concentration [%]	Solvent [*v*/*v*%]	Voltage [kV]	Needle [Gauge]	Flow Rate [mL/h]	Tip to Collector Distance [cm]	Temperature, Humidity
21%	75% DMF: 25% Ac	30	18	0.5	15	30 °C, 45%

The 21% PVDF solution was stirred at 80 °C for 3 h to achieve a homogenous mixture. Different concentrations of $AgNO_3$ (0%, 0.3%, and 3% *w*/*w*) were then added to the solution, followed by stirring for an additional 12 h. The mixture was subsequently dispersed using an ultrasonicator and allowed to cool at 28 °C. The solution turned grey, signifying the formation of AgNPs [2].

The prepared solutions were transferred into a 5 mL syringe fitted with an 18-gauge needle. Electrospinning was performed under a voltage of 30 kV, with a flow rate of 0.5 mL/h and a needle-to-collector distance of 15 cm. The process was conducted under controlled conditions of 45% humidity and a temperature of 30 °C. After electrospinning, the nanofibers were cautiously collected onto aluminum foil and stored in a dark, room-temperature environment to allow the residual solvent to evaporate fully.

3.2. SEM and EDS Analysis

The nanofiber mat morphology was examined with a scanning electron microscope (SEM) (FEI Scios2 Dual Beam System, Hillsboro, OR, USA) at the Institute of Nuclear Sciences Vinča, University of Belgrade, Serbia. Square samples measuring 2 cm by 2 cm were prepared and gold-coated for 30 s to facilitate SEM analysis. The gold-coated samples were then carefully placed into an SEM operated at 10 kV. Additionally, energy-dispersive spectroscopy (EDS) was employed for elemental analysis to confirm the presence of Ag-loaded nanofibers.

3.3. FTIR Spectroscopy Analysis

The presence of crystalline phases in the electrospun PVDF samples was analyzed using transmission infrared spectroscopy with a portable FTIR/FT-NIR spectrometer (Interspec 301-X, Toravere, Estonia). The measurements were conducted in the range of 400 cm^{-1} to 1600 cm^{-1}, with a resolution of 4 cm^{-1}.

3.4. Electromechanical Tests at Macro Scale

To measure the thickness of electrospun materials, Trimos Labconcept Premium (Renens, Switzerland), samples were measured ten times each, in order to determine average thickness. In addition, specimen tensile testing was performed using a mechanical uniaxial testing machine, Ametek Brookfield CT3 Texture Analyzer (Devon–Bervin, PA, USA), equipped with 500 N load cells. Testing was performed with a testing speed of 10 mm/min, following the literature [95,96], due to a lack of standards for tensile testing of such materials [97]. Before testing, samples were cut to 45 × 20 mm dimensions. Since electrospun specimens are thin and sensitive, a paper frame with a 10 × 10 mm opening was introduced in which specimens were inserted using Scotch tape. The structure formed in this way was then inserted into tensile testing machine grips, after which frame sides were removed using scissors, as shown in figure 9.0, which ensures that tensile testing is performed on the electrospun specimen only [96]. Three samples of every material type were tested to ensure repeatability, and the average values are given.

To determine the piezoelectric d_{33} coefficient, a modified Sawyer–Tower circuit (Figure 1b) was employed [32]. This was coupled with load measurements through load cells, ensuring simultaneous load and electrical voltage measurements, as described in Figure 1a.

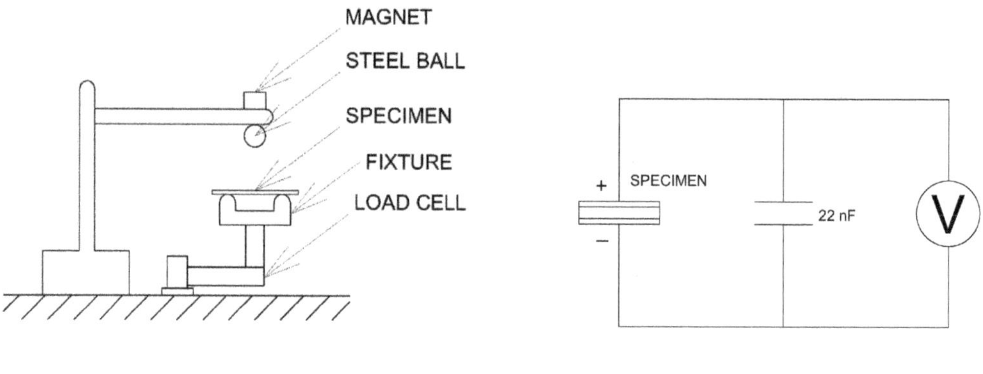

Figure 1. Piezoelectric testing fixture (**a**) and modified Sawyer–Tower circuit (**b**).

Testing was conducted by dropping a steel ball from a fixed height (50 mm) onto test specimens, measuring both load and electrical voltage [34] across a 22 nF capacitor; the voltage measured across the capacitor is later used to calculate the piezoelectric coefficient:

$$d_{33} = \frac{Q}{F} = \frac{CV}{F} \tag{1}$$

where d_{33} is the piezoelectric coefficient, Q is the charge in the capacitor, F is the applied load (measured), C is the capacitance of the capacitor (22 nF) and V is the voltage across the capacitor (measured). For voltage and load measurements, the HBM (Virum, Denmark) QuantumX A410 and CATMAN 3.4.1 Software were used. Three different samples were prepared for every nanocomposite configuration and examined, with the average value considered.

3.5. Finite Element Analysis (FEA)

3.5.1. Piezoelectric Constitutive Relations

The direct piezoelectric effect can be expressed with the following equation [98,99]:

$$\mathbf{P}_{\text{piezo}} = \mathbf{dT} \tag{2}$$

where $\mathbf{P}_{\text{piezo}}$ represents the mechanically induced polarization, **d** represents piezoelectric charge constants matrix and **T** is the applied stress vector. The total electric displacement field **D**, including $\mathbf{P}_{\text{piezo}}$, is defined by the following equation [98]:

$$\mathbf{D} = \varepsilon^{\mathrm{T}}\mathbf{E} + \mathbf{dT} \tag{3}$$

where ε^{T} represents the dielectric permittivity matrix under constant stress and E is the electric field vector.

The inverse piezoelectric effect can be expressed with the following equation [99]:

$$\mathbf{S} = \mathbf{d}^{\mathrm{T}}\mathbf{E} \tag{4}$$

where S is elastic strain vector and $\mathbf{d}^{\mathrm{T}}$ represents a piezoelectric charge constant matrix under constant stress. The elastic strain vector, including the inverse piezoelectric effect, can then be expressed as [98]

$$\mathbf{S} = \mathbf{s}^{\mathrm{E}}\mathbf{T} + \mathbf{d}^{\mathrm{T}}\mathbf{E} \tag{5}$$

where $\mathbf{s}^{\mathrm{E}}$ represents compliance for the constant electric field. By combining Equations (3) and (5), the following equations are obtained [100]:

$$\begin{aligned} \mathbf{S} &= \mathbf{s}^{\mathrm{E}}\mathbf{T} + \mathbf{d}^{\mathrm{T}}\mathbf{E} \\ \mathbf{D} &= \mathbf{dT} + \varepsilon^{\mathrm{T}}\mathbf{E} \end{aligned} \tag{6}$$

Previous equations represent the constitutive relations for linear piezoelectric materials in the strain–charge form. This equation can be expressed in other forms in which different coefficients are used. Accordingly, in addition to the strain–charge form, the stress–charge form is also often used, which is given by the following equations:

$$\begin{aligned} \mathbf{T} &= \mathbf{c}^{\mathrm{E}}\mathbf{S} - \mathbf{e}^{\mathrm{t}}\mathbf{E} \\ \mathbf{D} &= \mathbf{eS} + \varepsilon^{\mathrm{S}}\mathbf{E} \end{aligned} \tag{7}$$

where $\mathbf{c}^{\mathrm{E}}$ is the elastic stiffness coefficient matrix evaluated at a constant electric field, e is the piezoelectric stress coefficient matrix, et is the transpose of **e** and ε^{S} is the dielectric matrix evaluated at constant strain.

3.5.2. FE Model of PVDF/AgNPs Nanocomposite

The 2D FE model of PVDF/AgNPs nanocomposites was created in COMSOL software version 6.3 in order to prove the concept and set the foundations for future research and calibration. Numerical analysis was performed within the electromagnetics–structure interaction module which couples linear-elastic statics with the constitutive relations of piezoelectric materials given in Section 3.5.1 and is used to investigate the behaviour of piezoelectric materials at the macro scale, capturing the global structural response without going into the microstructure of the material. In this paper, three configurations were considered: pure PVDF and two nanocomposite configurations—PVDF with 0.3% and 3%

AgNPs. Silver particles were not physically modelled, but their influence was considered through material parameters.

FEM-based simulation material properties were determined from the experimental tests performed (details in Section 3.4) with the prepared piezoelectric samples and entered as input values to the COMSOL software version 6.3.

The piezoelectric coefficient d_{33} was calculated based on the experimental tests using the modified Sawyer–Tower circuit based on the measured voltage across the capacitor connected to the sample (details in Section 3.4). In order to calculate the contact area between the steel ball and the specimen, the Hertzian contact theory was employed [101] to calculate its radius as

$$a = \sqrt[3]{\frac{3}{8}FD(k_1 + k_2)} \tag{8}$$

where F is the measured load, D is the steel ball diameter and k_i is calculated as

$$k_i = \frac{1 - v_i^2}{E_i} \tag{9}$$

where v is Poisson's ratio, E is the tensile modulus and subscripts $i = 1, 2$ denote the sphere and plate. The calculated contact diameter is imported as geometry in COMSOL software version 6.3.

The values of material parameter d_{33} were measured for different variants of PVDF/AgNPs nanocomposites (0%, 0.3% and 0.5% AgNPs), and they were used in numerical simulations. Material relative partitivities were calculated from experimental data for equivalent capacitance of both the specimen and reference 22 nF capacitor. Other material characteristics, which were unable to be obtained from experimental work, were adopted from COMSOL's material library and were kept the same for every nanocomposite configuration.

The user-defined mapped quad mesh was generated with 38 elements. The geometry, which represents contact area radius and specimen thickness, and the FE model are given in Figure 2.

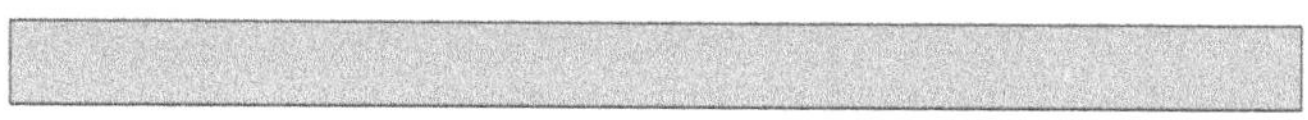

(**a**)

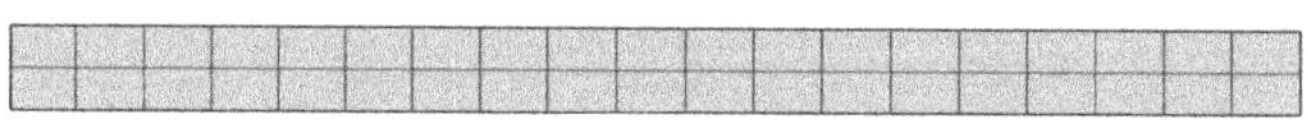

(**b**)

Figure 2. Geometry (**a**) and FE models (**b**) of PVDF and PVDF/AgNPs nanocomposite configurations.

Boundary conditions (Figure 3) in the solid mechanics module were set as follows: the lower and left edge have an applied roller boundary condition, which limits displacement in the direction perpendicular to the boundary, while a compressive force of 1 N acts on the upper edge. The force value used within numerical simulations corresponds to the peak value of experimentaltesting results. In the electrostatics module, grounding is defined on the lower edge, which corresponds to the connection of the testing specimen to an electrode at ground potential, with zero charge on the side edges, while charge conservation and the initial value of the electric potential of 0 V are defined on all surfaces.

The processing hardware included 64 GB of RAM and an Intel Core i9-13900KF CPU running at 3.00 GHz (32 CPUs), while computational time was about 3–4 s. The results of the numerical analysis are given in the next chapter.

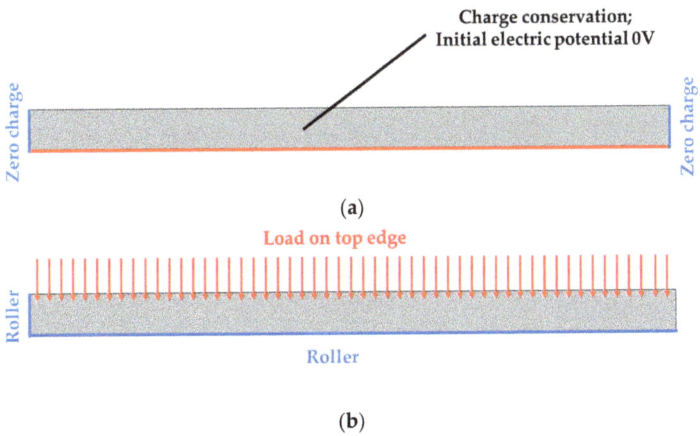

Figure 3. Loads and boundary conditions: electrical (**a**) and mechanical (**b**).

4. Results and Discussion

4.1. SEM and EDS Results

SEM images of PVDF and PVDF/AgNP nanofiber mats are shown in Figure 4. The addition of silver nanoparticles to PVDF nanofibers shows significant differences in morphologies. However, due to a very low concentration of $AgNO_3$, silver nanoparticles are not visible in SEM images. The addition of nanoparticles decreased the presence of beads which are notable in pure PVDF nanofiber mats. The reduction of beads is observed for both 0.3% AgNPs and 3% AgNPs specimens, although it is still present in structures. Also, Figure 5 shows EDS mapping of pure PVDF, PVDF + 0.3% AgNP and PVDF + 3% AgNP nanofiber mats. The existence of carbon (C), oxygen (O), nitrogen (N), fluor (F) and silver (Ag) atoms can be seen, confirming the successful integration of the silver nanoparticles in PVDF nanofibers. Additionally, Figure 5 shows the EDS spectrum, which confirms the existence and demonstrates the homogenous distribution of silver.

The selection of AgNPs proportion in nanocomposites was based on our previous experimental work experience and literature review [23,24,102]. In our previously published study [2], we tested PVDF composites with AgNP concentrations of 0.1%, 0.3%, 0.5%, 1%, 3% and 5%. We concluded that higher concentrations (1%, 3% and 5%) show strong cytotoxicity against both healthy MRC-5 and MDA-MB-231 cancer cells. In addition, testing lower concentrations (0.1%, 0.3% and 0.5%) showed that 0.3% AgNPs have the greatest β-phase fraction and the best nanofiber morphology compared to the 0.1% and 0.5% concentrations in terms of the lowest bead formation and nanofiber diameters.

One of the main reasons for bead formation during the electrospinning process is low solution viscosity, leading to inadequate chain entanglement, causing the polymer jet to break up in droplets rather than forming continuous fibres. A higher solution concentration increases viscosity, which promotes the formation of bead-free fibres [103]. In addition, high surface tension can promote bead formation, so increasing the electrical field increases electrostatic forces that should overcome the surface tension and initiate jet [104]. Also, tip-to-collector distance can have an impact on the prevention of bead formation because complete solvent evaporation at optimal distance is what should be pursued [104];

however, this weakens the electric force, which can cause jet instability [103,105]. Further, insufficient charge density can be responsible for jet instability, hence promoting bead formation. Because of all of this, optimizing both the solution and process parameters is of peak importance. The addition of AgNPs to the PVDF solution can alter all of the beforementioned properties by increasing the solution conductivity [2,106], allowing jet stabilization, but only to some extent, as AgNPs can lead to aggregation, which increases surface tension, which again promotes bead formation.

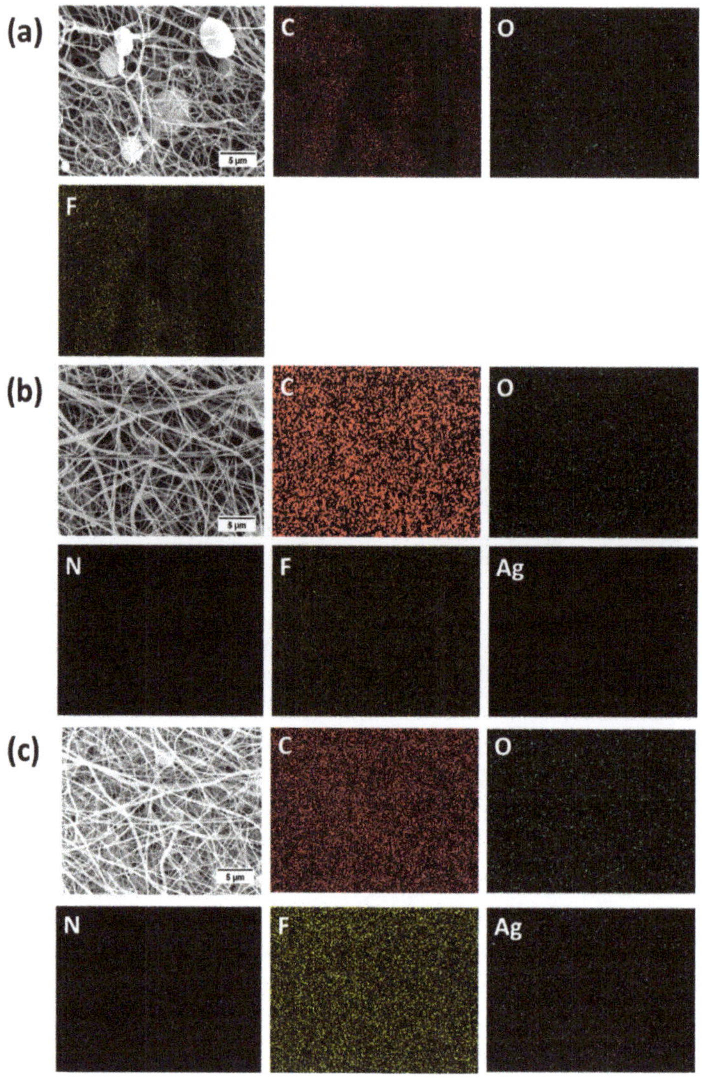

Figure 4. EDS analysis of (**a**) pure PVDF; (**b**) PVDF + 0.3% AgNPs; and (**c**) PVDF + 3% AgNPs nanofibers related to the distribution of C (red), N (dark blue), O (green), F (yellow) and Ag (white). The scale bar is 5 μm [2].

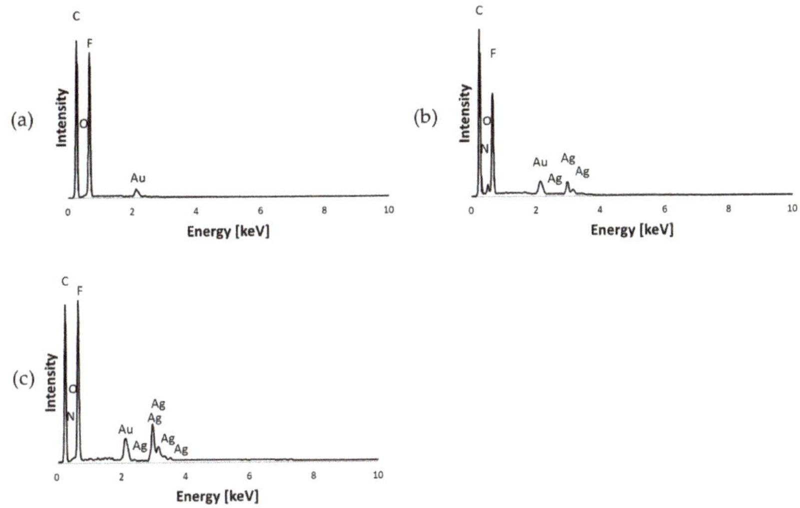

Figure 5. EDS diagrams of the elemental distribution of specimens: (**a**) pure PVDF; (**b**) PVDF + 0.3% AgNPs; (**c**) PVDF + 3% AgNPs.

4.2. FTIR Spectroscopy Analysis Results

The FTIR spectra of pure PVDF and PVDF + AgNP nanofiber mats are shown in Figure 6. The absorption peak near 1400 cm^{-1} corresponds to the CH_2 wagging vibration in PVDF. The band at 1180 cm^{-1} is associated with the asymmetric stretching vibration of the CF_2 group, while the 1068 cm^{-1} band is attributed to the CH_2 wagging mode [107,108]. The β-phase band at 874 cm^{-1} corresponds to CF_2 symmetric stretching, while the α phase is identified by characteristic peaks at 763 cm^{-1} and 610 cm^{-1}, related to CF_2 bending. Peaks at 880 cm^{-1} and 841 cm^{-1} are linked to C-C-C and CF stretching vibrations of PVDF [107]. Unique peaks for the electroactive, piezoelectric β phase are observed around 445, 473, and 1275 cm^{-1} [29]. The chemical structure of the nanocomposite mat preserves the characteristic absorption peaks of pure PVDF. However, the absorption peak at 1662 cm^{-1}, which indicates the presence of silver, was absent—likely due to the low concentration of AgNPs [2,109]. Nevertheless, the presence of silver was confirmed using EDS analysis.

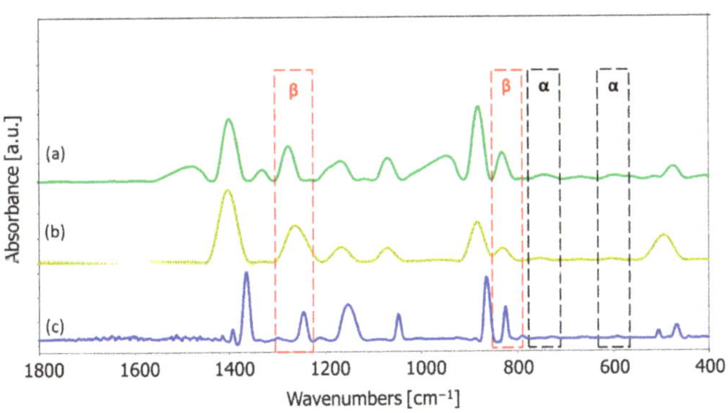

Figure 6. FTIR absorbance spectra of the specimens: (a) pure PVDF, (b) PVDF + 0.3% AgNPs, (c) PVDF + 3% AgNPs [2].

Peaks at 840 cm^{-1} and 510 cm^{-1} provide specific insights into the phases of PVDF. When the 1275 cm^{-1} peak is present, but the 1234 cm^{-1} peak is absent, the 840 cm^{-1} and 510 cm^{-1} bands are attributed to the β phase. Conversely, if the 1275 cm^{-1} peak is absent while the 1234 cm^{-1} peak is present, these two bands correspond to the γ phase. If both 1275 cm^{-1} and 1234 cm^{-1} peaks appear together, the 840 cm^{-1} and 510 cm^{-1} bands are considered indicative of both β and γ phases [29].

According to the FTIR spectra in Figure 6, with the omission of 1234 cm^{-1} bands, 510 cm^{-1} and 840 cm^{-1} bands can be considered as the β phase.

The relative fraction of the electroactive phase, in this case the β phase, can be quantified using the following formula, according to the Lambert–Beer law [2,29,110]:

$$F_{\beta} = \frac{A_{\beta}}{\left(\frac{K_{\beta}}{K_{\alpha}}\right)A_{\alpha} + A_{\beta}} \times 100 \quad (10)$$

where A_{α} and A_{β} are the absorbencies at 763 and 840 cm^{-1}, respectively, and K_{α} and K_{β} are the absorption coefficients at their respective wave numbers, obtained from the literature [2,29,110]. Therefore, the calculated fraction of the β phase in specimens is shown in Figure 7.

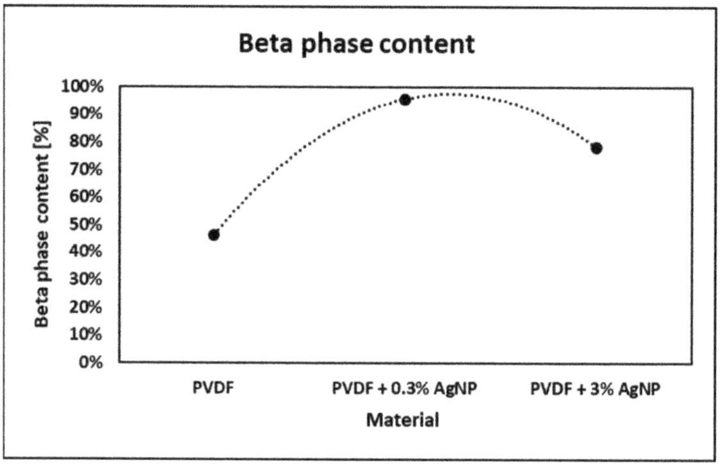

Figure 7. Beta-phase content in different materials.

Observing obtained FTIR data, it can be concluded that the addition of silver nanoparticles to electrospinning solution promotes PVDF β-phase formation, but only to a certain extent, because further addition of AgNPs led to β-phase content decreasing, as previously confirmed [2,23,102]. Silver nanoparticles have electron-rich surfaces, so their dipole field promotes polymer chain alignment on the AgNPs' surfaces to form a crystalline polar β phase [111]. In addition. this can facilitate the nucleation of the β phase. However, when the concentration of AgNPs increases further, too many β-phase spherulites are formed which tend to squeeze together, which allows the α phase to be formed in the inter region between these spherulites [111]. Jet instability promotes bead formation, which reduces β-phase content as it prevents adequate fibre stretching [103]. Also, a high concentration of AgNPs can lead to nanoparticle aggregation, which can interfere with the polymer chain alignment necessary for β-phase formation.

4.3. Mechanical Testing

Mean sample thickness values, as well as their standard deviations, are presented in Figure 8.

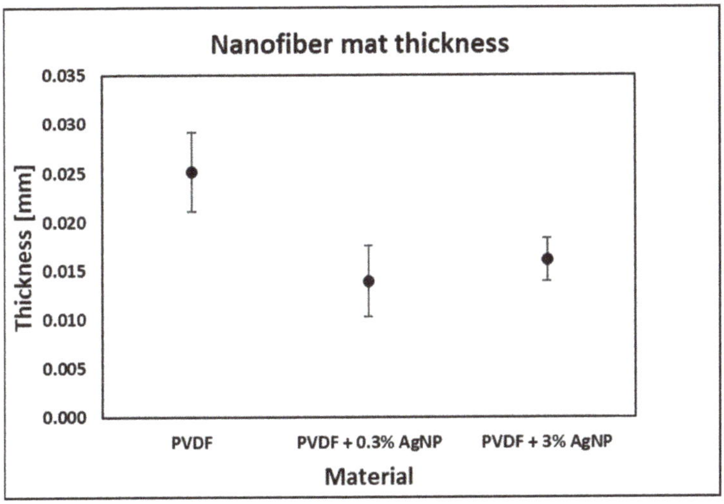

Figure 8. Mean sample thickness values.

The variation in the thickness of the tested samples is due to the random nature of the electrospinning process. We did not measure the fibre volume variations considering the thickness of the tested samples. However, results from the literature indicate that thicker nanofibers exhibit reduced strain under deformation and, therefore, lower output voltages because the fibre flexibility is decreased, leading to lower mechanical deformation and reduced electrical output [112].

The nanofiber mat tensile testing process is described in Figure 9, showing the specimen at the beginning of the process inside a paper frame which was cut after placing the sample in the grips, the specimen in the middle of the testing process, and at the end.

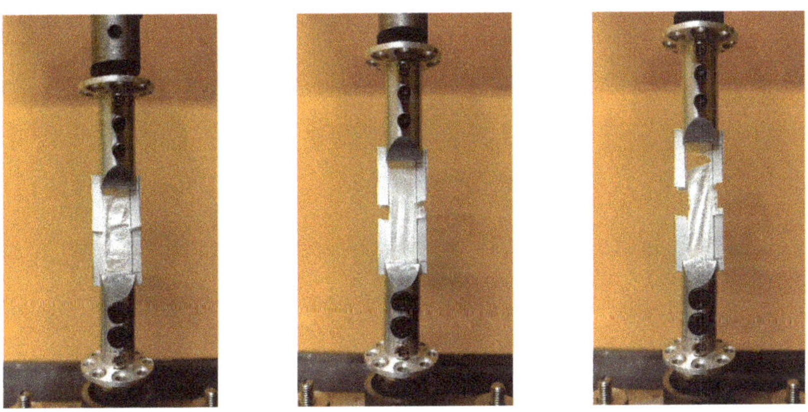

(a) (b) (c)

Figure 9. Nanofiber mat tensile testing process: (**a**) beginning; (**b**) middle; and (**c**) at the end of testing.

Tensile strength increases with the addition of 0.3% of Ag, as shown in Figures 10 and 11, from around 3 MPa to around 5 MPa for the pure PVDF specimen and PVDF + 0.3% AgNPs, respectively, as similarly reported in [113].

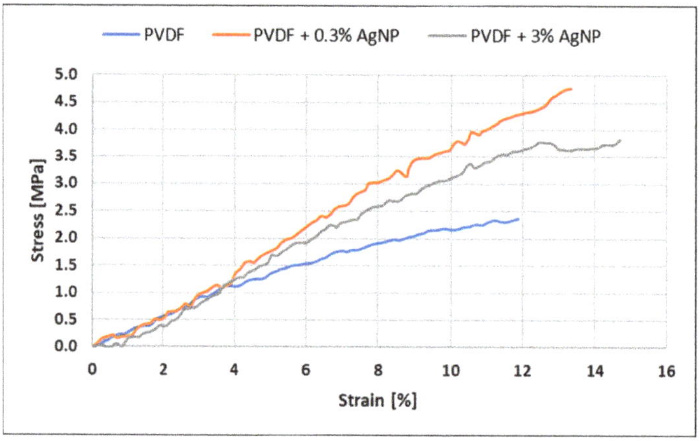

Figure 10. Stress–strain diagrams for different nanofiber mats.

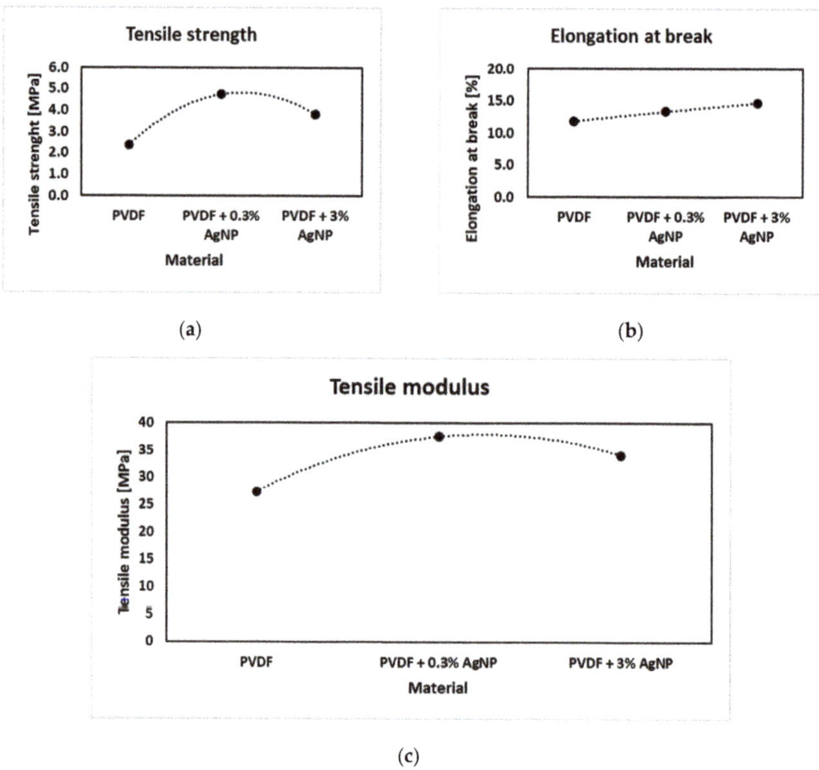

Figure 11. Nanofiber mat tensile testing results: (**a**) tensile strength; (**b**) elongation at break; and (**c**) tensile modulus.

But, further increases in the percentage of AgNPs increase the heterogeneity in the PVDF matrix, which seems to introduce premature flaws in the system at some point which lowers the stress required to fracture [24]. Still, the addition of AgNPs into PVDF leads to a decrease in impurities such as bead formation, as shown in SEM images (Figure 5), which led to improved mechanical properties compared to the pure PVDF specimen. The Ag nanoparticles help hinder crack propagation, even at low concentrations, resulting in increased elongation at break compared to pure PVDF nanofiber mats [24].

4.4. Measurements Related to Piezoelectric Outputs

Results for pure PVDF electrospun nanofiber mats are given in Figure 12. The measured load was around 1 N and the generated voltage in this case was around 0.2 V; similar results were reported by references [22,23].

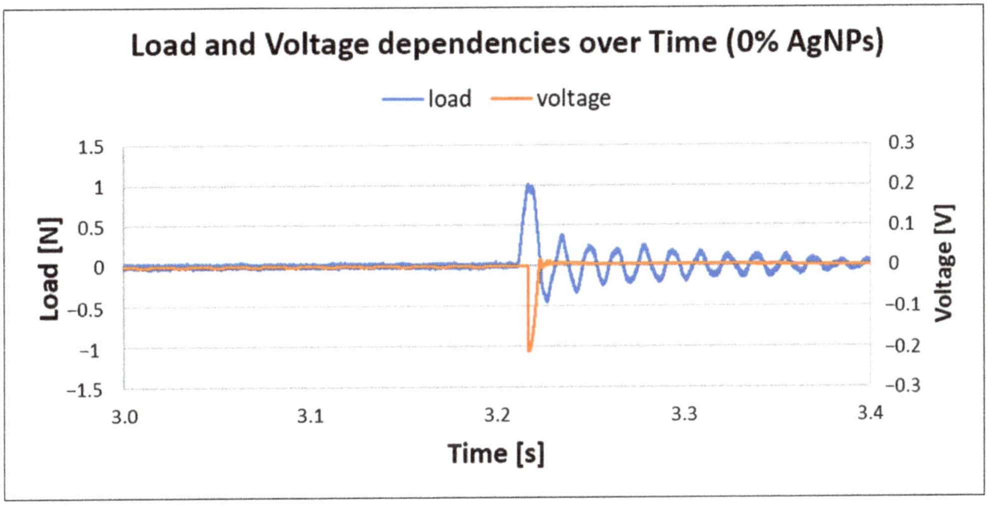

Figure 12. Load and voltage dependency in pure PVDF without AgNPs.

The results for PVDF nanofiber mats with 0.3% AgNPs are given in Figure 13. The generated load is the same as in the previous case, but with a noted increase in voltage, up to 0.48 V, which suggests that the incorporation of AgNPs into PVDF nanofiber mats increases output voltage; this is in correlation with previously reported work [23]. However, with a further increase in silver nanoparticles in composite material, the voltage output dropped, as shown in Figure 14.

The output voltage corresponds to the beta phase content in the PVDF material, so as the β-phase content increases with the addition of AgNPs, the piezoelectric coefficient increases [102]; hence, the output voltage also increases. However, with higher AgNP concentration, the β-phase content decreases, so the output voltage decreases as well. In addition, it is proven that higher AgNP wt.% leads to a decrease in nanocomposite surface potential, which corresponds to the lower piezoelectric coefficients [19].

Differences in material composition, microstructure and fibre orientations, because of the random nature of the electrospinning process, can cause differ fibre orientations from sample to sample, and can alter mechanical coupling and damping properties. In addition, thickness variations, fibre density and fibre diameters affect how the samples deform or absorb impact, which also affects the attenuation. For instance, the damping behaviour of fibre-reinforced composites is highly dependent on fibre content and orientation [114].

These differences can also lead to different sample conductivity and internal resistance, which also can affect voltage attenuation. Also, if one sample is closer to a resonant frequency of the system, it might exhibit sustained oscillations compared to others because piezoelectric materials generate maximum power at their resonance frequency [115].

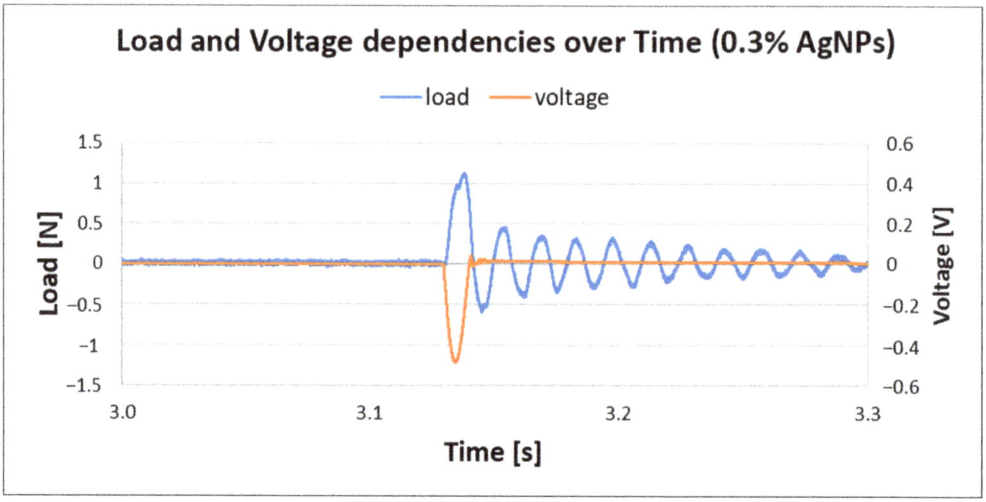

Figure 13. Load and voltage dependency in PVDF with 0.3% AgNPs.

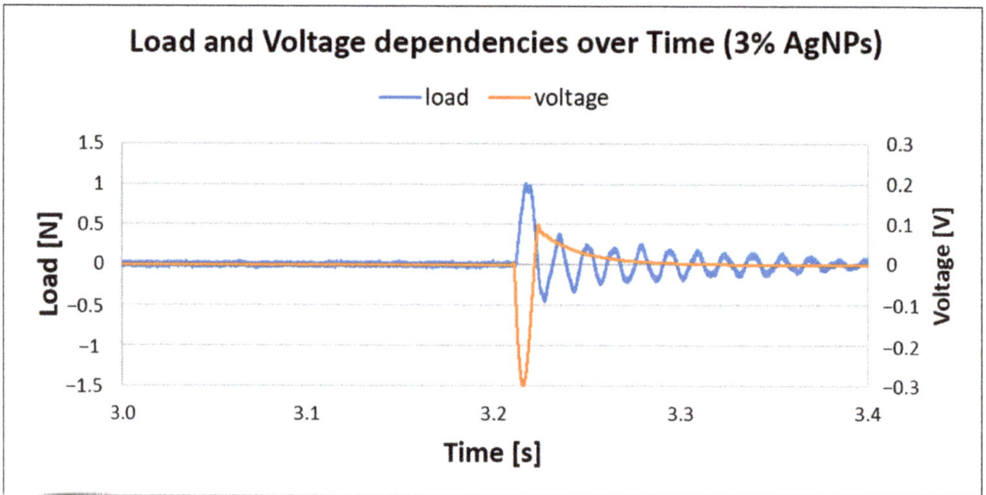

Figure 14. Load and voltage dependency in PVDF with 3% AgNPs.

4.5. Finite Element Analysis

Previously described experimental voltage measurements were used to calculate piezoelectric coefficient d_{33} using Equation (1). This calculated coefficient was implemented in numerical FEA software COMSOL Multiphysics software version 6.3 where electric potential output is simulated.

In Figure 15, numerical results are shown that were obtained through Finite Element Analysis for pure PVDF electrospun nanofiber mats.

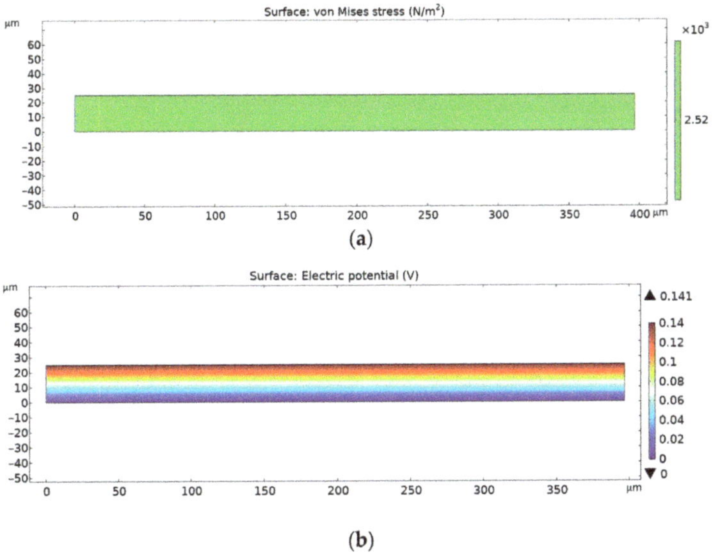

Figure 15. FEA results for pure PVDF nanofibers without (**a**) von Misses stress and (**b**) electric potential.

In Figure 16, numerical results are shown that were obtained through Finite Element Analysis for PVDF electrospun nanofiber mats with 0.3% AgNPs.

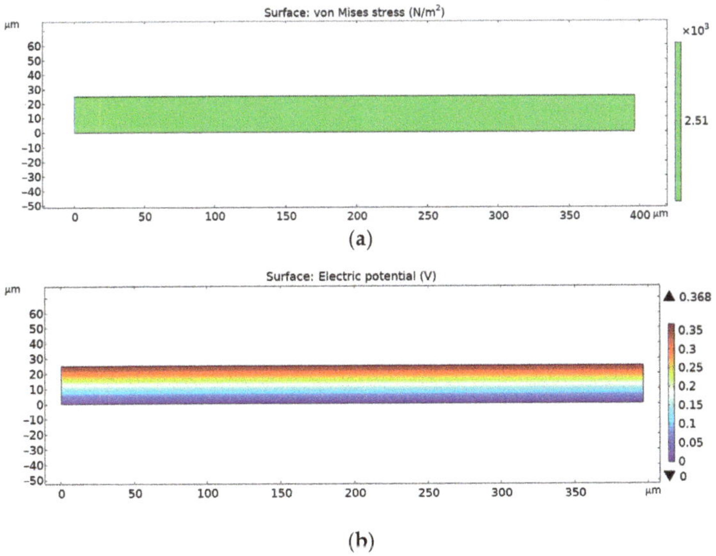

Figure 16. FEA results for PVDF nanofibers with 0.3% AgNPs: (**a**) von Misses stress and (**b**) electric potential.

Figure 17 denotes FEA results for PVDF + 3% AgNP electrospun nanomats.

As can be observed from the figures above, the von Mises stress is the same for every given case, but the electrical potential is different, due to different piezoelectric coefficient d_{33} values. The pure PVDF nanomats showed a maximum electrical potential of 0.141 V, while those with the addition of 0.3% of AgNPs to PVDF electrical potential exhibited an increase to 0.368 V. However, with further increases in AgNP content in nanomats, to 3%, a decrease in generated voltage was noted, of 0.227 V, which corresponds to the FTIR data and calculated β-phase content.

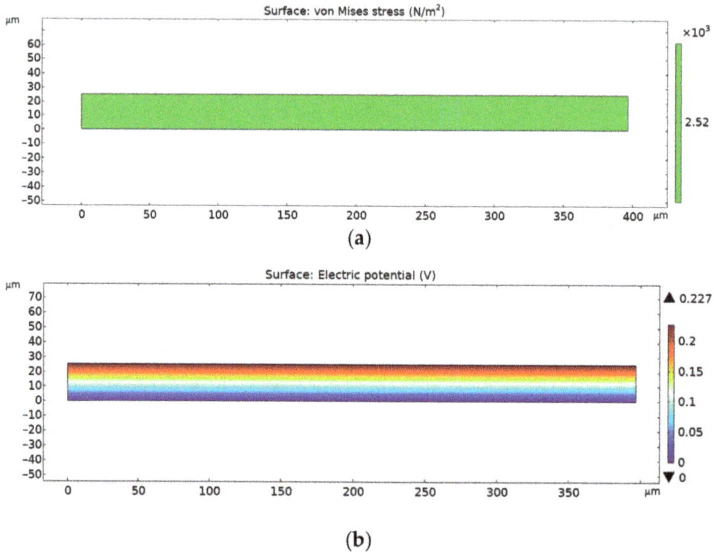

Figure 17. FEA results for PVDF nanofibers with 3% AgNPs: (**a**) von Misses stress and (**b**) electric potential.

Figure 18 describes the comparison of experimental and simulated data for generated voltage where good correlation is found, with minor differences. Pure PVDF and PVDF + 3% AgNP specimens showed greater simulated results compared to experimental specimens, while the PVDF + 0.3% AgNP specimen showed greater experimental output compared to the simulated one, where similar results are reported in the literature, with different filler types and concentrations [32].

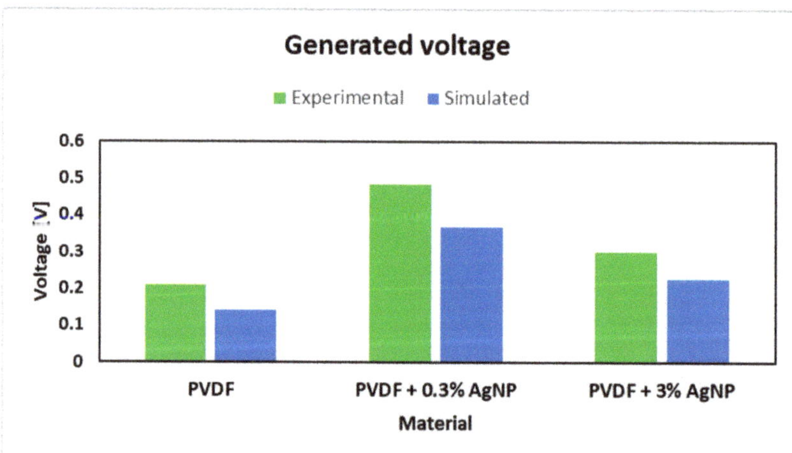

Figure 18. Generated voltage comparison between experimental and simulated data.

We used experimentally determined values of coefficients and material parameters in the 2D FE model and FE analysis, in combination with custom-designed meshing in the FE model with 38 elements (as shown in Figure 2), which showed good correspondence between the experiments and FEM models.

5. Conclusions

PVDF with incorporated AgNPs showed enhanced piezoelectricity up to a certain AgNP percentage (0.3%), while, with further increases in AgNP content in the composite, piezoelectricity dropped, compared to the 0.3% composite, but still showed increased piezoelectric response compared to the pure PVDF electrospun nanofiber mats.

SEM analysis proved that the addition of silver nanoparticles to the PVDF electrospun nanofibers provided better homogeneity of the material, removing PVDF beads and increasing overall fibre morphology, while EDS confirmed the presence of AgNPs in the composite due to SEM's inability to locate nanoparticles because of their size.

FTIR characterization has proven that electrospinning is an effective way to promote the electroactive phase in PVDF nanofiber composites, promoting their piezoelectric properties.

Mechanical tests showed an increase in tensile strength with the addition of 0.3% of AgNPs compared to the specimens without silver particles, while a further increase in AgNPs, to 3%, showed lower tensile strength compared to PVDF with 0.3% AgNPs, probably due to the increase in heterogeneity and the inhibition of premature flaws within the material.

A custom-built measurement setup allows one to tailor and adapt the fixture to specific experimental needs but has disadvantages due to signal instability during tests, so extensive preparations are needed in order to achieve measurement repeatability and reliability. These measurements confirmed that PVDF nanofibers exhibit piezoelectric response, and that the incorporation of AgNPs promotes voltage generation over the same amount of load, which is in agreement with the increase in the electroactive phase in PVDF.

Experimentally obtained results were imported and implemented in numerical modelling software, where FEM analysis showed good agreement with experimental results, thus pointing out that computational models can be used for further virtual experiments focusing on the development of this material for specific applications aiming at better tailoring of the composite piezoelectric outputs.

Author Contributions: Conceptualization, F.Ž., S.M., A.B., K.V. and N.G.; methodology, F.Ž., S.M., A.B. and K.V.; software, S.M. and A.B.; validation, F.Ž., S.M., A.B., K.V. and N.G.; formal analysis, F.Ž., S.M., A.B. and D.P.; investigation, F.Ž., S.M., A.B., K.V. and D.P.; resources, F.Ž., N.G., S.M. and A.B.; data curation, S.M. and A.B.; writing—original draft preparation, S.M., A.B., F.Ž., K.V. and D.P.; writing—review and editing, S.M., A.B., F.Ž. and N.G.; visualization, S.M., A.B. and K.V.; supervision, F.Ž. and N.G.; funding acquisition, F.Ž. All authors have read and agreed to the published version of the manuscript.

Funding: This paper was supported through the EIT's HEI Initiative SMART-2M and DEEPTECH-2M projects, supported by EIT Digital, coordinated by EIT RawMaterials, and funded by the European Union.

Institutional Review Board Statement: Not applicable.

Informed Consent Statement: Not applicable.

Data Availability Statement: The original contributions presented in the study are included in the article, further inquiries can be directed to the corresponding author.

Conflicts of Interest: The authors declare no conflicts of interest.

References

1. Ueberschlag, P. PVDF Piezoelectric Polymer. *Sens. Rev.* **2001**, *21*, 118–126. [CrossRef]
2. Milenković, S.; Virijević, K.; Živić, F.; Radojević, I.; Grujović, N. Composite Nanoarchitectonics of Electrospun Piezoelectric PVDF/AgNPs for Biomedical Applications, Including Breast Cancer Treatment. *Materials* **2024**, *17*, 3872. [CrossRef] [PubMed]
3. Anic, M.; Prodanovic, M.; Milenkovic, S.; Filipovic, N.; Grujovic, N.; Zivic, F. The Review of Materials for Energy Harvesting. In Proceedings of the 2021 IEEE 21st International Conference on Bioinformatics and Bioengineering (BIBE), Kragujevac, Serbia, 25 October 2021; IEEE: Kragujevac, Serbia, 2021; pp. 1–6.
4. Anic, M.; Prodanovic, M.; Milenkovic, S.; Filipovic, N.; Grujovic, N.; Zivic, F. Electrospinning as the fabrication technology for the energy harvesting composites. In Proceedings of the 38th International Conference on Production Engineering of Serbia ICPE-S 2021, Čačak, Serbia, 14 October 2021; Fakultet Tehničkih Nauka u Čačku, Univerziteta u Kragujevcu: Čačak, Serbia, 2021; pp. 167–178.
5. Fukada, E. History and Recent Progress in Piezoelectric Polymers. *IEEE Trans. Ultrason. Ferroelectr. Freq. Control* **2000**, *47*, 1277–1290. [CrossRef] [PubMed]
6. Ruan, L.; Yao, X.; Chang, Y.; Zhou, L.; Qin, G.; Zhang, X. Properties and Applications of the β Phase Poly(Vinylidene Fluoride). *Polymers* **2018**, *10*, 228. [CrossRef]
7. Martins, P.; Lopes, A.C.; Lanceros-Mendez, S. Electroactive Phases of Poly(Vinylidene Fluoride): Determination, Processing and Applications. *Prog. Polym. Sci.* **2014**, *39*, 683–706. [CrossRef]
8. Gao, Y.; Zhu, W.; Wang, T.; Yilmaz, D.E.; Van Duin, A.C.T. C/H/O/F/Al ReaxFF Force Field Development and Application to Study the Condensed-Phase Poly(Vinylidene Fluoride) and Reaction Mechanisms with Aluminum. *J. Phys. Chem. C* **2022**, *126*, 11058–11074. [CrossRef]
9. Marmolejo-Tejada, J.M.; De La Roche-Yepes, J.; Pérez-López, C.A.; Taborda, J.A.P.; Ávila, A.; Jaramillo-Botero, A. Understanding the Origin of Enhanced Piezoelectric Response in PVDF Matrices with Embedded ZnO Nanoparticles, from Polarizable Molecular Dynamics Simulations. *J. Chem. Inf. Model.* **2021**, *61*, 4537–4543. [CrossRef]
10. Guo, H.; Zhang, Z.; Zhu, Y. Study on the Effect of Stretching Behavior on the Properties of PVDF Films and Mechanism. *Polymer* **2024**, *297*, 126884. [CrossRef]
11. Hartono, A.; Satira, S.; Djamal, M.; Ramli, R.; Bahar, H.; Sanjaya, E. Effect of Mechanical Treatment Temperature on Electrical Properties and Crystallite Size of PVDF Film. *Adv. Mater. Phys. Chem.* **2013**, *3*, 71–76. [CrossRef]
12. Patil, Y.; Zhao, J.; Ameduri, B.; Rastogi, S. Tailoring Electroactive β- and γ-Phases via Synthesis in the Nascent Poly(Vinylidene Fluoride) Homopolymers. *Macromolecules* **2024**, *57*, 616–627. [CrossRef]
13. Mukherjee, A.; Dasgupta Ghosh, B.; Ghosh, A.; Roy, S. Polyvinylidene Fluoride Nanocomposites as Piezoelectric Nanogenerator: Properties, Fabrication and Market Applications. *Adv. Eng. Mater.* **2024**, *26*, 2400445. [CrossRef]
14. Bindhu, A.; Arun, A.P.; Pathak, M. Review on Polyvinylidene Fluoride-Based Triboelectric Nanogenerators for Applications in Health Monitoring and Energy Harvesting. *ACS Appl. Electron. Mater.* **2024**, *6*, 47–72. [CrossRef]
15. Virijević, K.; Živanović, M.N.; Nikolić, D.; Milivojević, N.; Pavić, J.; Morić, I.; Šenerović, L.; Dragačević, L.; Thurner, P.J.; Rufin, M.; et al. AI-Driven Optimization of PCL/PEG Electrospun Scaffolds for Enhanced In Vivo Wound Healing. *ACS Appl. Mater. Interfaces* **2024**, *16*, 22989–23002. [CrossRef] [PubMed]
16. Virijević, K.; Živanović, M.; Pavić, J.; Dragačević, L.; Ljujić, B.; Miletić Kovačević, M.; Papić, M.; Živanović, S.; Milenković, S.; Radojević, I.; et al. Electrospun Gelatin Scaffolds with Incorporated Antibiotics for Skin Wound Healing. *Pharmaceuticals* **2024**, *17*, 851. [CrossRef] [PubMed]
17. Purushothaman, S.M.; Tronco, M.F.; Kottathodi, B.; Royaud, I.; Ponçot, M.; Kalarikkal, N.; Thomas, S.; Rouxel, D. A Review on Electrospun PVDF-Based Nanocomposites: Recent Trends and Developments in Energy Harvesting and Sensing Applications. *Polymer* **2023**, *283*, 126179. [CrossRef]
18. Liu, X.; Zhang, M.; Jiang, B.; Zhang, Q.; Chen, H.; Shen, Y.; Wang, Z.; Yin, X. Process Investigation on Robust Electrospinning of Non-Aligned and Aligned Polyvinylidene Fluoride Nanofiber Mats for Flexible Piezoelectric Sensors. *Polymers* **2024**, *16*, 816. [CrossRef]
19. Chen, Z.; Guan, M.; Bian, Y.; Yin, X. Multifunctional Electrospun Nanofibers for Biosensing and Biomedical Engineering Applications. *Biosensors* **2023**, *14*, 13. [CrossRef]
20. Kumar, M.; Kulkarni, N.D.; Kumari, P. Piezoelectric Performance Enhancement of Electrospun Functionally Graded PVDF/$BaTiO_3$ Based Flexible Nanogenerators. *Mater. Res. Bull.* **2024**, *174*, 112739. [CrossRef]
21. Salama, M.; Hamed, A.; Noman, S.; Magdy, G.; Shehata, N.; Kandas, I. Boosting Piezoelectric Properties of PVDF Nanofibers via Embedded Graphene Oxide Nanosheets. *Sci. Rep.* **2024**, *14*, 16484. [CrossRef]

22. Pan, C.-T.; Dutt, K.; Yen, C.-K.; Kumar, A.; Kaushik, A.C.; Wei, D.-Q.; Kumar, A.; Wen, Z.-H.; Hsu, W.-H.; Shiue, Y.-L. Characterization of Piezoelectric Properties of Ag-NPs Doped PVDFNanocomposite Fibres Membrane Prepared by Near Field Electrospinning. *Comb. Chem. High Throughput Screen.* **2022**, *25*, 720–729. [CrossRef]
23. Issa, A.; Al-Maadeed, M.; Luyt, A.; Ponnamma, D.; Hassan, M. Physico-Mechanical, Dielectric, and Piezoelectric Properties of PVDF Electrospun Mats Containing Silver Nanoparticles. *C* **2017**, *3*, 30. [CrossRef]
24. Chae, D.W.; Hwang, S.S.; Hong, S.M.; Hong, S.P.; Cho, B.G.; Kim, B.C. Influence of High Contents of Silver Nanoparticles on the Physical Properties of Poly(Vinylidene Fluoride). *Mol. Cryst. Liq. Cryst.* **2007**, *464*, 233/[815]–241/[823]. [CrossRef]
25. Xi, Y.; Pan, W.; Xi, D.; Liu, X.; Yu, J.; Xue, M.; Xu, N.; Wen, J.; Wang, W.; He, H.; et al. Optimization, Characterization and Evaluation of ZnO/Polyvinylidene Fluoride Nanocomposites for Orthopedic Applications: Improved Antibacterial Ability and Promoted Osteoblast Growth. *Drug Deliv.* **2020**, *27*, 1378–1385. [CrossRef] [PubMed]
26. Zaszczyńska, A.; Zabielski, K.; Gradys, A.; Kowalczyk, T.; Sajkiewicz, P. Piezoelectric Scaffolds as Smart Materials for Bone Tissue Engineering. *Polymers* **2024**, *16*, 2797. [CrossRef]
27. Liu, T.; Xie, F.; Geng, L.; He, R.; Sun, M.; Ni, T.; Xu, P.; Xing, C.; Peng, Y.; Chen, K.; et al. Micro-Electro Nanofibrous Dressings Based on PVDF-AgNPs as Wound Healing Materials to Promote Healing in Active Areas. *Int. J. Nanomed.* **2025**, *20*, 771–789. [CrossRef]
28. Mokhtari, F.; Samadi, A.; Rashed, A.O.; Li, X.; Razal, J.M.; Kong, L.; Varley, R.J.; Zhao, S. Recent Progress in Electrospun Polyvinylidene Fluoride (PVDF)-Based Nanofibers for Sustainable Energy and Environmental Applications. *Prog. Mater. Sci.* **2025**, *148*, 101376. [CrossRef]
29. Cai, X.; Lei, T.; Sun, D.; Lin, L. A Critical Analysis of the α, β and γ Phases in Poly(Vinylidene Fluoride) Using FTIR. *RSC Adv.* **2017**, *7*, 15382–15389. [CrossRef]
30. Jiang, Y.; Gong, L.; Hu, X.; Zhao, Y.; Chen, H.; Feng, L.; Zhang, D. Aligned P(VDF-TrFE) Nanofibers for Enhanced Piezoelectric Directional Strain Sensing. *Polymers* **2018**, *10*, 364. [CrossRef]
31. Zeyrek Ongun, M.; Oguzlar, S.; Doluel, E.C.; Kartal, U.; Yurddaskal, M. Enhancement of Piezoelectric Energy-Harvesting Capacity of Electrospun β-PVDF Nanogenerators by Adding GO and rGO. *J. Mater. Sci. Mater. Electron.* **2020**, *31*, 1960–1968. [CrossRef]
32. Kloster, J.T.; Danley, M.J.; Lai, V.K.; Zhao, P. Effects of Porosity on Piezoelectric Characteristics of Polyvinylidene Fluoride Films for Biomedical Applications. *BME Front.* **2023**, *4*, 0009. [CrossRef]
33. Gusarov, B.; Gusarova, E.; Viala, B.; Gimeno, L.; Cugat, O. PVDF Piezoelectric Voltage Coefficient *in Situ* Measurements as a Function of Applied Stress. *J. Appl. Polym. Sci.* **2016**, *133*, app.43248. [CrossRef]
34. Nair, R.; Mahdi, J.; Amjad, O.; Maree, K.; Jain, A.; Al-Dubai, A.; Jaradat, S.; Shyha, I.; Trabelsi, M.; Alothamn, B.; et al. Parametric Study on PVDF Electrospun Nanofibers: Optical Characteristics, Piezoelectric Analysis, and Correlated Applications. *J. Electron. Mater.* **2024**, *53*, 4155–4172. [CrossRef]
35. Crespo-Miguel, J.; Lucarini, S.; Arias, A.; Garcia-Gonzalez, D. Thermo-Electro-Mechanical Microstructural Interdependences in Conductive Thermoplastics. *NPJ Comput. Mater.* **2023**, *9*, 134. [CrossRef]
36. Dunn, M.L.; Taya, M. Micromechanics Predictions of the Effective Electroelastic Moduli of Piezoelectric Composites. *Int. J. Solids Struct.* **1993**, *30*, 161–175. [CrossRef]
37. Mishra, N.; Das, K. A Mori–Tanaka Based Micromechanical Model for Predicting the Effective Electroelastic Properties of Orthotropic Piezoelectric Composites with Spherical Inclusions. *SN Appl. Sci.* **2020**, *2*, 1206. [CrossRef]
38. Ferreira, P.M.; Machado, M.A.; Vidal, C.; Carvalho, M.S. Modelling Electro-Mechanical Behaviour in Piezoelectric Composites: Current Status and Perspectives on Homogenisation. *Adv. Eng. Softw.* **2024**, *193*, 103651. [CrossRef]
39. Cañamero, F.J.; Buroni, F.C.; Rodríguez-Tembleque, L. Connectivity Patterns in Lead-Free Piezocomposites: A Critical Analysis for 0-3 and 1-3 Configurations. *Compos. Struct.* **2024**, *337*, 118062. [CrossRef]
40. Mackerle, J. Smart Materials and Structures—A Finite Element Approach—An Addendum: A Bibliography (1997–2002). *Model. Simul. Mater. Sci. Eng.* **2003**, *11*, 707–744. [CrossRef]
41. Newnham, R.E.; Skinner, D.P.; Cross, L.E. Connectivity and Piezoelectric-Pyroelectric Composites. *Mater. Res. Bull.* **1978**, *13*, 525–536. [CrossRef]
42. Mori, T.; Tanaka, K. Average Stress in Matrix and Average Elastic Energy of Materials with Misfitting Inclusions. *Acta Metall.* **1973**, *21*, 571–574. [CrossRef]
43. Jiménez Segura, N.; Pichler, B.L.A.; Hellmich, C. A Green's Function-Based Approach to the Concentration Tensor Fields in Arbitrary Elastic Microstructures. *Front. Mater.* **2023**, *10*, 1137057. [CrossRef]
44. Yang, Z.; Yabansu, Y.C.; Jha, D.; Liao, W.; Choudhary, A.N.; Kalidindi, S.R.; Agrawal, A. Establishing Structure-Property Localization Linkages for Elastic Deformation of Three-Dimensional High Contrast Composites Using Deep Learning Approaches. *Acta Mater.* **2019**, *166*, 335–345. [CrossRef]

45. Akinpelu, A.; Bhullar, M.; Yao, Y. Discovery of Novel Materials through Machine Learning. *J. Phys. Condens. Matter* **2024**, *36*, 453001. [CrossRef]
46. Guinovart-Díaz, R.; Bravo-Castillero, J.; Rodríguez-Ramos, R.; Sabina, F.J.; Martínez-Rosado, R. Overall Properties of Piezocomposite Materials 1–3. *Mater. Lett.* **2001**, *48*, 93–98. [CrossRef]
47. Derraz, M.; Ennawaoui, C.; Mastouri, H.; El Hmamssy, Y.; Abouricha, N.; Rjafallah, A.; Laadissi, E.M.; Hajjaji, A. Mathematical Modeling for Predicting Electrical Energy Harvested Using Piezoelectric Composite Materials for Smart System Applications. *Micro Nano Eng.* **2024**, *23*, 100253. [CrossRef]
48. Lv, S.; Li, B.; Zhang, Q.; Shi, Y.; Gao, C. A Finite Deformation Phase Field Model for Electromechanical Fracture of Flexible Piezoelectric Materials. *Appl. Math. Model.* **2024**, *130*, 728–748. [CrossRef]
49. Zhang, J.; Chadwick, A.F.; Chopp, D.L.; Voorhees, P.W. Phase Field Modeling with Large Driving Forces. *NPJ Comput. Mater.* **2023**, *9*, 166. [CrossRef]
50. Shen, J.; Yang, X. Numerical Approximations of Allen-Cahn and Cahn-Hilliard Equations. *Discret. Contin. Dyn. Syst.* **2010**, *28*, 1669–1691. [CrossRef]
51. Wight, C.L.; Zhao, J. Solving Allen-Cahn and Cahn-Hilliard Equations Using the Adaptive Physics Informed Neural Networks. *arXiv* **2020**, arXiv:2007.04542.
52. Berger, H.; Kari, S.; Gabbert, U.; Rodriguez-Ramos, R.; Bravo-Castillero, J.; Guinovart-Diaz, R.; Sabina, F.J.; Maugin, G.A. Unit Cell Models of Piezoelectric Fiber Composites for Numerical and Analytical Calculation of Effective Properties. *Smart Mater. Struct.* **2006**, *15*, 451–458. [CrossRef]
53. Wang, S.Y. A Finite Element Model for the Static and Dynamic Analysis of a Piezoelectric Bimorph. *Int. J. Solids Struct.* **2004**, *41*, 4075–4096. [CrossRef]
54. Kumar, A.; Mishra, N.; Das, K.; Roy, A. Effective Electromechanical Properties and Energy Harvesting Response in PMN-0.3PT/PDMS Flexible Piezoelectric Composites: A Combined Experimental and Theoretical Study. *J. Phys. Appl. Phys.* **2023**, *56*, 365503. [CrossRef]
55. Mishra, N.; Das, K. A Comparative Study of Incremental Selfconsistent and Eshelby–Mori–Tanaka Models for Estimating the Electroelastic Properties of Piezoelectric Polymer Composites with an Orthotropic Matrix. *Mech. Compos. Mater.* **2022**, *58*, 657–672. [CrossRef]
56. Wang, S.Y.; Quek, S.T.; Ang, K.K. Vibration Control of Smart Piezoelectric Composite Plates. *Smart Mater. Struct.* **2001**, *10*, 637–644. [CrossRef]
57. Odegard, G. Constitutive Modeling of Nanotube–Reinforced Polymer Composites. *Compos. Sci. Technol.* **2003**, *63*, 1671–1687. [CrossRef]
58. Jagdish, A.K.; Buroni, F.C.; Melnik, R.; Rodriguez-Tembleque, L.; Sáez, A. Flexoelectric Anisotropy and Shear Contributions in Lead-Free Piezocomposites. *Mech. Res. Commun.* **2024**, *140*, 104321. [CrossRef]
59. Knott, E.F. Dielectric Constant of Plastic Foams. *IEEE Trans. Antennas Propag.* **1993**, *41*, 1167–1171. [CrossRef]
60. Furukawa, T.; Fujino, K.; Fukada, E. Electromechanical Properties in the Composites of Epoxy Resin and PZT Ceramics. *Jpn. J. Appl. Phys.* **1976**, *15*, 2119–2129. [CrossRef]
61. Velayutham, T.S.; Halim, N.A.; Majid, W.H.A. Enhanced Functionalities in Flexible Poly (Vinylidene Fluoride-Trifluoroethylene)/Cerium Oxide Doped Sodium Bismuth Titanate [P(VDF-TrFE)/NBT-CeO_2] Ceramic Polymer Composite Films: A Comprehensive Investigation of Piezoelectric, Pyroelectric, and Ferroelectric Properties. *Next Mater.* **2025**, *6*, 100416. [CrossRef]
62. Rodrigues-Marinho, T.; Tubio, C.R.; Lanceros-Mendez, S.; Costa, P.F.R. Tailoring the Electrical Response of Polyvinylidene Fluoride Nanocomposites with Electrically Conductive and Dielectric Fillers. *Adv. Eng. Mater.* **2024**, *26*, 2301596. [CrossRef]
63. Voigt, W. Ueber Die Beziehung Zwischen Den Beiden Elasticitätsconstanten Isotroper Körper. *Ann. Phys.* **1889**, *274*, 573–587. [CrossRef]
64. Hashin, Z.; Shtrikman, S. A Variational Approach to the Theory of the Elastic Behaviour of Multiphase Materials. *J. Mech. Phys. Solids* **1963**, *11*, 127–140. [CrossRef]
65. Affdl, J.C.H.; Kardos, J.L. The Halpin-Tsai Equations: A Review. *Polym. Eng. Sci.* **1976**, *16*, 344–352. [CrossRef]
66. Malikan, M.; Dastjerdi, S.; Rucka, M.; Kadkhodayan, M. On Nonlinear 3D Electro-Elastic Numerical Modeling of Two-Phase Inhomogeneous FG Piezocomposites Reinforced with GNPs. *Int. J. Eng. Sci.* **2025**, *207*, 104174. [CrossRef]
67. Eshelby, J.D. The Determination of the Elastic Field of an Ellipsoidal Inclusion, and Related Problems. *Proc. R. Soc. Lond. Ser. Math. Phys. Sci.* **1957**, *241*, 376–396. [CrossRef]
68. Mikata, Y. Determination of Piezoelectric Eshelby Tensor in Transversely Isotropic Piezoelectric Solids. *Int. J. Eng. Sci.* **2000**, *38*, 605–641. [CrossRef]
69. Kuo, W.-S.; Huang, J.H. On the Effective Electroelastic Properties of Piezoelectric Composites Containing Spatially Oriented Inclusions. *Int. J. Solids Struct.* **1997**, *34*, 2445–2461. [CrossRef]

70. Gandarilla-Pérez, C.A.; Rodríguez-Ramos, R.; Sevostianov, I.; Sabina, F.J.; Bravo-Castillero, J.; Guinovart-Díaz, R.; Lau-Alfonso, L. Extension of Maxwell Homogenization Scheme for Piezoelectric Composites Containing Spheroidal Inhomogeneities. *Int. J. Solids Struct.* **2018**, *135*, 125–136. [CrossRef]
71. Guinovart-Díaz, R.; Rodríguez-Ramos, R.; López-Realpozo, J.C.; Bravo-Castillero, J.; Otero, J.A.; Sabina, F.J.; Lebon, F.; Dumont, S. Analysis of Fibrous Elastic Composites with Nonuniform Imperfect Adhesion. *Acta Mech.* **2016**, *227*, 57–73. [CrossRef]
72. Odegard, G.M. Constitutive Modeling of Piezoelectric Polymer Composites. *Acta Mater.* **2004**, *52*, 5315–5330. [CrossRef]
73. Kar-Gupta, R.; Venkatesh, T.A. Electromechanical Response of 1–3 Piezoelectric Composites: An Analytical Model. *Acta Mater.* **2007**, *55*, 1093–1108. [CrossRef]
74. Tize Mha, P.; Maréchal, P.; Ntamack, G.E.; Kenmeugne, B. Phenomenological Model for Predicting Thermoelectromechanical Response of 2-2 Piezocomposite. *J. Compos. Mater.* **2023**, *57*, 2145–2159. [CrossRef]
75. Tize Mha, P.; Maréchal, P.; Ntamack, G.E.; Charif d'Ouazzane, S. Homogenized Electromechanical Coefficients and Effective Parameters of 1–3 Piezocomposites for Ultrasound Imaging Transducers. *Phys. Lett. A* **2021**, *408*, 127492. [CrossRef]
76. Chen, Q.; Xiao, C.; Yang, Z.; Tabet, J.; Chen, X. Deep Neural Network Homogenization of Multiphysics Behavior for Periodic Piezoelectric Composites. *Compos. Part Appl. Sci. Manuf.* **2024**, *186*, 108421. [CrossRef]
77. Zheng, B.; Jin, Z.; Hu, G.; Gu, J.; Yu, S.-Y.; Lee, J.-H.; Gu, G.X. Machine Learning and Experiments: A Synergy for the Development of Functional Materials. *MRS Bull.* **2023**, *48*, 142–152. [CrossRef]
78. Yang, Z.; Park, K.; Nam, J.; Cho, J.; Choi, Y.J.; Kim, Y.; Kim, H.; Ryu, S.; Kim, M. Multi-Objective Bayesian Optimization for Laminate-Inspired Mechanically Reinforced Piezoelectric Self-Powered Sensing Yarns. *Adv. Sci.* **2024**, *11*, 2402440. [CrossRef]
79. Zhu, W.; Yang, C.; Huang, B. Searching for Optimal Patterns of Magnetoelectric Multi-Phase Composites with Machine Learning Method. In Proceedings of the 2020 15th Symposium on Piezoelectrcity, Acoustic Waves and Device Applications (SPAWDA), Zhengzhou, China, 16 April 2021; IEEE: Zhengzhou, China, 2021; pp. 85–89.
80. Zhu, W.; Yang, C.; Huang, B.; Guo, Y.; Xie, L.; Zhang, Y.; Wang, J. Predicting and Optimizing Coupling Effect in Magnetoelectric Multi-Phase Composites Based on Machine Learning Algorithm. *Compos. Struct.* **2021**, *271*, 114175. [CrossRef]
81. Qi, Z.; Zhang, N.; Liu, Y.; Chen, W. Prediction of Mechanical Properties of Carbon Fiber Based on Cross-Scale FEM and Machine Learning. *Compos. Struct.* **2019**, *212*, 199–206. [CrossRef]
82. Singh, K.; Adhikari, J.; Roscow, J. Prediction of the Electromechanical Properties of a Piezoelectric Composite Material through the Artificial Neural Network. *Mater. Today Commun.* **2024**, *38*, 108288. [CrossRef]
83. Matos, M.A.S.; Pinho, S.T.; Tagarielli, V.L. Application of Machine Learning to Predict the Multiaxial Strain-Sensing Response of CNT-Polymer Composites. *Carbon* **2019**, *146*, 265–275. [CrossRef]
84. Subeshan, B.; Atayo, A.; Asmatulu, E. Machine Learning Applications for Electrospun Nanofibers: A Review. *J. Mater. Sci.* **2024**, *59*, 14095–14140. [CrossRef]
85. Sarma, S.; Verma, A.K.; Phadkule, S.S.; Saharia, M. Towards an Interpretable Machine Learning Model for Electrospun Polyvinylidene Fluoride (PVDF) Fiber Properties. *Comput. Mater. Sci.* **2022**, *213*, 111661. [CrossRef]
86. Sukpancharoen, S.; Wijakmatee, T.; Katongtung, T.; Ponhan, K.; Rattanachoung, N.; Khojitmate, S. Data-Driven Prediction of Electrospun Nanofiber Diameter Using Machine Learning: A Comprehensive Study and Web-Based Tool Development. *Results Eng.* **2024**, *24*, 102826. [CrossRef]
87. Bai, C.-Y.; Yeh, F.-Y.; Shu, Y.-C. Physics-Informed Neural Network for Parameter Identification in a Piezoelectric Harvester. In Proceedings of the Active and Passive Smart Structures and Integrated Systems XVIII, Long Beach, CA, USA, 9 May 2024; Tol, S., Nouh, M.A., Shahab, S., Yang, J., Huang, G., Li, X., Eds.; SPIE: Long Beach, CA, USA, 2024; p. 39.
88. Chen, Q.; Wang, G.; Pindera, M.-J. Homogenization and Localization of Nanoporous Composites—A Critical Review and New Developments. *Compos. Part B Eng.* **2018**, *155*, 329–368. [CrossRef]
89. Mourched, B.; Hoxha, M.; Abdelgalil, A.; Ferko, N.; Abdallah, M.; Potams, A.; Lushi, A.; Turan, H.I.; Vrtagic, S. Piezoelectric-Based Sensor Concept and Design with Machine Learning-Enabled Using COMSOL Multiphysics. *Appl. Sci.* **2022**, *12*, 9798. [CrossRef]
90. Stojanovic, B.S.; Svicevic, M.R.; Kaplarevic-Malisic, A.M.; Ivanovic, M.R.; Nedic, D.M.; Filipovic, N.D.; Mijailovich, S.M. Coupling Finite Element and Huxley Models in Multiscale Muscle Modeling. In Proceedings of the 2015 IEEE 15th International Conference on Bioinformatics and Bioengineering (BIBE), Belgrade, Serbia, 2–4 November 2015; IEEE: Belgrade, Serbia, 2015; pp. 1–6.
91. Milićević, B.; Ivanović, M.; Stojanović, B.; Milošević, M.; Kojić, M.; Filipović, N. Huxley Muscle Model Surrogates for High-Speed Multi-Scale Simulations of Cardiac Contraction. *Comput. Biol. Med.* **2022**, *149*, 105963. [CrossRef]
92. Milicevic, B.; Simic, V.; Milosevic, M.; Ivanovic, M.; Stojanovic, B.; Kojic, M.; Filipovic, N. Integration of Surrogate Huxley Muscle Model into Finite Element Solver for Simulation of the Cardiac Cycle. In Proceedings of the 2022 44th Annual International Conference of the IEEE Engineering in Medicine & Biology Society (EMBC), Glasgow, UK, 11 July 2022; IEEE: Glasgow, UK, 2022; pp. 3943–3946.

93. Batra, R.; Song, L.; Ramprasad, R. Emerging Materials Intelligence Ecosystems Propelled by Machine Learning. *Nat. Rev. Mater.* **2020**, *6*, 655–678. [CrossRef]
94. Ramprasad, R.; Batra, R.; Pilania, G.; Mannodi-Kanakkithodi, A.; Kim, C. Machine Learning in Materials Informatics: Recent Applications and Prospects. *NPJ Comput. Mater.* **2017**, *3*, 54. [CrossRef]
95. Balan, K.K.; Sivanesan, V.; Moorthy, N.; Budhhan, D.; Jeyaseelan, S.; Sundaramoorthy, S. Effect of Thickness of Mat and Testing Parameters on Tensile Strength Variability of Electrospun Nanofibrous Mat. *Mater. Today Proc.* **2016**, *3*, 1320–1329. [CrossRef]
96. Molnar, K.; Vas, L.M.; Czigany, T. Determination of Tensile Strength of Electrospun Single Nanofibers through Modeling Tensile Behavior of the Nanofibrous Mat. *Compos. Part B Eng.* **2012**, *43*, 15–21. [CrossRef]
97. Maccaferri, E.; Cocchi, D.; Mazzocchetti, L.; Benelli, T.; Brugo, T.M.; Giorgini, L.; Zucchelli, A. How Nanofibers Carry the Load: Toward a Universal and Reliable Approach for Tensile Testing of Polymeric Nanofibrous Membranes. *Macromol. Mater. Eng.* **2021**, *306*, 2100183. [CrossRef]
98. Shams Nateri, M.; Azizollah Ganji, B. The Effect of Material Properties on Sensitivity of the Microelectromechanical Systems Piezoelectric Hydrophone. *Int. J. Eng.* **2017**, *30*, 1848–1855. [CrossRef]
99. Moheimani, S.O.R.; Fleming, A.J. Fundamentals of Piezoelectricity. In *Piezoelectric Transducers for Vibration Control and Damping*; Advances in Industrial Control; Springer: London, UK, 2006; pp. 9–35; ISBN 978-1-84628-331-4.
100. Erturk, A.; Inman, D.J. *Piezoelectric Energy Harvesting*, 1st ed.; Wiley: New York, NY, USA, 2011; ISBN 978-0-470-68254-8.
101. Lin, Y.-C.; Huang, Y.-H.; Ma, C.-C.; Chang, C.-K. Experimental and Theoretical Investigations on Sensing and Dynamic Characteristics of PVDF Thin Film. *J. Mech.* **2020**, *37*, 1–12. [CrossRef]
102. Wu, C.M.; Chou, M.H. Acoustic–Electric Conversion and Piezoelectric Properties of Electrospun Polyvinylidene Fluoride/Silver Nanofibrous Membranes. *Express Polym. Lett.* **2020**, *14*, 103–114. [CrossRef]
103. Tariq, A.; Behravesh, A.H.; Utkarsh; Rizvi, G. Statistical Modeling and Optimization of Electrospinning for Improved Morphology and Enhanced β-Phase in Polyvinylidene Fluoride Nanofibers. *Polymers* **2023**, *15*, 4344. [CrossRef]
104. Motamedi, A.S.; Mirzadeh, H.; Hajiesmaeilbaigi, F.; Bagheri-Khoulenjani, S.; Shokrgozar, M. Effect of Electrospinning Parameters on Morphological Properties of PVDF Nanofibrous Scaffolds. *Prog. Biomater.* **2017**, *6*, 113–123. [CrossRef]
105. Zhao, Z.; Li, J.; Yuan, X.; Li, X.; Zhang, Y.; Sheng, J. Preparation and Properties of Electrospun Poly(Vinylidene Fluoride) Membranes. *J. Appl. Polym. Sci.* **2005**, *97*, 466–474. [CrossRef]
106. Felix Swamidoss, V.; Bangaru, M.; Nalathambi, G.; Sangeetha, D.; Selvam, A.K. Silver-Incorporated Poly Vinylidene Fluoride Nanofibers for Bacterial Filtration. *Aerosol Sci. Technol.* **2019**, *53*, 196–206. [CrossRef]
107. Ismail, A.M.; El-Newehy, M.H.; El-Naggar, M.E.; Meera Moydeen, A.; Menazea, A.A. Enhancement the Electrical Conductivity of the Synthesized Polyvinylidene Fluoride/Polyvinyl Chloride Composite Doped with Palladium Nanoparticles via Laser Ablation. *J. Mater. Res. Technol.* **2020**, *9*, 11178–11188. [CrossRef]
108. Tommalieh, M.J.; Ismail, A.M.; Awwad, N.S.; Ibrahium, H.A.; Youssef, M.A.; Menazea, A.A. Investigation of Electrical Conductivity of Gold Nanoparticles Scattered in Polyvinylidene Fluoride/Polyvinyl Chloride via Laser Ablation for Electrical Applications. *J. Electron. Mater.* **2020**, *49*, 7603–7608. [CrossRef]
109. Abdul-Majeed, M.A. Preparation and Characterization of AgNp/PVDF Cmposite Ultrafiltration Membrane. *J. Eng.* **2018**, *24*, 50–63. [CrossRef]
110. Han, G.; Su, Y.; Feng, Y.; Lu, N. Approaches for Increasing the β-Phase Concentration of Electrospun Polyvinylidene Fluoride (PVDF) Nanofibers. *ES Mater. Manuf.* **2019**, *6*, 75–80. [CrossRef]
111. Li, B.; Xu, C.; Zheng, J.; Xu, C. Sensitivity of Pressure Sensors Enhanced by Doping Silver Nanowires. *Sensors* **2014**, *14*, 9889–9899. [CrossRef] [PubMed]
112. Pourbafrani, M.; Azimi, S.; Yaghoobi Nia, N.; Zendehdel, M.; Abolhasani, M.M. The Effect of Electrospinning Parameters on Piezoelectric PVDF-TrFE Nanofibers: Experimental and Simulation Study. *Energies* **2022**, *16*, 37. [CrossRef]
113. Sriyanti, I.; Ramadhani, R.F.; Almafie, M.R.; Ap Idjan, M.K.N.; Syafri, E.; Solihah, I.; Sanjaya, M.R.; Jauhari, J.; Fudholi, A. Physicochemical and Mechanical Properties of Polyvinylidene Fluoride Nanofiber Membranes. *Case Stud. Chem. Environ. Eng.* **2024**, *9*, 100588. [CrossRef]
114. Rahman, M.Z.; Xu, H. Damping under Varying Frequencies, Mechanical Properties, and Failure Modes of Flax/Polypropylene Composites. *Polymers* **2023**, *15*, 1042. [CrossRef]
115. Li, H.; Tian, C.; Deng, Z.D. Energy Harvesting from Low Frequency Applications Using Piezoelectric Materials. *Appl. Phys. Rev.* **2014**, *1*, 041301. [CrossRef]

materials

MDPI

Article

Organic Optocoupler with Simple Construction as an Effective Linear Current Transceiver

Jaroslaw Jung [1,*], Arkadiusz Selerowicz [1], Jacek Ulanski [1], Ruslana Udovytska [1], Beata Luszczynska [1], Artur Zawadzki [2] and Andrzej Rybak [2]

[1] Department of Molecular Physics, Faculty of Chemistry, Lodz University of Technology, 116 Zeromskiego Str., 90-924 Lodz, Poland; arkadiusz.selerowicz@dokt.p.lodz.pl (A.S.); jacek.ulanski@p.lodz.pl (J.U.); ruslana.udovytska@p.lodz.pl (R.U.); beata.luszczynska@p.lodz.pl (B.L.)

[2] ABB Corporate Technology Center, 13A Starowislna Str., 31-038 Krakow, Poland; artur.zawadzki@pl.abb.com (A.Z.); andrzej.rybak@pl.abb.com (A.R.)

* Correspondence: jaroslaw.jung@p.lodz.pl

Abstract: In this study, it is shown that an efficient organic optocoupler (OPC) can be fabricated using commercially available and solution-processable organic semiconductors. The transmitter is a single-active-layer organic light-emitting diode (OLED) made from a well-known polyparavinylene derivative, Super Yellow. The receiver is an organic light-emitting diode (OLSD) with a single active layer consisting of a mixture of the polymer donor PTB7-Th and the low-molecular-weight acceptor ITIC; the receiver operates without an applied reverse voltage. OLED and OLSD have the same geometry and simple structure without any interlayers: glass/ITO/PEDOT:PSS/(active layer)/Ca/Al; the OPC is formed by OLED and OLSD which adhere tightly to each other. Despite its simple structure, the OPC showed a current transfer ratio of 0.13%, good linearity, and good dynamic performance: a three-decibel cutoff frequency of 170 kHz and response times to a step change in current at the OPC input of 2 μs. Compared to most organic OPC devices with similar performance parameters, where the transmitter and receiver have complex structures with additional interlayers between the active layers and electrodes and the need to apply a reverse voltage to the receiver, the simple design of our OPC reduces the number of fabrication steps and greatly simplifies the device fabrication process.

Keywords: organic optocoupler; short circuited organic photodetector; DC and AC properties

1. Introduction

An optocoupler is a device that can be used to transmit alternating signals between galvanically separated electrical circuits [1,2]. In real organic optocouplers (OPCs), the ability to transmit an AC signal without distortion is limited [3,4]. These limitations result from external conditions (in the case of hermetically encapsulated optocouplers, it is mainly the operating temperature) and the materials and manufacturing technology used [5 10]. These limitations include the three-decibel cutoff frequency ($f_{-3dB,OPC}$) for the highest harmonic of the alternative current transmitted to the receiver without amplitude attenuation and the current transfer coefficient (CTR), which indicates the efficiency with which the optocoupler converts the current applied to the input into the current detected at the output.

It should be noted that even with a transmitter and receiver with very good operating parameters, it is not always possible to design an optocoupler that can transmit AC signals with low nonlinear distortion. To obtain an effective device, transmitters and receivers should be selected so that their light emission and absorption spectra and linear operating ranges overlap partially [11] or completely [12,13]. In addition, additional interlayers are often used in photodiodes to prevent current leakage when reverse bias is applied [14–17]. However, it is possible to use simpler photodetector designs and achieve good results, as we demonstrate in this paper.

Organic optocouplers are constructed using an organic light-emitting diode (OLED) as the transmitter and an organic light-sensitive diode (OLSD) as the receiver. OLSD devices can operate as photodetectors with short-circuited electrodes. The active layers in the OLED and OLSD are made of organic semiconductors that can be processed by solution methods, making them easy to manufacture. Encapsulated OPCs exhibit relatively good long-term stability and can transmit current signals in a wide frequency bandwidth while maintaining good linearity [18–21]. A diagram of an optocoupler with a photodetector with short-circuited electrodes in the receiver circuit is shown in Figure 1.

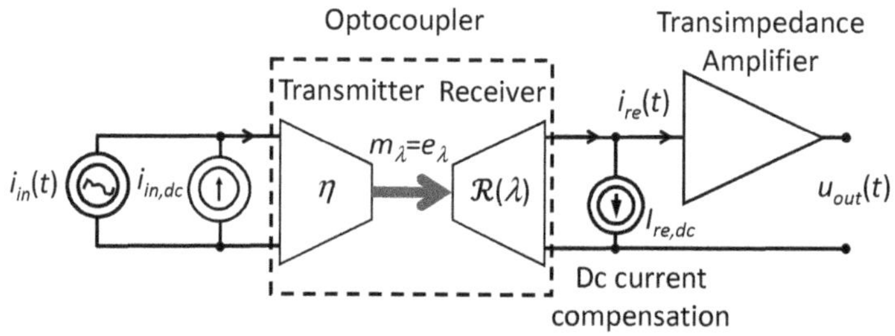

Figure 1. Schematic diagram of an optocoupler with dc $i_{in,dc}$ and ac $i_{in}(t)$ input current sources and a current source $i_{re,dc}$ designed to eliminate the DC component from the current at the output of the optocoupler.

In order to obtain the least distorted signals at the output of the optocoupler, it is necessary that in the frequency range corresponding to the frequency band of the transmitted signal, the current recorded at the output (i_{re}) is proportional to the current supplied to the input (i_{in}). For this purpose, a direct current ($i_{in,dc}$), is supplied to the input of the device, so that the device works in the linear range of the current transfer characteristic. An alternating current ($i_{in}(t)$), by means of which the signal is transmitted, is added to the constant component of current $i_{in,dc}$. The transmitter (i.e., OLED) emits alternating light with spectral emittance m_λ (where λ is the light wavelength) proportional to the variable component of current $i_{in}(t)$. The receiver (i.e., OLSD), exposed to radiation, converts the light energy into current i_{ph}. The output circuit contains a current source for eliminating the DC component of the receiver current ($i_{re,dc}$). The transimpedance amplifier (serving as an ideal ammeter with zero internal resistance) connected to the correction circuits converts the AC component of the receiver current ($i_{re}(t)$) into voltage ($u_{out}(t)$).

This paper presents an organic optocoupler with a simple structure consisting of the combination of an OLED as a transmitter and an OLSD as a receiver, wherein the active layers in the OLED and OLSD are fabricated by solution methods from well-known and widely studied organic semiconductors. The advantage of the device is the single-layer structure of both the OLED and OLSD, without interlayers. It is also important that, unlike most known organic optocouplers [9,22,23], the devices described in this paper do not require a reverse voltage on the receiver side. As the active layer in the OLED, a poly(para-phenylenevinylene) copolymer with the trade name SuperYellow (SY) (Figure 2a) [24–28] was used. SY is a well-known material that emits light in a yellow color and transports both holes and electrons. This material allows for the construction of OLEDs with a single active layer, which greatly simplifies the device fabrication process compared to most OLED devices with a complicated multi-layer structure. The active layer of the OLSD was a mixture of a donor and acceptor forming a bulk heterojunction. The donor was the polymer poly([2,6′-4,8-di(5-ethylhexylthienyl)benzo [1,2-b;3,3-b]dithiophene]{3-fluoro-2[(2-ethylhexyl)carbonyl]thieno [3,4-b]thiophenediyl}) with the commercial name PTB7-Th [29–31] (Figure 2c), and the acceptor was the low-molecular-weight, non-fullerene compound 3,9-bis(2-methylene-(3-(1,1-dicyanomethylene)-indanone))-5,5,11,11-tetrakis(4-hexylphenyl)-dithieno [2,3-d:2′,3′-d′]-s-indaceno [1,2-b:5,6-b′]dithiophene with the abbreviation ITIC [14,32–34] (Figure 2b). The OLED and OLSD had the same architecture, ITO/PEDOT:PSS/(active layer)/Ca/Al, and the same electrode geometry and were separated only by thin glass plates, i.e., glass supports of OLED and OLSD. It was therefore assumed that the spectral emittance of the emitter (m_λ) was equal to the spectral irradiance (e_λ) of the receiver.

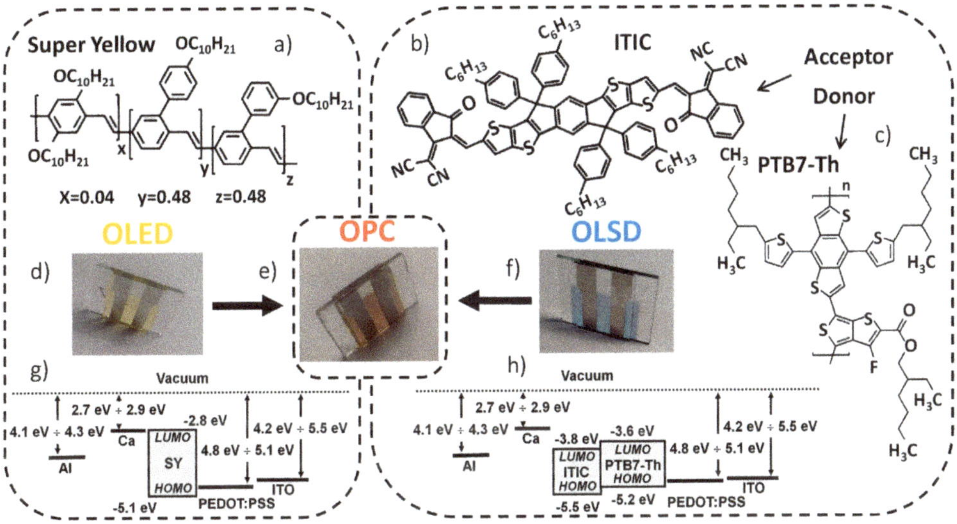

Figure 2. Materials and devices. Organic semiconductors used to produce active layer of (**a**) OLED (Super Yellow) and of OLSD ((**b**) acceptor ITIC and (**c**) donor PTB7-Th); photos of (**d**) OLED, (**e**) OPC, and (**f**) OLSD; (**g**,**h**) work functions for electrodes (Al, ITO), hole-transporting layer (PEDOT:PSS), and electron-transporting layer (Cu) and LUMO and HOMO energies of semiconductors [35,36].

Optocouplers manufactured in this way have good performance parameters, such as a high current transfer ratio (CTR) of 0.13%, a wide frequency band (three-decibel cutoff frequency of 170 kHz), and good linearity.

2. Materials and Methods

2.1. Sample Preparation and Characterization

Optocouplers were fabricated in two ways. One was to produce OLED devices (Figure 2d) and OLSD devices (Figure 2f) separately on glass substrates with six indium tin oxide (ITO) electrodes (anodes) (purchased from Ossila, S101, Sheffield, UK) and then bond the two devices closely together. The second method was to fabricate an optocoupler with ITO electrodes pre-patterned on both sides of the glass substrate (made by Instytut Fotonowy, Kraków, Poland) (Figure 2e). The geometry and area of the electrodes

(S = 4.5 × 10^{-6} m^2) for both one-sided and two-sided covered glass were identical. Substrates were ultrasonically washed twice in acetone and then twice in isopropanol for 5 min each time. Then, they were dried in a stream of nitrogen and then exposed to oxygen plasma for 150 s using a plasma surface treatment machine with a power of 200 W (Diener, Atto, Ebhausen, Germany).

The hole transport layers were prepared from a commercial PEDOT:PSS solution (purchased from Ossila, Al 4083) by spin coating them on ITO plates at a speed of 3000 rpm for 30 s. The substrates were then heated at 120 °C for 10 min in a nitrogen atmosphere. The active layer of the luminescent diodes was made by spin-coating (2000 rpm for 30 s) from a solution of Super Yellow (Sigma-Aldrich, St. Louis, MO, USA) in toluene (5 mg of polymer in 1 mL of solvent). The active layer of the OLSD was deposited from a mixture of ITIC and PTB7-Th in a mass ratio of 1:1. The total concentration of the solution was 20 mg of ITIC:PTB7-Th mixture per 1 mL of chloroform. Similar to OLED, the active layer of the OLSD device was prepared by spin coating (2500 rpm, 60 s). A scheme of the layer structure of the optocoupler is presented in Figure S1 in Supplementary Information.

The thicknesses of the prepared layers, measured using a Bruker DEKTAK XT needle profilometer (Billerica, MA, USA), were as follows: PEDOT:PSS-30 nm, SY-70 nm, and ITIC:PTB7-Th-150 nm. Finally, the metallic shared cathode was vacuum-evaporated on all six devices. This electrode was composed of a 10 nm layer of calcium evaporated directly on the surface of the active layer and a 100 nm thick protective layer of aluminum. The samples were prepared in a glovebox system with a controlled nitrogen atmosphere and then encapsulated to protect the organic layers from oxygen and moisture using an adhesive, non-destructive matrix encapsulation epoxy (Ossila, E132). Curing of epoxy resin was achieved by UV illumination at a wavelength of about 350 nm for 2 min. The resin in conjunction with a glass coverslip provided a robust barrier, ensuring an extended service life during the measurement and storage of samples.

2.2. Optical and Electrical Properties of Devices

DC Characterization

The DC current–voltage characteristics of OLSD and OLED were collected using a Keithley 2400 source measurement unit. The photocurrent was measured under OLSD illumination with the use of a solar simulator (AM1.5, 1000 Wm^{-2}, LOT Oriel, LS 0110-1000). The density of radiant flux per unit area of OLED were controlled with homemade equipment consisting of an Ocean Optics spectrometer (365–900 nm), the Keithley 2400 sourcemeter with a computer, and appropriate software. The external quantum efficiency (*EQE*) spectra for OPDs were determined using a photoelectric spectrometer (Instytut Fotonowy). The OLEDs were powered by a Matrix MPS-3005L-3 DC power supply and the current was measured with the use of a Fluke 8808A multimeter. The bias voltage applied to the OLSD and the photocurrent control were performed using a Keithley 2400 source meter.

AC Characterization

The OLEDs were powered by the sum of the DC and AC voltage from the sinusoidal- or square-wave generator (Matrix 8216A). The DC component of the bias voltage determined the operating conditions of the OLED devices (the operating point on the current–voltage characteristics). The AC analysis of the photocurrent was performed by illuminating the OLSD with an inorganic white light-emitting diode. Two Rigol DS1052E l oscilloscopes, electrically isolated from each other, were used to measure the dynamics of the signal changes. One was used to measure the LED or OLED supply voltage and the other to monitor the output voltage of the high-speed transimpedance amplifier (Femto, DHPCA-100) that converts the OLSD photocurrent into voltage. The output voltage of the transimpedance amplifier was proportional to the measured current over the full range of the frequency spectrum. A bias voltage applied to the photodiode was supplied by the voltage source (Matrix MPS-3005L-3). A silicon photodiode with a dedicated amplifier (ThorLabs, PDA36A-EC) was used to analyze the changes in the light emitted by the OLED over time. A scheme of the DC and AC control systems is presented in Figure S2 in the Supplementary Information.

3. Results and Discussion

An example of the current–voltage characteristics of an OLED based on SY is shown in Figure 3a. For a supply voltage (u_{OLED}) lower than the turn-on voltage u_{ON} = 2.1 V, a current flows in the diode, for which the density (j_{OLED}) is a linear function of the voltage. When the turn-on voltage is exceeded, the current density suddenly increases. According to the literature [36,37], the value of u_{ON} corresponds to the voltage of the built-in field, which, after multiplying by the elementary charge (q), is equal to the energy difference between the work functions of electrons exiting the anode and cathode (made of PEDOT:PSS and calcium layers, respectively). For voltages above this, light emission occurs.

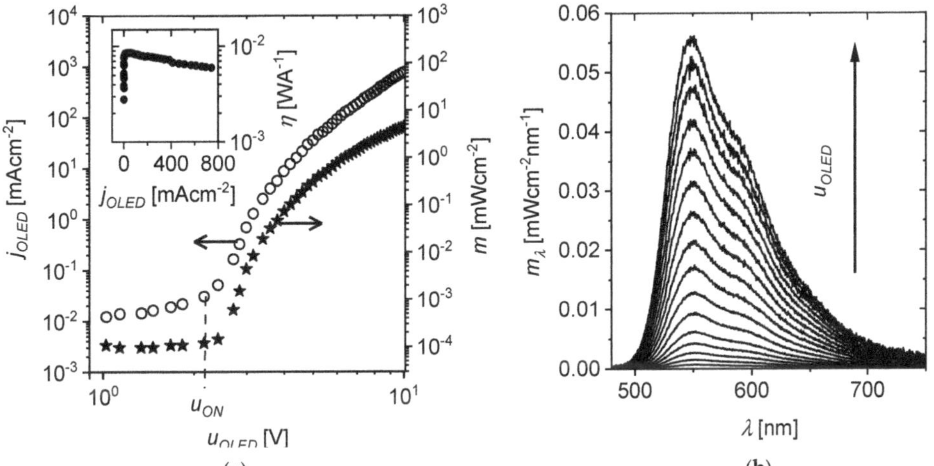

Figure 3. (**a**) Current–voltage characteristics $j_{OLED}(u_{OLED})$ (circles) and the dependence of the total radiant emittance on the voltage $m(u_{OLED})$ (stars) for OLED based on SY; the symbol u_{ON} denotes the threshold voltage; the inset shows the dependence of the current efficiency η on the current density j_{OLED}; (**b**) spectral emittance m_λ for different voltage values u_{OLED}.

3.1. Characteristics of the OLED Emitter

In optocouplers, the detector is an electronic device and not the human eye, so instead of photometric quantities (such as luminance), spectroscopic quantities such as the spectral density of radiation emittance m_λ and radiation emittance m were used to describe the operation of the OLED transmitter properties. The spectra m_λ measured for different values of voltage u_{OLED} are shown in Figure 3b. The dependence of the total radiant emittance on voltage $m(j_{OLED})$ shown in Figure 3a was determined from the spectral emittance $m_\lambda(j_{OLED})$ measured using Equation (1):

$$m(j_{OLED}) = \int_0^\infty m_\lambda(j_{OLED})d\lambda \tag{1}$$

Based on the *I-V* characteristic and the $m(u_{OLED})$ dependence, current efficiency was determined as a function of current density ($\eta(j_{OLED}) = m(j_{OLED})/j_{OLED}$). The graph in the inset of Figure 3a shows that as current density increases from 0.1 to 800 mA/cm, current efficiency slowly decreases from 85 to 60 A/W.

An important parameter of the transmitter is the speed at which the OLED emits light after the voltage is applied. To test the effectiveness of the emission of alternating light, the OLED devices were supplied with a voltage containing a constant time component, $u_{OLED,dc}$, and a harmonic variable component: $u_{OLED\sim}(t) = u_{0,OLED}\sin(2\pi ft)$. For the value of the constant voltage component $u_{OLED,dc}$ = 7 V and the amplitude of the variable component $u_{0,OLED}$ = 0.1 V, the current density varied from 190 to 210 mAcm^{-2}. The operating point of the OLED was the current density $j_{OLED,dc}$ = 200 mAcm^{-2} and the corresponding radiation emittance m_{dc} = 16 Wm^{-2}.

Based on the series of measured oscillograms for different frequencies, the relative amplitude of the radiation emittance, $m_{0,rel}(f)$, expressed in decibels, was calculated:

$$m_{0,rel}(f) = 20\log[m_0(f)/m_0(10\text{Hz})] \tag{2}$$

where $m_0(f)$ is the amplitude of the alternating component of the radiation emittance. Then, the amplitude spectrum of the radiation emittance was prepared and on its basis, the three-decibel cutoff frequency, $f_{-3dB,OLED}$, was determined to be $f_{-3dB,OLED} \approx 640$ kHz.

The speed of OLED devices was limited by contact resistance, $R_{s,OLED}$, differential resistance of the OLED, $R_{diff,OLED}$ (equal to the inverse derivative of the I-V characteristic for a fixed voltage value), and geometric capacitance $C_{OLED} = S\varepsilon\varepsilon_0/L$ (where ε, ε_0, S, and L are the vacuum permittivity, the relative permittivity of SY, the OLED area, and the SY layer thickness, respectively).

From the I-V characteristic presented in Figure 3a, it follows that for a voltage of 7 V, resistance $R_{diff,OLED}$ was equal to 220 Ω, and based on the electrical equivalent model of OLED (made on the basis of impedance tests, described in [36]), contact resistance $R_{s,OLED} = 100$ Ω was calculated. The three-decibel cutoff frequency is

$$f_{-3dB,OLED} = 1/(2\pi\tau_{RC,OLED}) \tag{3}$$

where $\tau_{RC,OLED} = 2\pi C_{OLED}R_{diff,OLED}R_{s,OLED}/\left(R_{diff,OLED}+R_{s,OLED}\right)$ is the time constant for the simplified OLED equivalent circuit shown in Figure S3 in the Supplementary Information. Based on the above formula and the experimentally obtained results, the value of the dielectric constant of SY, ε, was calculated from the following formula:

$$\varepsilon = \frac{L\left(R_{diff,OLED}+R_{s,OLED}\right)}{2\pi f_{-3dB,OLED}S\varepsilon_0 R_{diff,OLED}R_{s,OLED}} \tag{4}$$

The estimated ε value was 6.3.

3.2. Characteristics of the OLSD Photodetectors

Five different OLSD photodetectors were tested as a receiver in optocouplers with the OLED transmitter based on SY. It was assumed that the optocouplers should fulfill three basic conditions: (a) they should transmit signals from the transmitter to the photodetector with high efficiency, (b) they should be characterized by a high signal transmission rate while maintaining good DC parameters, and (c) they should not lose their electro-optical properties after encapsulation and operate stably under normal atmospheric conditions for at least several weeks. The characteristics of the tested OLSD devices with active layers made of mixtures of different semiconductors are given in the Supplementary Information (Figure S4 and Table S1). Based on the measured spectra of emittance spectral density m_λ for the OLED emitter (measured for j_{OLED} = 0.11 mAcm^{-2}) and the OLSD response $\mathcal{R}(\lambda)$, the overlap integrals were calculated. These were used to determine the ability of each transmitter/receiver pair to effectively convert light energy into photocurrent. The highest values of the relative overlap integral ξ (see Figure 4) were found for the OLED/IPT1:1chb and OLED/IPT1:1chf systems.

$$\xi = \frac{\int_{300\ \text{nm}}^{800\ \text{nm}} m_\lambda \mathcal{R}(\lambda) d\lambda}{\int_{300\ \text{nm}}^{800\ \text{nm}} m_\lambda \mathcal{R}_{IPT1:1chr}(\lambda) d\lambda} \tag{5}$$

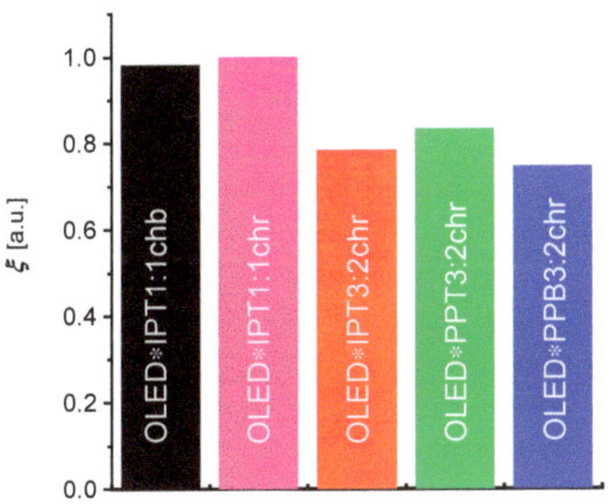

Figure 4. Comparison of relative overlap integrals ξ of spectral emittance m_λ of OLED based on SY transmitter and responsivity $\mathcal{R}(\lambda)$ of OLSD photodetectors based on different donor–acceptor mixtures.

The semiconductors that make up the active layer of OLSD devices are unstable in air, so OLSD devices were fabricated and characterized in glove boxes filled with nitrogen. The IPT1:1chb system was found to have the best stability in air after encapsulation of the OLSD devices tested. It was also found that this OLSD converted light signals to photocurrent at a much higher rate than the other OLSD devices and showed good DC properties without an applied reverse voltage. Therefore, this system was selected for use in the optocoupler together with SY-based OLED.

3.3. *Characteristics of the Optocoupler*

Constant current measurements were performed on the constructed OLED/OLSD devices and the dependence of the photocurrent density on input current $j_{ph,dc}(j_{in,dc})$ was determined. The current generated by light in the receiver, j_{re}, was measured using a transimpedance amplifier without reverse-biasing the OLSD photodetector. It was assumed that the current density in the circuit at the receiver output was equal to the photocurrent density j_{ph} in OLSD; according to the optocoupler model shown in Figure 1, $j_{re} = j_{ph} R_{sh,re}/(R_{sh,re} + R_{s,re})$, which for $R_{sh,re} \gg R_{s,re}$ gives $j_{re} \cong j_{ph}$. Figure 5 shows the measured current characteristic for an example optocoupler. The characteristic is linear for the current density at the OPC input in the range from 55 to 580 mAcm^{-2}.

The key parameter that characterizes an optocoupler is the current transfer ratio (*CTR*), which is defined as the ratio of the current detected at the receiver output to the current supplied to the transmitter input:

$$CTR = \frac{i_{re}}{i_{in}} \approx \frac{i_{ph}}{i_{in}} = \frac{j_{ph}}{j_{in}} \frac{S_{ph}}{S_{em}} \tag{6}$$

where S_{em} and S_{ph} are the surfaces of the emitter and photodetector, respectively (for the tested OLED/OLSD, both surfaces were the same; $S_{ph}/S_{em} = 1$).

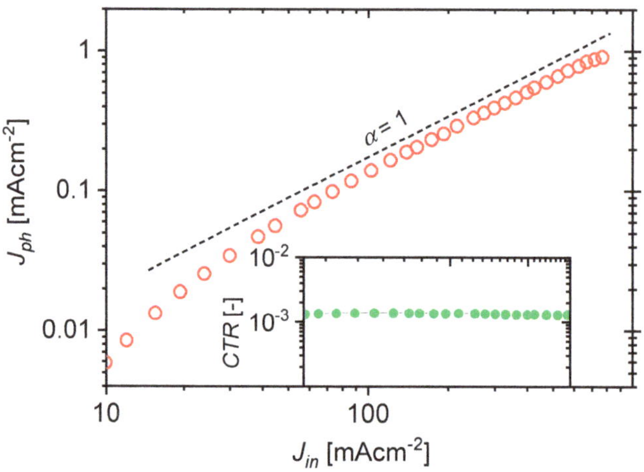

Figure 5. Dependence of photocurrent density at photodetector output j_{ph} on current density at transmitter input j_{in} for the OLED/OLSD optocoupler. The inset shows the dependence of current ratio CTR on j_{in} in the linear range of the optocoupler operation.

An important requirement is that the range of variations in the current applied to the device input, for which the value of CRT coefficient remains constant, should be as wide as possible. For a current density range j_{in} from 55 to 580 mAcm^{-2}, where the dependence $j_{ph,dc}(j_{in,dc})$ could be considered linear, the CRT value was about 0.13% and was similar to those found in the literature [4,8,38]. However, in the tested systems, the CTR value was only approximately constant over a limited range of current density $j_{in,dc}$. Moreover, the current transfer ratios for the DC and AC components, $CTR_{dc} = i_{ph,dc}/i_{in,dc}$ and $CTR_{ac} = i_{ph,rms}/i_{in,rms}$, respectively (where the rms subscript denotes the root mean square value), were not equal for all current density values.

For time-varying signals, for frequencies above the three-decibel optocoupler cutoff frequency ($f_{-3dB,OPC}$), CTR_{ac} decreased with increasing f according to the formula

$$CTR_{ac}(f) = CTR_0\left(1 + \frac{f^2}{f^2_{-3dB,SY}}\right)^{-1/2} \tag{7}$$

where CTR_0 is the current transfer ratio for $f \ll f_{-3dB,OPC}$.

In Figure 6a, it is visible that the experimentally determined CTR_{dc} and CTR_0 values decrease with current density j_{in} below a certain threshold value of current density $j_{in,t}$. This threshold value $j_{in,t}$ could be reduced by biasing the OLSD photodetector with a reverse voltage (u_{rev}). In order to check what effect the reverse voltage polarization of the OLSD photodetector may have on the linearity of current transfer, the characteristics $j_{ph,dc}(j_{in,dc})$ were measured for $u_{rev} = -1$ V and $u_{rev} = -2$ V. The photocurrent density $j_{ph,dc}$ was calculated as the difference between receiver current density ($j_{re,dc}$) for $j_{in,dc} > 0$ A and the leakage current ($j_{re,l}$) of the receiver at $j_{in,dc} = 0$ A ($j_{ph,dc} = j_{re,dc} - j_{re,l}$).

To estimate the values of CTR_{dc} and CTR_0, two approaches were used. In the first one, the coefficients were estimated based on independent tests performed separately for the receiver and transmitter (these coefficients are marked with the upper index *e*—estimated), and the second way was to determine these coefficients according to the definition as the ratio of the measured current at the OPC output and the current supplied to its input (these coefficients are marked with the upper index *d*—determined).

(a)

(b)

(c)

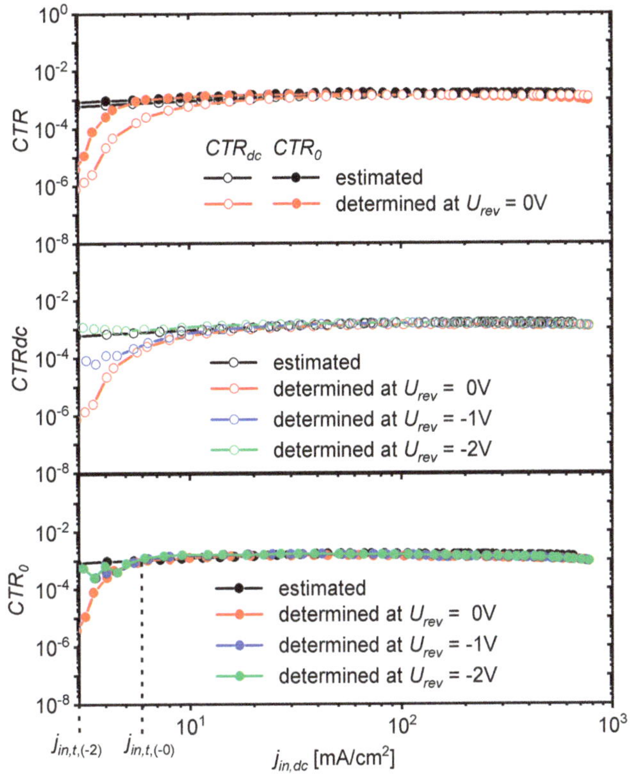

Figure 6. Comparison of the variations in direct CTR_{dc} and alternating CTR_0 current transfer ratios as a function of the density of direct current component $j_{in,dc}$ for (**a**) the experimentally determined DC and AC current transfer ratios for the OLED/OLSD optocoupler at reverse voltage $u_{rev} = 0$ and their equivalents estimated based on the experimental results obtained for the OLED transmitter and the OLSD photodetector; (**b**) direct CTR_{dc}, and (**c**) alternating CTR_0 current transfer ratios determined for the photodetector biased with the reverse voltage u_{rev} of two different values.

First, according to the definition of responsivity, the total photocurrent density was determined according to the formula $j_{ph} = \int_{\lambda_1}^{\lambda_2} m_\lambda \mathcal{R}(\lambda) d\lambda$. The dependence of the spectral density of emittance of the transmitter radiation m_λ on the current density at the transmitter input $j_{in,dc} = j_{OLED,dc}$ for OLED (Figure 3) and the responsivity spectrum ($\mathcal{R}(\lambda)$) of the receiver) for OLSD were determined experimentally (see Supplementary Information Figure S4). Then, according to Equation (6), $CTR_{dc}^{(e)}$ was calculated. It was assumed that for small signals of the variable component, the approximate equation holds $j_{ph,rms}/j_{pin,rms} \approx \mathrm{d}j_{ph,rms}/\mathrm{d}j_{pin,rms}$. Based on this, $CTR_0^{(e)}$ was calculated as follows:

$$CTR_0^{(e)}(j_{in,dc}) = \int_{\lambda_1}^{\lambda_2} \frac{dm_\lambda(j_{in,dc})}{dj_{in,dc}} \mathcal{R}(\lambda) d\lambda \quad (8)$$

The determined dependencies $CTR_{dc}^{(e)}(j_{in,dc})$ and $CTR_0^{(e)}(j_{in,dc})$ were treated as reference curves for coefficients $CTR_{dc}^{(d)}(j_{in,dc}) = j_{ph,dc}(j_{in,dc})/j_{in,dc}$ and $CTR_0^{(d)}(j_{in,dc}) =$

$dj_{ph,dc}(j_{in,dc})/dj_{in,dc}$ determined for the optocoupler created by assembling (i.e., simply by putting together) this transmitter–receiver pair.

Figure 6a compares the plots of the $CTR_{dc}(j_{in,dc})$ and $CTR_0(j_{in,dc})$ dependences determined by using the two methods described above for the case where the photodetector was not reverse-biased (u_{rev} = 0 V). For a current density higher than 20 mAcm^{-2}, all determined DC coefficients had very similar values in the range from 1.3×10^{-3} to 1.7×10^{-3}. However, as the current density decreased below 20 mAcm^{-2}, the coefficients $CTR_{dc}^{(d)}$ had lower values in relation to $CTR_{dc}^{(e)}$. A similar trend was observed for the AC coefficients, with significant differences between the estimated and experimentally determined CTR_0 values occurring only for the density of the direct current component $j_{in,dc}$ below 6 mAcm^{-2}.

As shown in Figure 6b, the values of coefficients $CTR_{dc}^{(e)}$ and $CTR_{dc}^{(d)}$ have approximately similar values for a current density $j_{in,dc}$ higher than 15 mAcm^{-2}, independently of the reverse voltage values applied to the OLSD photodetector. When the current density $j_{in,dc}$ is lower than 15 mAcm^{-2}, the values of coefficients $CTR_{dc}^{(e)}$ and $CTR_{dc}^{(d)}$ are different. However, after applying the reverse voltage to the photodetector, these differences decrease; the higher the absolute value of the $|u_{rev}|$, the smaller this difference is. A similar effect of applying the reverse voltage to the photodetectors was observed for the AC coefficients CTR_0 (Figure 6c).

The discrepancy between the estimated and experimentally determined CTR_{dc} and CTR_0 coefficients could be due to the phenomenon of electron trapping in the acceptor phase or/and holes in the donor phase in the bulk heterojunction in OLSD. Trapping reduces the population of mobile charge carriers and therefore reduces the photocurrent intensity. For a given trap concentration, the trapping effect will be particularly pronounced for low photocurrents because a relatively large fraction of the photogenerated charge carriers can get stuck in the traps. After applying the reverse voltage u_{rev} an additional external electric field is created in the heterojunction, which, according to the Poole–Frenkel mechanism, reduces the energy barrier for carriers to leave the traps [39,40]. At a sufficiently high voltage, the residence time of the carriers in the traps is significantly reduced, and the trapping of the charge carriers has little effect on the current flow.

In the case where $f < f_{-3dB,OPC}$ and $j_{in} > j_{in,t}$, the DC and AC current transfer ratios have similar values, $CTR_{dc} \approx CTR_0$, which apparently do not depend on the current intensity, as can be seen in Figure 6. However, the DC and AC CTR values are not exactly equal and constant in the OPC input current density ranges described above. The CTR coefficients changed slightly with the change in current density $j_{in,dc}$, resulting in the deterioration of linearity of the transmitting of high-amplitude AC currents (j_{in0}). In order to estimate the degree of this nonlinearity, we assumed that the range of linear operation of the example OLED/OLSD optocoupler for u_{rev} = 0 V was within the current density limits $j_{in,dc}$ from the threshold value $j_{in,t}$ = 6 mAcm^{-2} to the maximum value $j_{in,m}$ = 500 mAcm^{-2} (Figure 6c). For the assumed range of the constant component of current density, the maximum possible amplitude of the variable component of current density ($j_{in0,m}$) depended on the value of the constant component $j_{in,dc}$ and was $j_{in0,m} = (j_{in,dc} - j_{in,t})/2$ for $j_{in,dc} \leq (j_{in,m} - j_{in,t})/2$ or $j_{in0,m} = (j_{in,m} - j_{in,dc})/2$ for $j_{in,dc} \geq (j_{in,m} + j_{in,t})/2$, as shown in Figure 7a.

For $j_{in,dc} \in (j_{in,t}, j_{in,m})$, the coefficients CTR_0 changed slightly with the changing current, and their deviation from the mean value ($\overline{CTR_o}$) was expressed by coefficient δ_0 defined as

$$\delta_0(j_{in,dc}, j_{in0}) = \frac{\sqrt{\int_{j_{in,dc}-j_{in0}}^{j_{in,dc}+j_{in0}} \left(CTR_o(\xi) - \overline{CTR_o}(j_{in,dc}, j_{in0})\right)^2 d\xi}}{\overline{CTR_o}(j_{in,dc}, j_{in0})} \tag{9}$$

where

$$\overline{CTR_o}(j_{in,dc}, j_{in0}) = \frac{1}{2j_0} \int_{j_{in,dc}-j_{in0}}^{j_{in,dc}+j_{in0}} (CTR_o(\xi) - CTR_o(j_{in,dc}))^2 d\xi \tag{10}$$

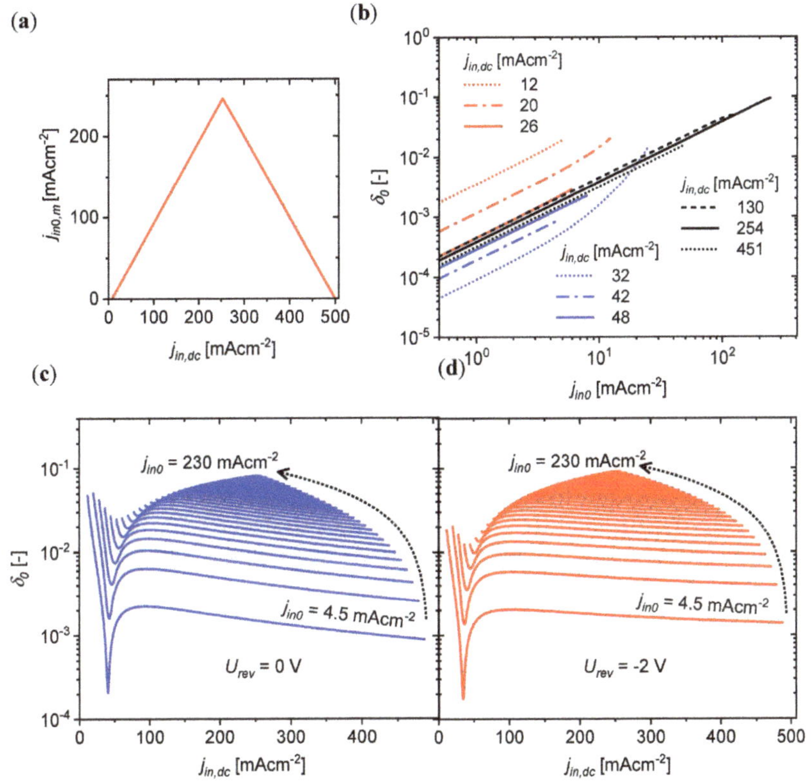

Figure 7. (**a**) The highest possible value of the amplitude of the variable component of current density $j_{in0,m}$ as a function of $j_{in,dc}$ for the assumed range of linear operation of OLED/OLSD for a photodetector unpolarized by reverse voltage; (**b**) dependence of coefficient δ_0 on $j_{in,0}$ for different values of the constant component of current $j_{in,dc}$ in the range 12–451 mAcm^{-2}; (**c**,**d**) coefficient δ_0 as a function of the constant component of current density $j_{in,dc}$ supplied to the OPC input for different values of the amplitude of variable component $j_{in,0}$, for two cases: (**c**) a non-polarized photodetector and (**d**) a photodetector polarized with a reverse voltage $u_{rev} = -2$ V.

The value of δ_0 depended on the choice of the constant component of current density $j_{in,dc}$ (OPC operating point) and the amplitude of the variable component of signal j_{in0}. Figure 7b shows the dependencies of the δ_0 coefficient on the amplitude of the variable signal for different values of the constant component of current $j_{in,dc}$ in the range

12–451 mAcm^{-2}. In each case, regardless of the value of the constant component of the current adopted, the nonlinearity increased as the amplitude of the variable component $j_{in0,m}$ increased. This is intuitive, since one might expect that the smaller the signal amplitude relative to the DC component of the input current, the better the linear function that can be used to approximate the current characteristic of the optocoupler in the immediate vicinity of the operating point (Figure 5).

Plots of the dependence of the δ_0 coefficient on the DC component of the current density for different values of the AC signal amplitude have been drawn for an optocoupler with an unpolarized photodetector (u_{rev} = 0 V—Figure 7c) and a polarized one (u_{rev} = −2 V—Figure 7d). After exceeding the current density $j_{in,dc}$ (e.g., 100 mAcm^{-2}), which is characteristic for the amplitude $j_{in,0}$ (e.g., 4.5 mAcm^{-2}), the δ_0 coefficient decreased, which was caused by the slightly decreasing current efficiency η with increasing current characteristic of OLED devices (see the inset in Figure 3a). In the whole range of current densities tested at the OPC input, δ_0 had a higher value for u_{rev} = −2 V, which indicates a deterioration in the linearity of light conversion into photocurrent by the optocoupler after polarizing the receiver with a voltage.

To operate in the third quadrant of the I-V characteristic and to broaden the frequency band photodiodes working as the photodetectors in optocoupler require reverse polarization [7,41–45]. However, from the above analysis, it is clear that in order to minimalize linearity degradation, the photodetector should be polarized with the lowest possible reverse voltage or should not be polarized at all. In addition, the following rules must be followed: (a) small values of direct current from the range in which δ_0 reaches the lowest value should be used for alternating signals with low amplitude, or (b) for higher-amplitude alternating signals, the highest possible values of direct current should be used. On the other hand, one should be aware that too small an amplitude of AC current and/or too large a DC component can hinder transmission. The limitation of a small AC amplitude is the noise occurring in the transmitter, receiver, and amplifier, as well as the fluctuations of the power of the emitted light. It is often necessary to use a preamplifier to increase the AC amplitude and improve the signal-to-noise ratio before feeding the signal to the OPC input. However, this may preclude the use of variant a), as small values of direct current are only suitable for small AC signal amplitudes (Figure 7c,d). In turn, the use of large values of direct current (variant b) will contribute to an increase in shot noise in the photodetector, which can also worsen the signal-to-noise ratio at the OPC output. In the case of polarization of the OLSD photodetector with reverse voltage, the leakage current can increase significantly with the increasing voltage u_{rev} and then the detectivity [44,46,47] of the receiver is drastically worsened.

Time-varying signals transmission tests

Taking into account the above analysis, in order to ensure low nonlinear distortions during the transmission of time-varying signals through the OLED(SY)/OLSD(IPT) optocoupler, the operating point determined by the input voltage $u_{in,dc}$ = 7 V (current density $j_{in,dc}$ = 200 mAcm^{-2}) was chosen. The corresponding constant component of the radiation emittance was m_{dc} = 16 Wm^{-2}. The variable component consisted of sinusoidal voltage waves $u_{in\sim}(t) = u_{in0}\sin(2\pi f t)$ and rectangular $u_{in\sqcap}(t) = u_{in0}[1(t) - 1(t-0.5/f) + 1(t-1/f) - t - 1.5/f + \cdots]$, where t is time, f is frequency, and $1(t)$ is the Heaviside step function. The amplitude of the variable component u_{in0} was 0.1 V. It corresponded to the range of transmitter current changes from 190 to 210 mAcm^{-2}. For the adopted operating point and the range of transmitter current variation, both the emitter radiative emittance values and the corresponding photodetector photocurrent were within the ranges of linear operation of the OPC (see the inset in Figure 5).

Figure 8 shows sample waveforms of the variable components of the voltage applied to the device input $u_{in\sqcap}$ and $u_{in\sim}$ and the corresponding current densities at the photodetector output $j_{ph\sqcap}$ and $j_{ph\sim}$ developed based on the measured oscillograms (see Supplementary Information Figure S5). The results are presented for two selected frequency values of 100 Hz and 158 kHz. At 100 Hz, there was no attenuation of the signal at the output of the device. There was no phase shift in the output current relative to the input voltage, and the input current increased immediately after the voltage jump (on a time scale corresponding to the period of a rectangular voltage wave). However, at a frequency of 158 kHz, the response to a rectangular wave did not resemble the shape of the input voltage. The rise and fall of the output current corresponded to the rise (τ_{rise}) and relaxation (τ_{fall}) times of 2 µs, respectively.

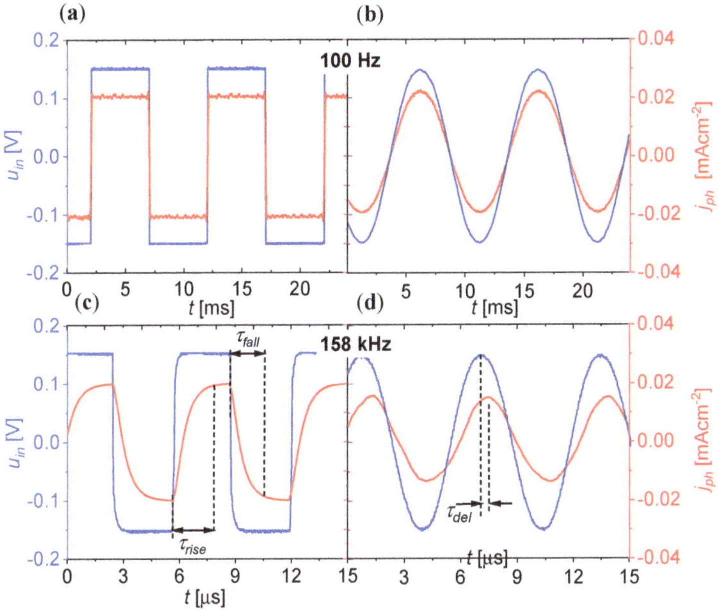

Figure 8. Variable components of current density at the output of the OLED/OLSD optocoupler $j_{ph\sqcap}$ and $j_{ph\sim}$ as a function of time t (red) after applying a voltage wave (blue) to the transmitter input. (**a**,**c**) Rectangular $u_{in\sqcap}$ and (**b**,**d**) sinusoidal $u_{in\sim}$ of two frequencies of 100 Hz and 158 kHz; (**c**) the symbols τ_{rise} and τ_{fall} denote the times of the rise and fall of the current at the output, respectively, after applying a rectangular voltage wave to the input; (**d**) τ_{del} is the delay time of the receiver photocurrent after applying a voltage harmonic component to the input.

After the deconvolution of the waveform of the variable component shown in the oscillogram in Figure S5, we obtained two time constants $\tau_1 = 249$ ns and $\tau_2 = 440$ ns. For the former, the three-decibel cutoff frequency ($1/(2\pi\tau_1)$) was 660 kHz, which roughly corresponded to the transmitter cutoff frequency $f_{-3\text{dB,OLED}}$. The second time constant τ_2 indicated that the three-decibel receiver cutoff frequency should be about 360 kHz, which is higher than $f_{-3\text{dB,OLSD}} = 250$ kHz for OLSD(IPT) (see Supplementary Information, Table S1).

For a sinusoidal signal with a frequency of 158 kHz, the output current amplitude decreased by $\sqrt{2}$ compared to the amplitude for lower frequencies (100 Hz). This would indicate that 158 kHz is equal to the three-decibel cutoff frequency $f_{-3\mathrm{dB,OPC}}$, but the phase shift between the input voltage and the output current was less than $\pi/4$ ($\tau_{rise} = 1/(8f)$). This in turn would mean that the three-decibel cutoff frequency was higher than 158 kHz. In order to determine $f_{-3\mathrm{dB,OPC}}$, the frequency dependence of the relative transadmittance $TA_{rel}(f)$ was investigated, calculated as the quotient of the amplitudes of the variable components of the current at the photodetector output, and the voltage applied to the transmitter input for frequency f was studied by referring to the same quotient, but for the undamped frequency of 10 Hz: $TA_{rel}(f) = [i_{in0}(f)u_{in0}(10\mathrm{Hz})]/[u_{in0}(f)i_{in0}(10\mathrm{Hz})]$. Figure 9 shows a comparison of the frequency spectra for the relative amplitude of radiant emittance ($m_{rel}(f) = m_0(f)/m_0(10\mathrm{Hz})$) for OLED, relative amplitude of photocurrent density $j_{ph,rel}(f) = j_{ph,0}(f)/j_{ph,0}(10\mathrm{Hz})$ recorded by OLSD, product $m_{rel}(f)j_{ph,rel}(f)$, and relative transadmittance $TA_{rel}(f)$. As expected, the spectrum obtained by multiplying the transmitter and receiver spectra ($m_{rel}(f)j_{ph,rel}(f)$) completely overlapped with the relative transadmittance spectrum of the optocoupler in a wide range of considered frequencies (from 10 Hz to 640 kHz). The determined three-decibel cutoff frequency of the OLED/OLSD optocoupler was 170 kHz.

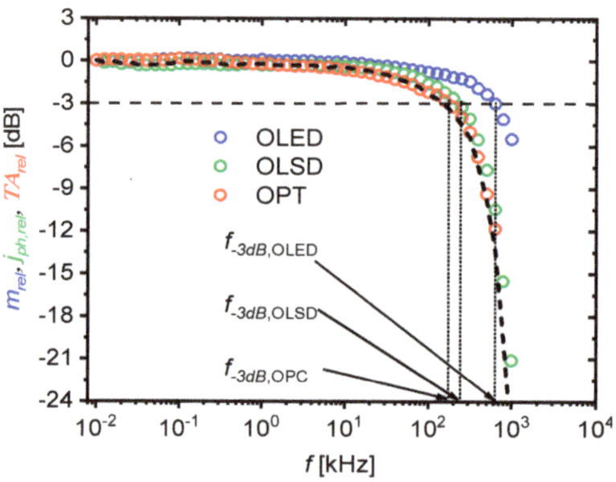

Figure 9. Comparison of frequency spectra for relative amplitude of radiant emittance m_{rel} of the OLED (blue circles), relative amplitude of photocurrent density $j_{ph,rel}$ of the OLSD photodiode (red circles), and relative transadmittance TA_{rel} of the OLED/OLSD optocoupler (green circles). Dashed curve—spectrum resulting from multiplication of m_{rel} and $j_{ph,rel}$ spectra. Vertical dotted lines indicate three-decibel cutoff frequencies of OLED $f_{-3\mathrm{dB},SY}$, OLSD $f_{-3\mathrm{dB},IPT}$ and OPC $f_{-3\mathrm{dB},OPT}$.

The spectrum obtained by mathematically combining the frequency spectra of the transmitter and photodetector matched the transadmittance spectrum measured for the optocoupler almost perfectly. Considering the considerable t time interval (several weeks) between the time of conducting a complete set of experiments, first performed separately for the transmitter and photodetector, and then the production of the optocoupler and its characterization, the overlap of the spectra indicated the stability of the fabricated devices. It could be expected that applying a reverse voltage to the photodetector would increase the cutoff frequency of the device. This broadening of the bandwidth also follows from the analysis of the OPC equivalent circuit (see Figure S3 in Supplementary Information), where capacitor C_{re} symbolizes the photodiode capacitance, which is the resultant of the geometric capacitance, the capacitance of the electrode–semiconductor junctions, and the capacitance of the bulk heterojunction. The application of an external electric field causes a decrease in junction capacitance [48–50] and this in turn causes an increase in limit frequency $f_{-3\mathrm{dB,OPC}}$. As a result, the frequency band for signals that can be transmitted without distortion is extended. However, for the OLED/OLSD optocouplers tested, no effect of increasing the limit frequency $f_{-3dB,OPC}$ was observed after applying a reverse voltage to the photodetector. This may be explained by the low leakage resistance and the small number of junctions in the simple structure of the OLSD photodetector tested compared to typical multilayer photodiodes described in the literature. In these photodiodes, the leakage current reduction is achieved by introducing additional interlayers [14,41] to block the injection of holes from the cathode and electrons from the anode. However, the additional junctions created by the interlayers hinder the charge carrier transport for alternating signals, because charges are accumulated near these junctions. In order to accelerate the operation of the device, these space-charges should be removed, which is achieved by applying a reverse voltage to the photodiode. In the case of our single-layer OLSD, the charges from the active layer and the areas near the junctions are quickly removed even with shorted electrodes due to the low leakage resistance; in this case, the application of a reverse voltage does not improve the photodiode dynamics.

It should be emphasized that it is difficult to produce an optocoupler that has both a high *CTR* and fast response time. In general, it can be seen that optocouplers with high *CTR* values tend to operate slower than optocouplers with low *CTR* [8,41,51–53]. The improvement in receiver speed is usually achieved by biasing the OPDs with the voltage u_{rev}. The assumption is that an increase in the effective electric field leads to less carrier trapping and higher effective mobility, resulting in faster carrier extraction from the active layer. However, in our devices, we obtained a high value of $f_{-3dB,OPC}$ = 170 kHz, comparable to other OPC devices fabricated using similar organic semiconductors, but without the need to apply u_{rev}. To the best of our knowledge, the three-decibel frequency for the photodetectors with an ITIC:PTB7-Th mixture layer has not yet been determined. The published data for OPD with different mixtures of the PTB7-Th/acceptor [29,30] or with ITIC/donor [14,32] show that only photodetectors with the mixture PTB7-Th:CO1-4Cl had cutoff frequencies $f_{-3dB,OPC}$ = 240 kHz comparable to our results [54,55].

The results presented in this article allow us to formulate several detailed summary statements. An organic light-emitting diode with a single active layer made of the well-known Super Yellow can be used as an efficient transmitter. An organic light-sensitive diode with a single active layer made of a mixture of a polymer donor PTB7-Th and a low-molecular-weight non-fullerene acceptor ITIC can be used as an efficient receiver. An OPC can be easily formed by inducing a tight adhesion of the glass substrates of OLED and OLSD, since the OLED and OLSD have the same geometry and structure: glass/ITO/PEDOT:PSS/(active layer)/Ca/Al. Despite their simplified structure, our OPC devices showed the ability to transmit current signals over a wide frequency band with a good linearity of current transmission. The current transfer ratio (CTR) was 0.13%; the three-decibel cutoff frequency was 170 kHz for harmonic signals, and the response time to a step change in the current at the OPC input was 2 μs.

4. Conclusions

- We have shown that it is possible to produce an optocoupler that meets the application requirements using commercially available organic semiconductors that can be processed using solution techniques.
- The organic light-sensitive diode can efficiently operate as a receiver with short-circuited electrodes, i.e., without any reverse voltage applied. It is often stated in the literature that it is necessary to apply a reverse voltage to the photodiode to make it operate in the third quadrant of its I-V characteristic and to extend the frequency bandwidth of the device. However, from our analysis, it is clear that in order to

minimize linearity degradation, the photodetector should not be biased or should be biased with the lowest possible reverse voltage.

- Compared to most OPC devices with complex multi-layer transmitter and receiver structures and the need to apply reverse voltage to the receiver, the simplest possible design of our OPC reduces the number of manufacturing steps and greatly simplifies the device manufacturing process.
- The simple structure of our OPC, if the vacuum-deposited Ca/Al electrodes are replaced with a conductive ink (e.g., zinc oxide and nanosilver suspensions), enables the production of optocouplers by printing on flexible double-sided ITO-coated substrates. This structure will also reduce light losses resulting from reflections at interfaces.

Supplementary Materials: The following supporting information can be downloaded at https://www.mdpi.com/article/10.3390/ma18010152/s1.

Author Contributions: Conceptualization, J.J., J.U. and B.L.; Methodology, A.S., A.Z. and A.R.; Formal analysis, J.J., A.Z. and A.R.; Investigation, A.S. and R.U.; Resources, A.S. and R.U.; Writing—original draft, J.J., J.U. and B.L.; Writing—review & editing, J.J., J.U., R.U. and B.L.; Supervision, J.J. All authors have read and agreed to the published version of the manuscript.

Funding: This work was financially supported by the National Science Centre (Poland) under grant 2020/37/B/ST5/03929.

Institutional Review Board Statement: Not applicable.

Informed Consent Statement: Not applicable.

Data Availability Statement: The original contributions presented in this study are included in the article/Supplementary Material. Further inquiries can be directed to the corresponding author.

Acknowledgments: The authors would like to thank Allison Proszowska for her support in proofreading the English language of the article.

Conflicts of Interest: Authors Artur Zawadzki and Andrzej Rybak were employed by ABB Corporate Technology Center. The remaining authors declare that the research was conducted in the absence of any commercial or financial relationships that could be construed as a potential conflict of interest.

References

1. Lai, P.; Gonzalez, D.; Madhusoodhanan, S.; Sabbar, A.; Ahmed, S.; Dong, B.; Wang, J.; Mantooth, H.A.; Yu, S.-Q.; Chen, Z. Development of LTCC-packaged optocouplers as optical galvanic isolation for high-temperature applications. *Sci. Rep.* **2022**, *12*, 11685. [CrossRef] [PubMed]
2. Dimitrov, B.; Collier, G.; Cruden, A. Design and Experimental Verification of Voltage Measurement Circuits Based on Linear Optocouplers with Galvanic Isolation for Battery Management Systems. *World Electr. Veh. J.* **2019**, *10*, 59. [CrossRef]
3. Ren, H.; Chen, J.; Li, Y.; Tang, J. Recent Progress in Organic Photodetectors and their Applications. *Adv. Sci.* **2021**, *8*, 2002418. [CrossRef] [PubMed]
4. Sun, Q.J.; Dong, G.F.; Wang, L.D.; Qiu, Y. Organic optocouplers. *Sci. China Chem.* **2011**, *54*, 1017–1026. [CrossRef]
5. Cantatore, E. (Ed.) Applications of Organic and Printed Electronics. In *Integrated Circuits and Systems*; Springer: Berlin/Heidelberg, Germany; New York, NY, USA; Dordrecht, The Netherlands; London, UK, 2013.
6. Buga, C.S.; Viana, J.C. A Review on Materials and Technologies for Organic Large—Area Electronics. *Adv. Mater. Technol.* **2021**, *6*, 2001016. [CrossRef]
7. Pierre, A.; Deckman, I.; Lechêne, P.B.; Arias, A.C. High Detectivity All—Printed Organic Photodiodes. *Adv. Mater.* **2015**, *27*, 6411–6417. [CrossRef] [PubMed]
8. de Arquer, F.P.G.; Armin, A.; Meredith, P.; Sargent, E.H. Solution-processed semiconductors for next-generation photodetectors. *Nat. Rev. Mater.* **2017**, *2*, 16100. [CrossRef]
9. Strobel, N.; Seiberlich, M.; Eckstein, R.; Lemmer, U.; Hernandez-Sosa, G. Organic photodiodes: Printing, coating, benchmarks, and applications. *Flex. Print. Electron.* **2019**, *4*, 043001. [CrossRef]

10. Liu, J.; Gao, M.; Kim, J.; Zhou, Z.; Chung, D.S.; Yin, H.; Ye, L. Challenges and recent advances in photodiodes-based organic photodetectors. *Mater. Today.* **2021**, *51*, 475–503. [CrossRef]
11. Li, D.; Dong, G.; Duan, L.; Zhang, D.; Wang, L.; Qiu, Y. High-Performance Organic Optocouplers Based on an Organic Photodiode With High Blue Light Sensitivity. *IEEE Electron Device Lett.* **2013**, *34*, 1295–1297. [CrossRef]
12. Li, D.; Kim, K.; Zhang, S.; Dong, G.; Kuo, Y. High-performance organic–inorganic hybrid optocouplers based on organic light-emitting diodes and *a-Si:H* Photodiodes. *Sens. Actuators A Phys.* **2015**, *236*, 364–368. [CrossRef]
13. De-Zhi, Y.; Heng-Da, S.; Jiang-Shan, C.; Dong-Ge, M. High Current Transfer Ratio Organic Optocoupler Based on Tandem Organic Light-Emitting Diode as the Input Unit. *Chin. Phys. Lett.* **2012**, *29*, 114216.
14. Klab, T.; Luszczynska, B.; Ulanski, J.; Wei, Q.; Chen, G.; Zou, Y. Influence of PEIE interlayer on detectivity of red-light sensitive organic non-fullerene photodetectors with reverse structure. *Org. Electron.* **2020**, *77*, 105527. [CrossRef]
15. Simone, G.; Dyson, M.J.; Meskers, S.C.J.; Janssen, R.A.J.; Gelinck, G.H. Organic Photodetectors and their Application in Large Area and Flexible Image Sensors: The Role of Dark Current. *Adv. Funct. Mater.* **2020**, *30*, 1904205. [CrossRef]
16. Ishii, H.; Sugiyama, K.; Ito, E.; Seki, K. Energy Level Alignment and Interfacial Electronic Structures at Organic/Metal and Organic/Organic Interfaces. *Adv. Mater.* **1999**, *11*, 605–625. [CrossRef]
17. Tan, Z.; Li, S.; Wang, F.; Qian, D.; Lin, J.; Hou, J.; Li, Y. High performance polymer solar cells with as-prepared zirconium acetylacetonate film as cathode buffer layer. *Sci. Rep.* **2014**, *4*, 4691. [CrossRef] [PubMed]
18. Planes, E.; Juillard, S.; Matheron, M.; Charvin, N.; Cros, S.; Qian, D.; Zhang, F.; Berson, S.; Flandin, L. Encapsulation Effect on Performance and Stability of Organic Solar Cells. *Adv. Mater. Interfaces* **2020**, *7*, 2000293. [CrossRef]
19. Zheng, H.; Tian, R.; Zhou, X.; Chen, J.; Liu, X.; Zhou, Y. Encapsulation of flexible organic solar cells via parylene and alumina dyads. *Org. Electron.* **2024**, *128*, 107035. [CrossRef]
20. Song, S.; Lu, J.; Ye, W.; Zhang, B.; Liu, X.; Xing, G.; Zhang, S. Stability, encapsulation and large-area fabrication of organic photovoltaics. *Sci. China Chem.* **2021**, *64*, 1441–1459. [CrossRef]
21. Estrada, R.; Luo, D.; Lee, C.-C.; Iskandar, J.; Biring, S.; Al Amin, N.R.; Akbar, A.K.; Chen, C.-H.; Yu, C.-W.; Pham, T.M.D.; et al. Developing efficient small molecule-based organic photo-couplers by optimizing the cathode interfacial layer in the photodetector. *J. Mater. Chem. C* **2023**, *11*, 5378. [CrossRef]
22. Jiang, B.-H.; Lin, D.-W.; Shiu, M.-N.; Su, Y.-W.; Tsai, T.-H.; Tsai, P.-H.; Shieh, T.-S.; Chan, K.; Lu, J.-H.; Chen, C.-P. Efficient, Ambient-Stable, All-Polymer Organic Photodetector for Machine Learning-Promoted Intelligent Monitoring of Indoor Plant Growth. *Adv. Opt. Mater.* **2023**, *11*, 2203129. [CrossRef]
23. Shekhara, H.; Lami, V.; Solomeshch, O.; Fenigstein, A.; Tomer, L.; Becky, L.; Vaynzof, Y.; Tessler, N. Doping induced performance enhancement in inverted small molecule organic photodiodes operating below 1V reverse bias—Towards compatibility with CMOS for imaging applications. *Org. Electron.* **2019**, *67*, 1–9. [CrossRef]
24. Szymański, M.C.A.; Łuszczyńska, B.; Ulański, J. Inkjet Printing of Super Yellow: Ink Formulation, Film Optimization, OLEDs Fabrication, and Transient Electroluminescence. *Sci. Rep.* **2019**, *9*, 8493.
25. Schlisske, S.; Rosenauer, C.; Rödlmeier, T.; Giringer, K.; Michels, J.J.; Kremer, K.; Lemmer, U.; Morsbach, S.; Daoulas, K.C.; Hernandez-Sosa, G. Ink Formulation for Printed Organic Electronics: Investigating Effects of Aggrega tion on Structure and Rheology of Functional Inks Based on Conjugated Polymers in Mixed Solvents. *Adv. Mater. Technol.* **2020**, *6*, 2000335. [CrossRef]
26. Snedden, E.W.; Cury, L.A.; Bourdakos, K.N.; Monkman, A.P. High Photoluminescence Quantum Yield Due to Intramolecular Energy Transfer in the Super Yellow Conjugated Copolymer. *Chem. Phys. Lett.* **2010**, *490*, 76–79. [CrossRef]
27. Höfle, S.; Lutz, T.; Egel, A.; Nickel, F.; Kettlitz, S.W.; Gomard, G.; Lemmer, U.; Colsmann, A. Influence of the Emission Layer Thickness on the Optoelectronic Properties of Solution Processed Organic Light-Emitting Diodes. *ACS Photonics* **2014**, *1*, 968–973. [CrossRef]
28. Burns, S.; MacLeod, J.; Trang Do, T.; Sonar, P.; Yambem, S.D. Effect of thermal annealing Super Yellow emissive layer on efficiency of OLEDs. *Sci. Rep.* **2017**, *7*, 40805. [CrossRef] [PubMed]
29. Huang, J.; Lee, J.; Nakayama, H.; Schrock, M.; Cao, D.X.; Cho, K.; Bazan, G.C.; Nguyen, T.-Q. Understanding and Countering Illumination-Sensitive Dark Current: Toward Organic Photodetectors with Reliable High Detectivity. *ACS Nano* **2021**, *15*, 1753–1763. [CrossRef] [PubMed]
30. Song, Y.; Zhong, Z.; He, P.; Yu, G.; Xue, Q.; Lan, L.; Huang, F. Doping Compensation Enables High-Detectivity Infrared Organic Photodiodes for Image Sensing. *Adv. Mater.* **2022**, *34*, 2201827. [CrossRef] [PubMed]
31. Tetreault, A.R.; Dang, M.-T.; Bender, T.P. PTB7 and PTB7-Th as universal polymers to evaluate materials development aspects of organic solar cells including interfacial layers, new fullerenes, and non-fullerene electron acceptors. *Synth. Met.* **2022**, *287*, 117088. [CrossRef]
32. Ban, D.; Yoo, K.; Kang, J. Characteristics of an organic photodetector with a conjugated donor and non-fullerene acceptor for indirect X-ray detection. *J. Inst.* **2019**, *14*, P12012. [CrossRef]

33. Baik, S.; Kim, D.W.; Kang, H.-S.; Hong, S.H.; Park, S.; An, B.-K.; Park, S.Y. TIC derivative acceptors for ternary organic solar cells: Fine-tuning of absorption bands, LUMO energy levels, and cascade charge transfer. *Sustain. Energy Fuels* **2022**, *6*, 110–120. [CrossRef]
34. Zhu, J.; Lv, L.; Dai, T.; Pan, R.; Wang, F.; Yang, J.; Do, H.; Zhou, E.; Chen, Z.-K.; Chen, F. Side-Chain Engineering in ITIC Skeleton Enabling As—Cast Organic Solar Cells with Reduced Energy Loss and Improved Vertical Phase Distribution. *ACS Mater. Lett.* **2024**, *6*, 2100–2110. [CrossRef]
35. Bi, P.; Zheng, F.; Yang, X.; Niu, M.; Feng, L.; Qin, W.; Hao, X. Dual Förster resonance energy transfer effects in non-fullerene ternary organic solar cells with the third component embedded in the donor and acceptor. *J. Mater. Chem. A* **2017**, *5*, 12120–12130. [CrossRef]
36. Udovytska, R.; Chulkin, P.; Wypych-Puszkarz, A.; Jung, J. Three Destinies of Solution-Processable Polymer Light-Emitting Diodes under Long-Time Operation. *Polymers* **2021**, *13*, 1853. [CrossRef]
37. Köhler, A.; Bässler, H. *Electronic Processes in Organic Semiconductors: An Introduction*; Wiley-VCH: Weinheim, Germany, 2015.
38. El Amrani, A.; Lucas, B.; Antony, R. Organic optocoupler consisting of an optimized blue organic light emitting diode and an organic photoconductor. *Superlattices Microstruct.* **2015**, *85*, 880–885. [CrossRef]
39. Frenkel, J. On Pre-Breakdown Phenomena in Insulators and Electronic Semi-Conductors. *Phys. Rev. B* **1938**, *54*, 647–648. [CrossRef]
40. Jung, J.; Ulanski, J. Charge Carrier Transport in Organic Semiconductor Composites—Models and Experimental Techniques. In *Solution—Processable Components for Organic Electronic Devices*; Luszczynska, B., Matyjaszewski, K., Ulański, J., Eds.; Wiley-VCH Verlag GmbH & Co. KGaA: Weinheim, Germany, 2019.
41. Armin, A.; Vuuren, R.D.J.-V.; Kopidakis, N.; Burn, P.L.; Meredith, P. Narrowband light detection via internal quantum efficiency manipulation of organic photodiodes. *Nat. Commun.* **2015**, *6*, 6343. [CrossRef] [PubMed]
42. Gong, X.; Tong, M.; Xia, Y.; Cai, W.; Moon, J.S.; Cao, Y.; Yu, G.; Shieh, C.-L.; Nilsson, B.; Heeger, A.J. High—Detectivity Polymer Photodetectors with Spectral Response from 300 nm to 1450 nm. *Science* **2009**, *325*, 1665–1667. [CrossRef]
43. Lai, L.-H.; Hsieh, C.-C.; Wu, J.-L.; Chang, Y.-M. Organic Photodiode Integration on Si Substrates beyond 1000 nm Wavelength. *ACS Appl. Electron. Mater.* **2022**, *4*, 168–176. [CrossRef]
44. Armin, A.; Hambsch, M.; Kim, I.K.; Burn, P.L.; Meredith, P.; Namdas, E.B. Thick junction broadband organic photodiodes. *Laser Photonics Rev.* **2014**, *8*, 924–932. [CrossRef]
45. Xia, Y.; Aguirre, L.E.; Xu, X.; Inganäs, O. All—Polymer High—Performance Photodetector through Lamination. *Adv. Electron. Mater.* **2020**, *6*, 1901017. [CrossRef]
46. Sze, S.M. *Physics of Semiconductor Devices*, 3rd ed.; John Wiley and Sons Inc.: Hoboken, NJ, USA, 2007.
47. Jones, R.C. A Method of Describing the Detectivity of Photoconductive Cells. *Rev. Sci. Instrum.* **1953**, *24*, 1035–1040. [CrossRef]
48. Kim, J.-M.; Lee, K.H.; Lim, J.; Lee, J.Y. Interface charge transport of multilayer devices for exact analysis of charge behavior in organic optoelectronic devices. *Appl. Phys. Lett.* **2021**, *118*, 203302. [CrossRef]
49. Guan, M.; Niu, L.; Zhang, Y.; Liu, X.; Li, Y.; Zeng, Y. Space charges and negative capacitance effect in organic light-emitting diodes by transient current response analysis. *RSC Adv.* **2017**, *7*, 50598–50602. [CrossRef]
50. Tsai, M.-N.; Chang, T.C.; Liu, P.-T.; Ko, C.-W.; Chen, C.; Lo, K. Short-diode like diffusion capacitance of organic light emission devices. *Thin Solid Films* **2006**, *498*, 244–248. [CrossRef]
51. Kato, K. Ultrawide-Band/High-Frequency Photodetectors. *IEEE Trans. Microw. Theory Tech.* **1999**, *47*, 1265–1281. [CrossRef]
52. Zhang, Z.-X.; Yao, J.-S.; Liang, L.; Tong, X.-W.; Lin, Y.; Liang, F.-X.; Yao, H.-B.; Luo, L.-B. High—Performance Photocoupler Based on Perovskite Light Emitting Diode and Photodetector. *ACS Appl. Mater. Interfaces* **2018**, *10*, 39441–39447. [CrossRef]
53. Li, D.; Dong, G.; Li, W.; Wang, L. High performance organic-inorganic perovskite-optocoupler based on low-voltage and fast response perovskite compound photodetector. *Sci. Rep.* **2016**, *5*, 7902. [CrossRef]
54. Lan, Z.; Lau, Y.S.; Wang, Y.; Xiao, Z.; Ding, L.; Luo, D.; Zhu, F. Filter—Free Band—Selective Organic Photodetectors. *Adv. Opt. Mater.* **2020**, *8*, 2001388. [CrossRef]
55. Huang, J.; Lee, J.; Vollbrecht, J.; Brus, V.V.; Dixon, A.L.; Cao, D.X.; Zhu, Z.; Du, Z.; Wang, H.; Cho, K.; et al. A High—Performance Solution—Processed Organic Photodetector for Near—Infrared Sensing. *Adv. Mater.* **2019**, *32*, 1906027. [CrossRef] [PubMed]

materials

MDPI

Article

Geopolymer Foam with Low Thermal Conductivity Based on Industrial Waste

Patrycja Bazan [1], Beata Figiela [1], Barbara Kozub [1], Michał Łach [1], Katarzyna Mróz [2], Mykola Melnychuk [3] and Kinga Korniejenko [1,*]

[1] Faculty of Materials Engineering and Physics, Cracow University of Technology, Jana Pawła II 37, 31-864 Cracow, Poland; patrycja.bazan@pk.edu.pl (P.B.); beata.figiela@pk.edu.pl (B.F.); barbara.kozub@pk.edu.pl (B.K.); michal.lach@pk.edu.pl (M.Ł.)

[2] Faculty of Civil Engineering, Cracow University of Technology, Warszawska 24, 31-155 Cracow, Poland; katarzyna.mroz@pk.edu.pl

[3] Department of Materials Science, Lutsk National Technical University, Lvivska 75, 43000 Lutsk, Ukraine; m.melnychuk@lntu.edu.ua

* Correspondence: kinga.korniejenko@pk.edu.pl; Tel.: +48-609974988

Abstract: Geopolymer materials are increasingly being considered as an alternative to environmentally damaging concrete based on Portland cement. The presented work analyzed waste from mines and waste incineration plants as potential precursors for producing geopolymer materials that could be used to make lightweight foamed geopolymers for insulation applications. The chemical and phase composition, radioactivity properties, and leachability of selected precursors were analyzed. Then, geopolymer materials were produced, and their strength properties were examined through compression and flexural tests. The results of the strength tests guided the material selection for foamed geopolymer materials. Next, geopolymer foams were foamed with hydrogen peroxide and aluminum powder. The produced foamed materials were subjected to strength and thermal conductivity tests. The results demonstrated the great potential of mine waste in the synthesis of geopolymers and the production of lightweight geopolymer foams with good insulating properties.

Keywords: mine waste; geopolymer; foaming process; mechanical properties; thermal conductivity

1. Introduction

Geopolymers are inorganic materials created by the alkaline activation of aluminosilicates [1,2]. The base materials to produce geopolymers are most often naturally occurring minerals from the group of silicates classified as clay minerals; namely, kaolin and metakaolin, rice husk ash, volcanic rock powders, fly ash, or slag are other precursor materials [3,4]. Nowadays, geopolymer research is gaining more and more interest, as it can be used as a cost-effective and economical alternative to organic polymers and inorganic cement in various applications, such as aviation, high-tech ceramics, thermal insulation foams, refractory building materials and adhesives, protective coatings, and hybrid organic-inorganic composites [5,6]. This budding interest is due to the physical, chemical, and mechanical properties of the geopolymers [7,8].

Research on geopolymer materials includes the analysis of the precursor selection, geopolymerization methods and modifications that tailor strength properties through additional components, time, annealing temperatures, and the type of activator [9,10] The basic properties and advantages of geopolymers over conventional concrete are their high compressive strength, thermal resistance, non-flammability, reduced energy demand, low carbon dioxide emission, increased durability, better workability, and reduced shrinkage [11,12]. Moreover, some geopolymers, especially those based on metakaolin or fly ash class F, have limited predisposition to cracking. Because of that, these materials are used in civil engineering for soil stabilization, slabs, insulation, cubes, etc. [13,14]. In addition, these materials are characterized by low permeability and chemical stability. Therefore, geopolymers are also considered advantageous materials for use in different civil engineering applications, including in marine infrastructure [15,16].

Thermal resistance is a critical property of geopolymers. A comparison of the fire resistance of conventional concrete and a fly ash-based geopolymer indicated that the concrete was limited to a temperature of 800 °C, beyond which the material undergoes intensive cracking. In contrast, at the same temperature, the geopolymer was characterized by small and few cracks [17–19].

The current literature on geopolymer materials focuses on solid materials, while scientific reports on geopolymer materials in the foamed form are receiving increased attention [20,21]. Foamed geopolymer materials are characterized by relatively low thermal conductivities. Similarly, the low weight of foamed geopolymers is considered regarding insulation materials, partitions, and void fillings [20,22,23]. Foamed geopolymer is obtained by chemical or mechanical foaming [24,25]. Research on foamed geopolymers has shown that these materials achieve thermal conductivities ~50% lower than concrete, ranging from 0.072–0.48 W/(m·K) [25–27]. The thermal properties of geopolymers are influenced by parameters such as size, distribution, pore shape, type and content of the foaming agent, type and content of the binder, activator content, temperature, and method of foaming [28–30].

The presented work demonstrates the enormous potential of geopolymer materials due to their advantageous properties. The indispensable advantages of geopolymer materials related to environmental protection, reducing the negative environmental impact of concrete based on Portland cement, and meeting commitments associated with the circular economy were highlighted. The work aims to synthesize a geopolymer foam with low thermal conductivity based on industrial waste as an innovative material for the circular economy. The work was divided into two stages. The first stage concerns the selection and preparation of waste materials (coal shale, fly ash, post-processing waste) for use as raw materials in the synthesis of inorganic polymers. In the second stage, the synthesis of waste-based geopolymers and the development of the foaming process were studied. Various foaming methods, in particular aluminum powder, hydrogen peroxide, and mixed methods, were utilized to prepare the foamed geopolymers. The presented research will also contribute to the knowledge of the risks associated with waste materials disposal. According to the circular economy, the methods of immobilization and the use of by-products stored in landfills to develop building materials are critical.

2. Materials and Methods

2.1. Materials

The first stage of research was the preparation of materials based on waste sources (fly ash, mining coal waste, and post-processing waste), and the following list presents applied materials:

- Ash from the municipal waste treatment plant in Białystok—a waste incineration plant that processes ~15 tons of municipal waste per hour [31].
- Fly ash from the combined heat and power plant in Skawina—this plant is a producer of electricity produced in combination with heat and is also a supplier of process steam, drinking water, and industrial water [31].
- Mining waste from the Tauron Wydobycie S.A. mine Zakład Górniczy Sobieski—a hard coal mine located in Jaworzno. The type of material used is flame coal.
- Mining waste from the Murcki-Staszic hard coal mine in Katowice—the exploited coal is described as low quality due to a large amount of sulfur and a small amount of chlorine.
- Mining waste from the Wieczorek II and Wujek hard coal mines—a mine located in Katowice.

The selected ashes showed sufficient reactivity in the synthesis process and did not require an additional calcination process. This was confirmed by the previous research [31].

Therefore, the activation process using mechanical methods was carried out using mine waste. Due to their large dimensions, the first stage of work with coal shale was fragmentation into smaller particles by crushing. Then, each of the crushed materials was ground in a wheel mill to obtain particles smaller than 200 μm. After each grinding process, the obtained type of fraction was examined. First, the particles larger than 200 μm were crushed in an automatic grinder on a sieve with a mesh size of 0.7 mm and then a 200 μm mesh sieve. In total, 4 kg of ground powders for each raw material were obtained. Then, the powders were heated (calcined) in a laboratory furnace at 700 °C for 24 h to facilitate alkaline activation, allowing for the appropriate microstructure.

The thermal treatment of aluminosilicate materials causes changes in their structure, resulting from the growth of the amorphous phase. Therefore, the mechanical strength of alkali-activated binders depends on the structural conditions of the aluminosilicate materials. Scientific research has shown that higher mechanical strength results from using calcined materials, such as metakaolin, or materials produced as by-products of high-temperature fly ash processes or blast furnace slag [32,33]. As with the pozzolanic reactivity, the alkali activation reactivity depends on the amorphous content of silica and aluminum. The reactivity is related to the structure of the material and increases with greater amorphous content [34].

2.2. Sample Preparation

Geopolymers were prepared using the sodium promoter and six waste materials acting as precursors. The activation process was performed using a 10 M solution of sodium hydroxide (NaOH) in combination with a solution of sodium silicate (and an aqueous solution of sodium silicate R-145 with a module of 2.5 molar and a density of approximately 1.45 g/cm^3) in the proportion of 1:2, which is the most used hydroxide activator in the synthesis of geopolymers. Furthermore, this solution is also the cheapest and most widely available of alkali hydroxides. The solution was mixed and allowed to stabilize to temperature and equilibrium concentrations, which took about 2 h, and then the solution was mixed with the precursors (in six compositions obtained from waste materials that are presented in Table 1). After obtaining a homogeneous mass with a thick, plastic consistency, the mixtures were transferred to molds and then vibrated using a vibrating table. The formed geopolymers were heated in a laboratory dryer (SLW 750 STD, POL-EKO-APARATURA, Wodzisław Śląski, Poland) for 24 h at 75 °C under atmospheric pressure. After 24 h, the samples were removed from the molds. The samples were designated according to the places and origin of the raw materials—Table 1.

Table 1. The designations of the tested samples—description of the selected sample index regarding the origin of the materials used in the production of geopolymers.

No	Designation	Origin of the Materials
1	Białystok	Municipal Waste Disposal Plant, Białystok
2	Skawina	Skawina Heat and Power Plant, Skawina
3	Sobieski	Tauron Wydobycie S.A. Z G. Sobieski, Jaworzno
4	Staszic	Polska Grupa Górnicza S.A. Branch of KWK Murcki-Staszic, Katowice
5	Wujek	Polska Grupa Górnicza S.A. KWK Wujek branch, Katowice
6	Wieczorek	SRK S.A. in Bytom, Branch of KWK "Wieczorek II", Katowice

A foaming process was carried out on selected precursors to obtain lightweight geopolymer materials for insulation purposes. The foaming process was carried out by adding hydrogen peroxide 36% and aluminum powder to the geopolymer paste (Table 2).

The detailed amounts are presented in Section 3.6.

Table 2. List of components investigated in the manufactured panels for future research.

Index	Solution	Foaming Agent	Stabilizer
Wieczorek	NaOH 10M	4.5 g Al	-
		1.5 g Al	
		30 mL Perhydrol 36%	20 mL polyglycol
		30 mL Perhydrol 36%	5 g cellulose
Staszic		4.5 g Al	
		30 mL Perhydrol 36%	20 mL polyglycol
		30 mL Perhydrol 36%	5 g cellulose

2.3. Methods

The mineral composition of the geopolymer precursors was analyzed with X-ray diffraction (XRD) (PANalytical, Almelo, The Netherlands) using the powder X-ray method (Debye-Scherrer). Phase analysis for all materials was performed on the PANalytical AERIS X-ray diffractometer using Cu-Kα radiation. The obtained distance between the planes was used to identify the phases. X-ray analysis was performed using the HighScore Plus software (v5.1) and the PDF4 + crystallographic database. Particle size analysis was made using the laser diffraction method on the Anton Paar PSA 1190 LD Graz particle analyzer (Anton Paar GmbH, Graz, Austria). Leachability and radioactivity tests were also commissioned for all analyzed materials. A leachability test was performed by GIG (Katowice, Poland). It was made according to PN EN 12457-2:2006 standard. Radioactivity tests were performed at the accredited AP Geotechnika Laboratory (Siemianowice Śląskie, Poland) following the PN EN 12457-2:2006 standard. Fourier-transform infrared spectroscopy (FT-IR) analysis of bonding in the material was carried out with the use Spektromer FT-IR Nicolet 6700 (Nicolet, Klapálkova, Czech Republic). It was performed by the Department of Medical Physics of the Jagiellonian University. Microscopic observations and analysis of the chemical composition in the micro-regions were performed using scanning electron microscopy (SEM, JEOL JSN5510LV) with an energy dispersive spectrometer (EDS), (JEOL, Tokyo, Japan). The density was determined using a geometric method. Density was determined as a ratio of mass to volume. Volume was calculated based on the arithmetic means of the measurements for five cuboidal samples for compressive strength tests for each of the analyzed compositions. Strength tests of the geopolymer materials were performed on cubic samples (50 mm × 50 mm × 50 mm) to determine compressive strength (PN-EN 12390-3:2019 Testing of concrete—Part 3: Compressive strength of test specimens). Flexural strength tests (PN-EN 12390-5:2019 Testing hardened concrete—Part 5: Flexural strength of test specimens) were carried out on prismatic samples (50 mm × 50 mm × 200 mm) with a 150 mm distance between the support points. Strength tests were carried out on a Matest 3000 kN testing machine (Matest, Treviolo, Italy); the speed of the tests was 0.05 MPa/s. The thermal conductivity was tested on the HFM 446 plate apparatus (Wittelsbacherstrasse, Germany). The average temperature of testing was 10 °C. The temperature of the upper plate was 20 °C and the lower plate was 0 °C. The distance between the plates was equal to the plate thickness. All tests were performed after 28 days of conditioning the geopolymeric samples at room temperature.

3. Results and Discussion

3.1. Summary XRD Analyses Performed and Results

Although geopolymers have an amorphous phase, the XRD technique helps analyze their chain formation. Figure 1 summarizes the obtained patterns from XRD analysis.

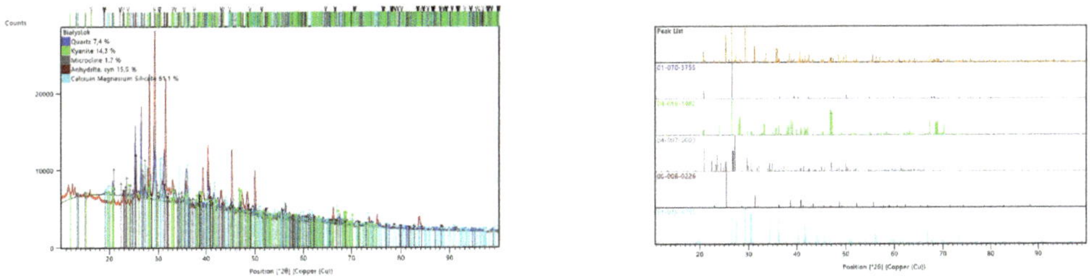

(a)

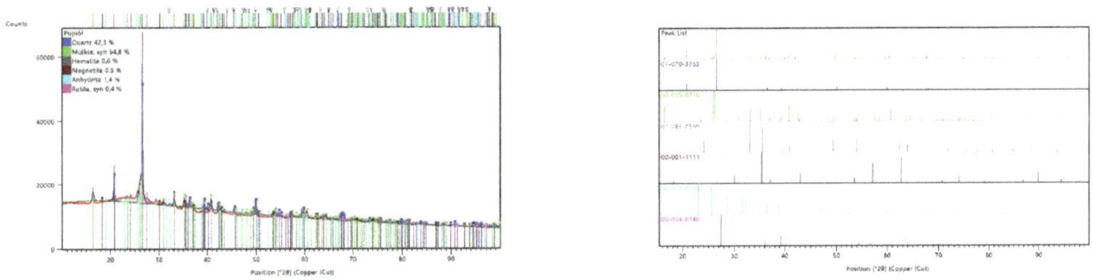

(b)

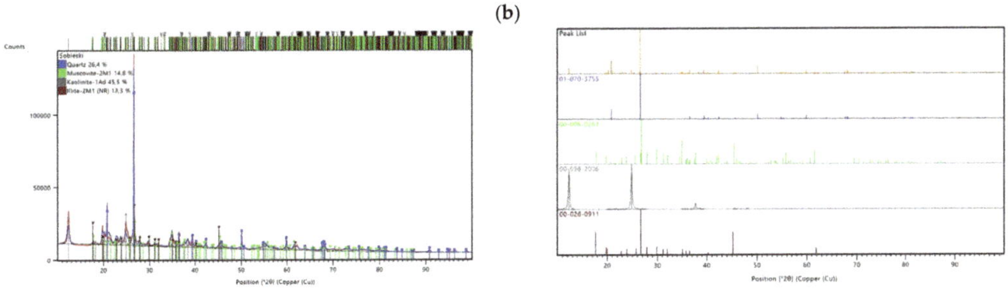

(c)

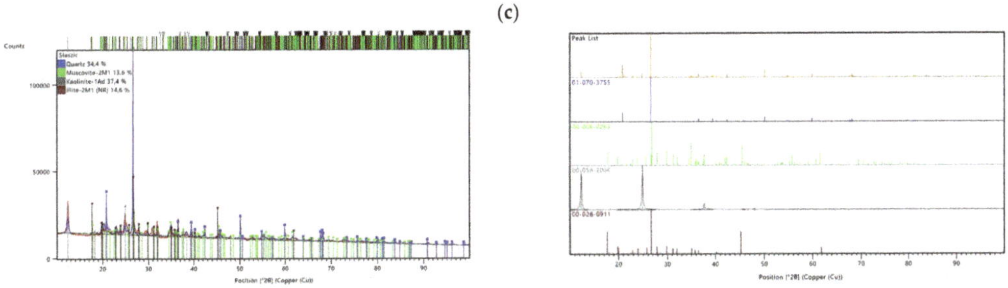

(d)

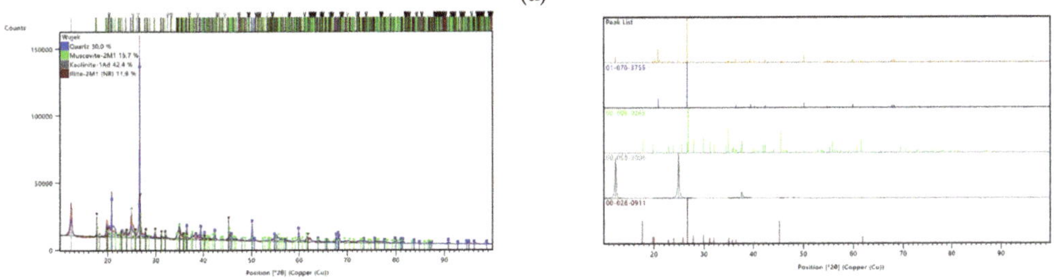

(e)

Figure 1. *Cont.*

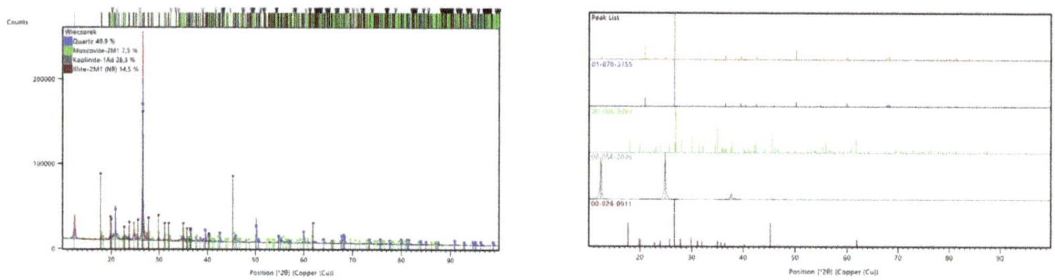

(**f**)

Figure 1. XRD patterns for raw materials (**a**) Białystok ash, (**b**) Skawina fly ash, (**c**) Sobieski, (**d**) Staszic, (**e**) Wujek, (**f**) Wieczorek.

Table 3 summarizes the percentages of individually identified phases for all tested materials to enable comparative analysis.

Table 3. Summary of the XRD phase analysis for all tested materials.

Index	Percentage [%]					
	Quartz SiO_2	Muscovite-2M1 $KAl_2(Si_3Al)O_{10}(OH,F)_2$		Kaolinite-1Ad $Al_2Si_2O_5(OH)_4$	Illite-2M1 $(K,H_3O)Al_2Si_3AlO_{10}(OH)_2$	
Sobieski	26.4	14.8		45.5	13.3	
Staszic	34.4	13.6		37.4	14.6	
Wieczorek	49.9	7.3		28.3	14.5	
Wujek	30.0	15.7		42.4	11.9	
Index	Percentage [%]					
	Quartz SiO_2	Cyanite Al_2SiO_5	Microcline $KAlSi_3O_8$	Anhydrite $CaSO_4$	Diopside $MgCaSi_2O_6$	
Białystok	7.4	14.3	1.7	15.5	61.1	
Skawina	Percentage [%]					
	Quartz SiO_2	Mullite $Al_6Si_2O_{13}$	Hematite Fe_2O_3	Magnetite Fe_3O_4	Anhydrite $CaSO_4$	Rutile TiO_2
	42.3	54.8	0.6	0.5	1.4	0.4

The analysis of the XRD techniques demonstrated that the tested carbon shales contain layered silicates such as quartz, kaolinite, muscovite, and illite. Waste material obtained from Katowice's Wieczorek II hard coal mine exhibited the highest quartz content (49.9%) followed by the Skawina mine (42.3%). However, material from the Staszic coal mine and Wujek contained almost equal silicon content. The greatest concentration of kaolinite, with the lowest range of illite, was exhibited in material from the Sobieski mine.

3.2. Summary of the Microscopic Examinations and EDS Analysis

The shapes and sizes of the raw materials' particles depend on many factors. The particle size of the tested materials ranged from 5–200 μm [31]. The smallest, regular-shaped particles were found in fly ash from the Skawina CHP Plant, which contained few impurities (Table 4). The remaining materials were characterized by a large particle size distribution. The chemical composition of the materials is characterized by the content of silicon, aluminum, potassium, titanium, sulfur, chlorine, sodium, calcium, and magnesium compounds. Minerals rich in iron, titanium, sulfur, phosphorus, boron, alkali metal, and alkaline earth metal cations are common, and their distribution in the material is very variable. Each coal particle may contain different amounts of inorganic substances, and thus the resulting ash may be highly heterogeneous.

The titanium present in the ashes can affect the formability of the material. Titanium is present as an impurity in kaolin and other clays. Moreover, titanium is an excellent nucleating agent for crystallization. Other particles may also be present in the coal fly ash, including reactive calcium silicate phases similar to those in Portland cement. These particles are the result of cement that fills the pores in the carbon. The presence of calcium in large amounts can interfere with the polymerization process and change the microstructure [34]. However, there are reports of a positive effect of calcium on the geopolymerization process. Previous research has shown that fly ash with a high concentration of calcium is suitable as a source material for producing good-quality geopolymer materials [35,36]. The calcium in the fly ash creates the hydrated phase of calcium silicate (C-S-H). The coexistence of this phase with geopolymer gel positively affects the strength properties of the obtained geopolymer materials. In addition, the high calcium content causes calcium atoms to bind in the geopolymer lattice and act as a charge-balancing cation [37,38]. The microscopic images and EDS analyses are shown in Figures 2–4.

(a)

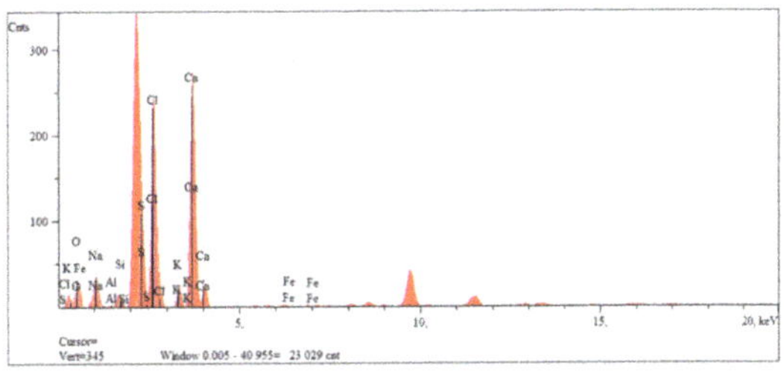

Element	IntensityKa [c/s]	atom.[%]	mass[%]	error 2-sig
O	10.51	44.226	26.680	2.041
Na	17.60	7.252	6.287	0.422
Mg	1.49	0.410	0.376	0.223
Al	1.69	0.317	0.323	0.190
Si	6.56	0.958	1.014	0.175
S	78.74	9.013	10.896	0.266
Cl	135.01	16.160	21.602	0.391
K	13.64	1.613	2.378	0.198
Ca	169.56	19.818	29.951	0.445
Fe	1.70	0.234	0.493	0.179

(b)

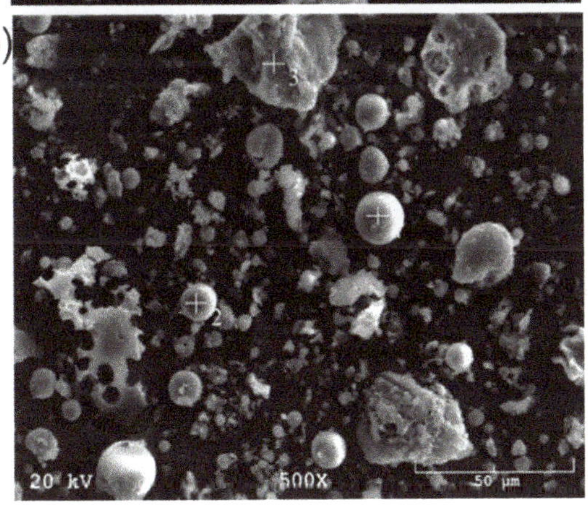

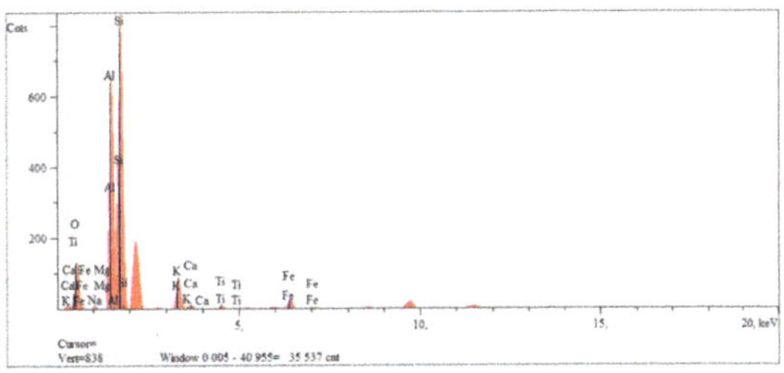

Element	IntensityKa [c/s]	atom.[%]	mass[%]	error 2-sig
O	55.84	42.996	29.558	0.973
Na	3.83	0.466	0.461	0.137
Mg	1.48	0.115	0.120	0.107
Al	368.57	20.751	24.057	0.272
Si	467.36	31.598	38.132	0.370
K	52.80	2.519	4.232	0.137
Ca	5.70	0.262	0.451	0.088
Ti	7.75	0.333	0.686	0.107
Fe	18.78	0.960	2.304	0.131

Figure 2. SEM micrographs and EDS analyses for (**a**) Białystok ash and (**b**) Skawina fly ash.

Table 4. Summary of the particle size and chemical composition of the tested materials.

Index	Particle Size, [µm]	Chemical Composition
Białystok	5–100	Compounds of calcium, chlorine, sulfur, sodium, potassium, and silicon
Skawina	<20	Compounds of silicon, aluminum, calcium, and iron
Sobieski	20–100	Compounds of silicon, aluminum, potassium, iron, and titanium
Staszic	2–100	Compounds of silicon, aluminum, potassium, and iron
Wujek	10–70	Compounds of silicon, aluminum, potassium, and iron
Wieczorek	30–200	Compounds of silicon, aluminum, potassium, iron, chlorine, calcium, and magnesium

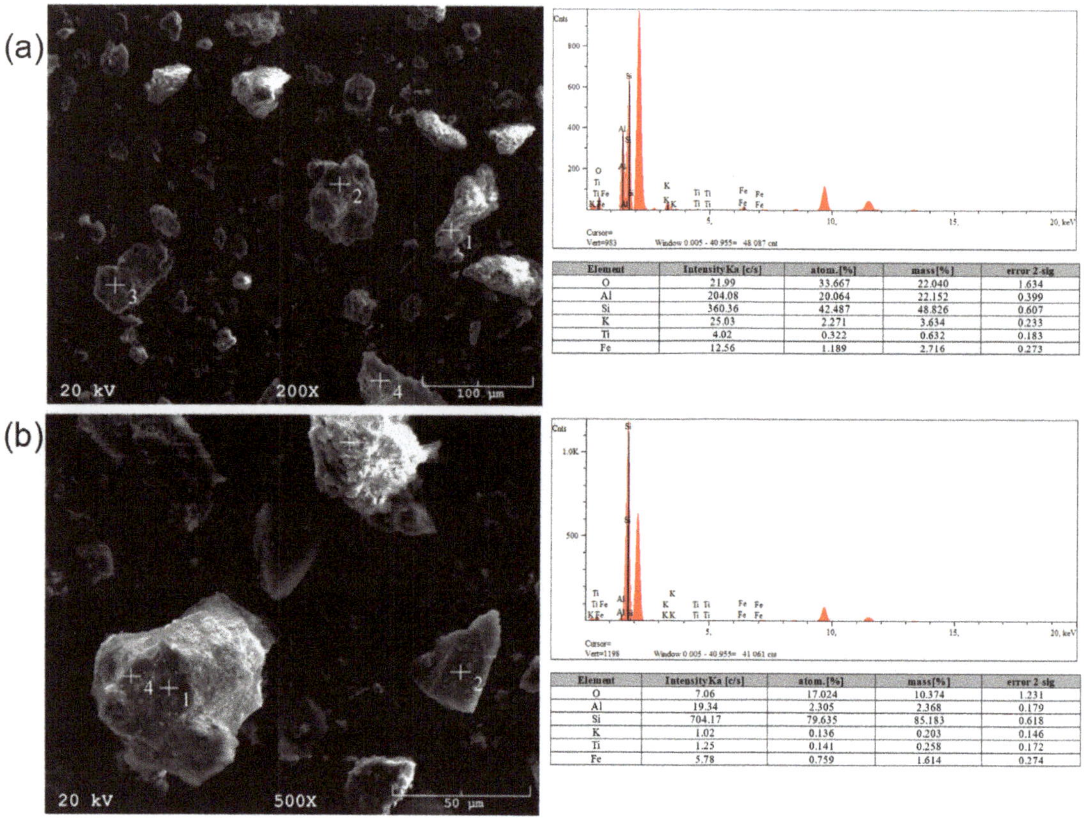

Figure 3. SEM micrographs and EDS analyses for (**a**) Sobieski and (**b**) Staszic coal shell.

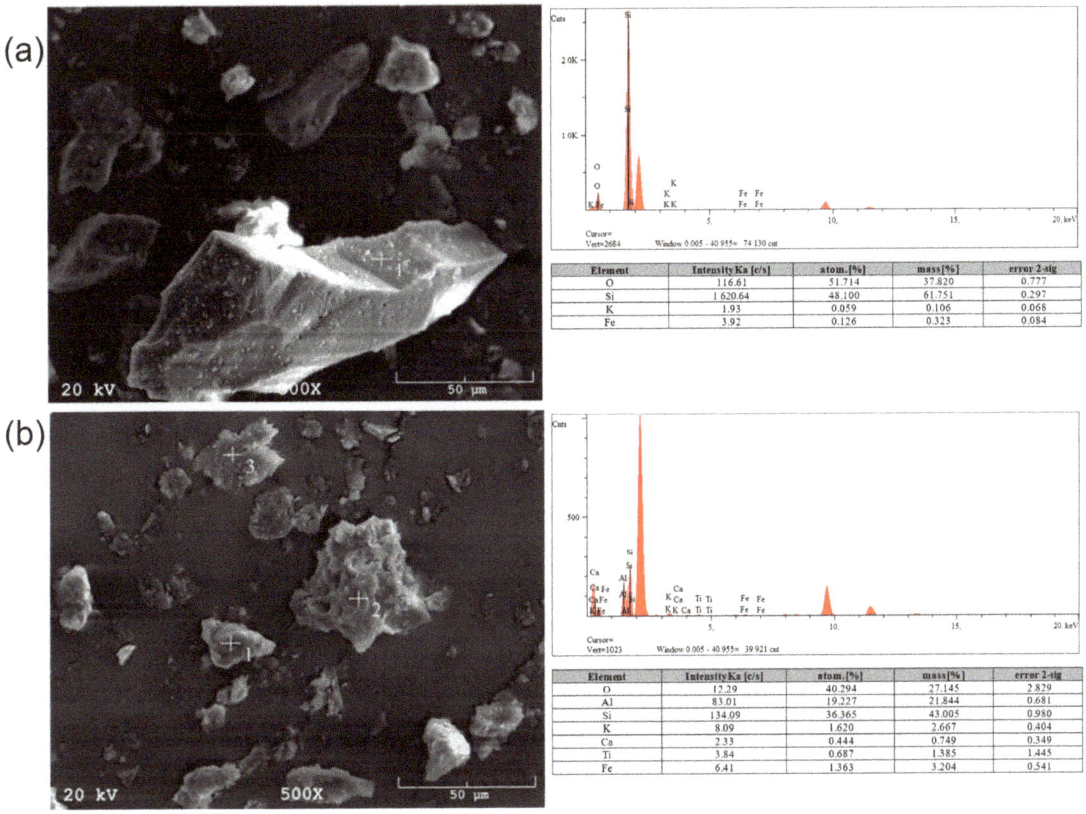

Element	IntensityKa [c/s]	atom.[%]	mass[%]	error 2-sig
O	116.61	51.714	37.820	0.777
Si	1 620.64	48.100	61.751	0.297
K	1.93	0.059	0.106	0.068
Fe	3.92	0.126	0.323	0.084

Element	IntensityKa [c/s]	atom.[%]	mass[%]	error 2-sig
O	12.29	40.294	27.145	2.829
Al	83.01	19.227	21.844	0.681
Si	134.09	36.365	43.005	0.980
K	8.09	1.620	2.667	0.404
Ca	2.33	0.444	0.749	0.349
Ti	3.84	0.687	1.385	1.445
Fe	6.41	1.363	3.204	0.541

Figure 4. SEM micrographs and EDS analyses for (**a**) Wieczorek and (**b**) Wujek coal shell.

3.3. *Analysis of Radioactivity and Leachability*

Leachability and radioactivity tests were carried out for all analyzed materials (Tables 5 and 6).

Table 5. Leachability results for the analyzed materials.

Element	Index						
	Białystok	Skawina	Sobieski	Staszic	Wujek	Wieczorek	Standard
pH	10.3	11.4	10.9	11.9	11.0	11.0	--- [1]
Antimony [mg/L]	0.13	0.35	0.005	<0.005	<0.005	<0.005	PN-EN ISO 11885:2009
Arsenic [mg/L]	0.019	1.31	0.32	1.07	0.081	0.24	PN-EN ISO 11885:2009
Bar [mg/L]	0.24	0.033	0.23	0.022	0.079	0.14	PN-EN ISO 11885:2009
Chromium [mg/L]	0.19	0.032	0.19	3.96	0.19	0.023	PN-EN ISO 11885:2009
Chromium (IV) [mg/L]	0.14	0.030	0.18	3.9	0.18	0.020	PN-EN ISO 18412:2007
Zinc [mg/L]	<0.01	0.022	0.069	0.012	0.041	0.070	PN-EN ISO 11885:2009

Table 5. *Cont.*

Element	Index						
	Białystok	**Skawina**	**Sobieski**	**Staszic**	**Wujek**	**Wieczorek**	**Standard**
Cadmium [mg/L]	<0.001	0.0023	<0.0005	0.0014	<0.0005	<0.0005	PN-EN ISO 11885:2009
Copper [mg/L]	<0.01	0.0051	0.0063	<0.005	<0.005	0.007	PN-EN ISO 11885:2009
Molybdenum [mg/L]	0.14	0.48	0.043	0.41	0.85	0.12	PN-EN ISO 11885:2009
Nickel [mg/L]	0.024	<0.005	0.0051	<0.005	<0.005	<0.005	PN-EN ISO 11885:2009
Lead [mg/L]	<0.005	0.0054	<0.005	<0.005	<0.005	<0.005	PN-EN ISO 11885:2009
Mercury [mg/L]	<0.001	<0.001	<0.001	<0.001	<0.001	<0.001	PN-EN ISO 12846:2012
Selenium [mg/L]	0.078	0.16	0.011	0.053	0.011	0.017	PN-EN ISO 11885:2009
Chlorides [mg/L]	12,500	56	5.3	45	5.5	5.5	PN-EN ISO 10304:2009
Fluorides [mg/L]	0.165	2.7	0.55	9.2	0.44	0.68	PN-EN ISO 10304:2009
Sulfates [mg/L]	5160	738	301	398	518	194	PN-EN ISO 10304:2009
RWO [2] [mg/L]	16	32	20	31	15	18	PN-EN 1484:1999
Solvent Subst.—TDS [mg/L]	31,600	4820	1480	3760	1900	1590	PN-EN 152162010

[1] Not applicable. [2] Organic carbon (solved).

Table 6. Summary of radioactivity results for the analyzed materials.

Index	Activity Index Values		Naturally Radioactive Elements' Concentrations		
	f_1 [-]	f_2 [Bq/kg]	Potassium: K^{40} [Bq/kg]	Radium: Ra^{226} [Bq/kg]	Thorium: Th^{228} [Bq/kg]
Białystok	0.25 (±0.02)	7.05 (±2.36)	625.95 (±38.30)	7.05 (±2.36)	4.08 (±1.45)
Skawina	0.9 (±0.06)	100.64 (±8.38)	573.57 (±42.97)	100.64 (±8.38)	75.25 (±5.38)
Sobieski	0.69 (±0.05)	43.00 (±5.19)	690.37 (±45.52)	43.00 (±5.19)	63.10 (±4.60)
Staszic	0.56 (±0.04)	43.41 (±5.00)	488.87 (±53.33)	43.41 (±5.00)	52.61 (±3.97)
Wujek	0.7 (±0.05)	44.85 (±5.3)	690.61 (±45.62)	44.85 (±5.3)	65.58 (±4.72)
Wieczorek	0.5 (±0.04)	28.40 (±3.97)	629.15 (±40.51)	28.40 (±3.97)	40.70 (±3.32)

Regarding the interpretation of the radioactivity results, the important values are f_1 and f_2. The most restrictive requirements are for the buildings to permanently house people and animals—for example, residential buildings. In this case, the values of f_1 and f_2 have to fulfill the following requirements: $f_1 < 1.2$. and $f_2 < 240$ Bq/kg. All materials meet these requirements.

The leachability and radioactivity tests for all studied materials showed compliance with the applicable standards and regulations [39,40]. Therefore, the materials can be safely used for construction products.

3.4. Summary FTIR Analyses Performed and Results

Infrared spectroscopy using attenuated total reflectance is a technique widely used to detect the presence of water and the soluble silicate ratio in geopolymers. Some of the infrared bands found in materials and geopolymers include an intense band located in the range of 1250–950 cm^{-1}, which corresponds to the asymmetric stretching of Si-O-R * (where R * may represent Si o Al) [41–43]. However, these vibrations could be found in the infrared spectra of aluminosilicates and aluminum. Some authors [44,45] suggest that the shift of this band towards lower or higher frequencies indicates the successful completion of the polycondensation reaction. The band located in the range of 3600–2700 cm^{-1} corresponds to the vibrations of -OH groups due to the presence of hydrogen bonds. Some authors reported weak Si-O-Al vibrations in the range of 460–470 cm^{-1} or 610–620 cm^{-1} [41,44]. The band located in the range of 1425–1450 cm^{-1} is associated with $-CO_3$ and appears as a result of atmospheric carbonation occurring during sample synthesis [46]. The detailed results are presented in Figure 5.

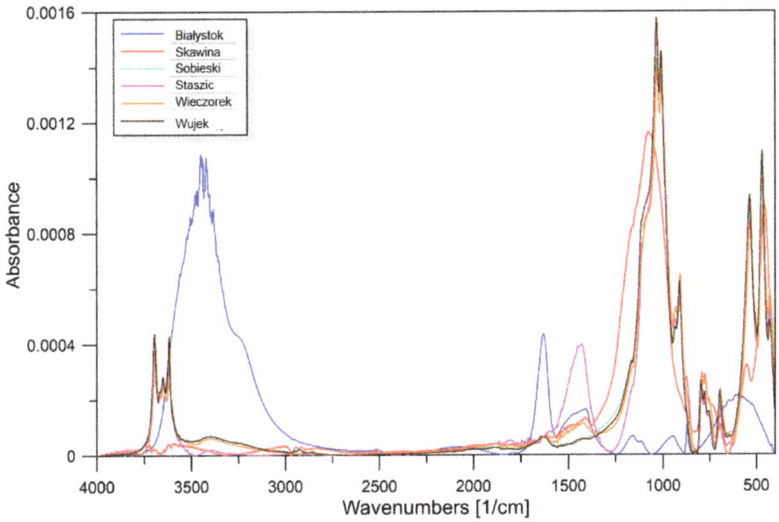

Figure 5. FTIR micrographs of the geopolymers made from the obtained waste materials.

3.5. Strength Tests of Manufactured Geopolymers

Before carrying out the strength tests, the density of the samples was measured using the geometric method. Figure 6 shows the average density values.

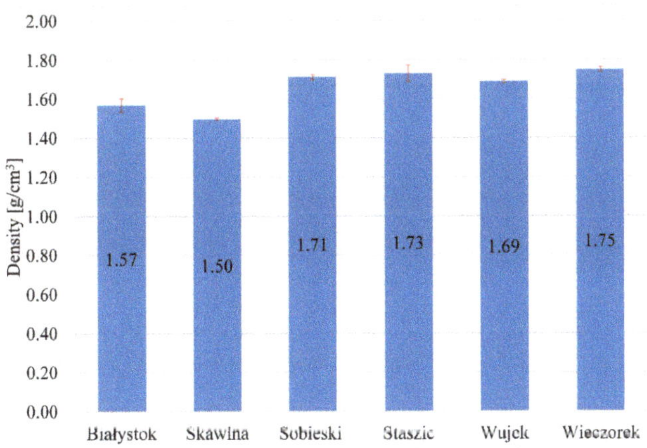

Figure 6. Densities of the geopolymers made from the obtained waste materials.

The density should correlate with the compressive and flexural strength of geopolymers. In the case of the tested samples, this tendency is slight. This tendency is more visible in the case of foamed materials, where more voids appear. Samples prepared using materials obtained from the mine have a density range of 1.69–1.75 g/cm^3. The obtained flexural and compressive strength results were presented graphically (Figures 7 and 8).

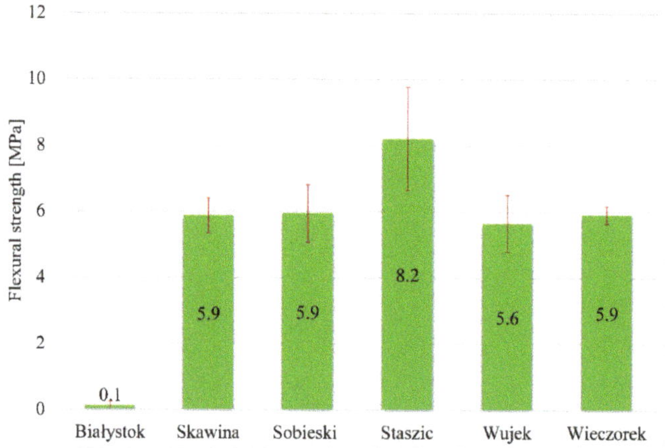

Figure 7. Flexural strength of the geopolymers made from the obtained waste materials.

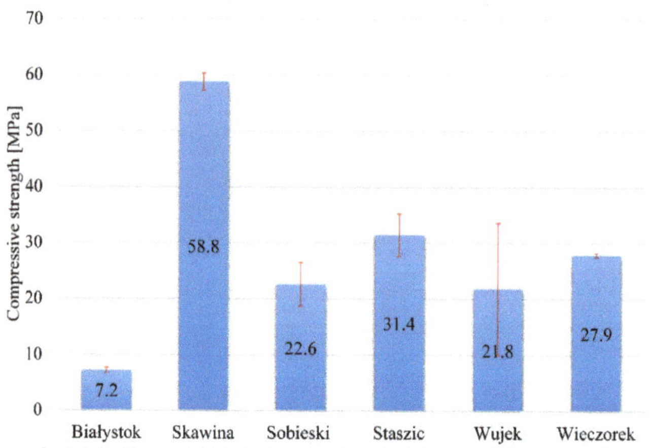

Figure 8. Compressive strength of the geopolymers made from the obtained waste materials.

The Skawina sample with Skawina fly ash content exhibited the greatest compressive strength (58.8 MPa). The Staszic sample had a compressive strength nearly two times lower (31.4 MPa) than the Skawina sample. The Wieczorek sample had a slightly weaker compressive strength (27.9 MPa) than the Staszic sample. The most insufficient compressive strength was for the sample prepared using waste material from Białystok (7.2 MPa).

In the case of flexural tests, the highest flexural strength was obtained for the Staszic sample (8.2 MPa). The remaining samples, comprised of coal shale and Skawina fly ash, exhibited similar flexural strength values of ~5.9 MPa. Samples from the waste incineration plant in Białystok had the lowest flexural strength (similar to the compression test results).

Summarizing the obtained compressive and flexural strength results, the highest values were obtained for geopolymers based on waste material from the Staszic and Skawina mine, which had a high content of quartz and kaolinite. On the other hand, samples obtained using the material from the waste incineration plant in Białystok demonstrated significantly worse results. It is worth noting that some studies of geopolymer materials indicate that even a small amount of quartz may positively affect mechanical properties. In contrast, the presence of other minerals may have opposite effects. For example, illite causes changes in geopolymer morphology and leaching behavior [47].

The results of the remaining samples, which are not significantly different, are the starting point for future studies regarding these materials, e.g., in variable proportions. The main difference between geopolymer concrete and Portland concrete is the binder. The silicon and aluminum oxides in the low calcium fly ash react with the alkaline liquid to form a geopolymer paste that binds loose, coarse aggregates, fine aggregates, and other unreacted materials to form geopolymer concrete. As with Portland concrete, coarse and fine aggregates account for approximately 75–80% of the weight of geopolymer concrete. This component of geopolymer concrete mixes can be designed using the tools currently available for Portland concrete. The compressive strength and workability of geopolymer concrete are influenced by the proportions and properties of the constituent materials composed by the geopolymer paste [34].

3.6. Microstructure Analysis of Manufactured Geopolymers

Figure 9a presents a geopolymer based on Białystok ash. In the case of this material, at higher magnifications, polygonal pores were observed. The material's structure was brittle and segmented, resulting in deficient strength properties. EDS analysis showed high calcium, sodium, and potassium content, and low levels of sulfur, silicon, aluminum, chromium, iron, nickel, and copper. In the micrographs of geopolymers based on Skawina fly ash (Figure 9b), a compact structure and visible pores with a maximum size of about 0.3 mm can be observed. However, most of the pores ranged from 10–30 μm. The structure is amorphous. The characteristic fine-grained structure of zeolites and undiluted ash particles were visible in the cavities. EDS analysis revealed the presence of elements such as silicon, aluminum, calcium, sodium, potassium, iron, and carbon.

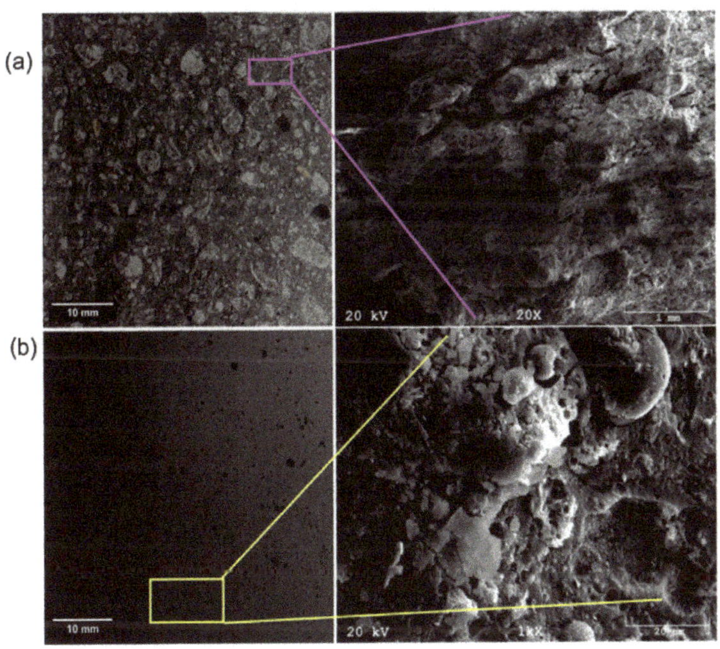

Figure 9. Optical and SEM images of (**a**) Białystok and (**b**) Skawina geopolymer samples.

Another tested material was a geopolymer based on coal shale from the Sobieski mine (Figure 10a). The pore size was ~25 μm. The structure is brittle but relatively contained, with no visible cracks. Undiluted particles 10–20 μm in size were also observed. EDS analysis showed the presence of silicon, aluminum, calcium, magnesium, and iron compounds. There is a lack of significant changes compared to raw materials because light elements such as sodium are not properly detected in this method. It is also worth mentioning that EDS analysis is a qualitative method, so it does not give representative results according to the amount of element in the whole material volume. The microstructure of the geopolymer made from the Staszic mine (Figure 10b) is significantly different compared to previous materials. A significant variation in the size of the structure can be noticed, ranging from a few micrometers to several dozen. Large amorphous areas and very fine crystalline areas are presented. The pore size is ~50 μm. EDS analysis showed the presence of silicon, aluminum, calcium, chlorine, iron, and potassium compounds.

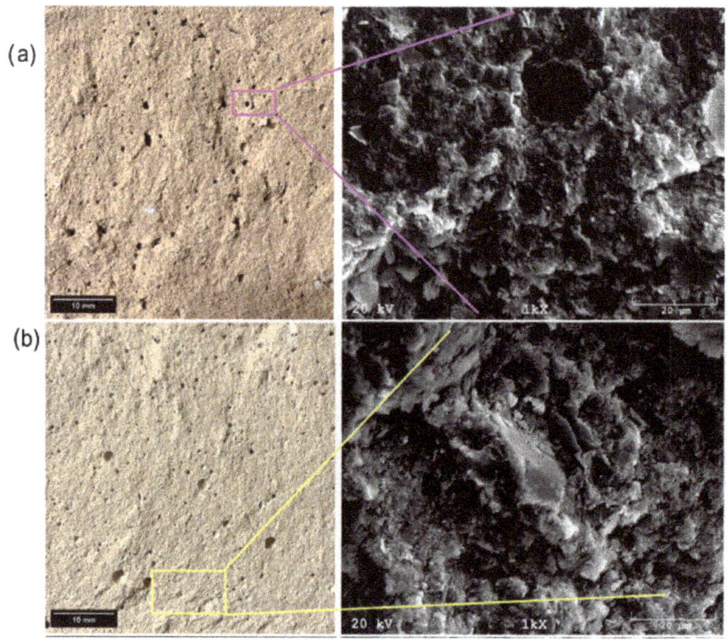

Figure 10. Optical and SEM images of (**a**) Sobieski and (**b**) Staszic geopolymer samples.

Geopolymers based on slate obtained from the Wieczorek mine (Figure 11a) are characterized by significant porosity, ranging from tiny pores in the range of 2–3 μm to pores as large as 0.5 mm. The EDS analysis identified silicon, aluminum, sodium, calcium, potassium, chlorine, and iron compounds, as well as trace amounts of carbon. The last tested material was a geopolymer based on ash from the Wujek mine (Figure 11b). The microstructure analysis showed the presence of pores in the 50 μm range. Again, this material consists of silicon, aluminum, calcium, chlorine, iron, and potassium compounds.

The present research facilitated material selection for future works connected with the forming process.

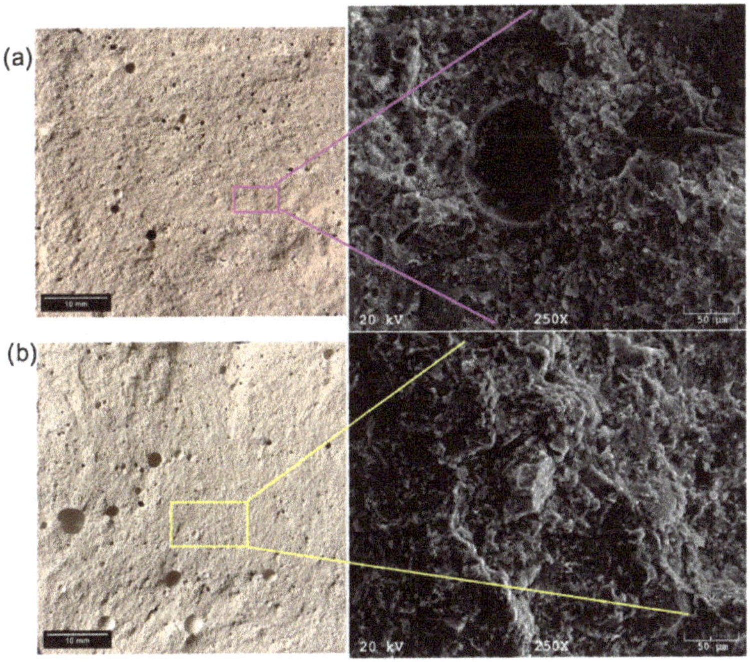

Figure 11. Optical and SEM images of (**a**) Wieczorek and (**b**) Wujek geopolymer samples.

3.7. Development of the Foaming Process of Waste-Based Geopolymers

The final stage of the presented results was the preparation of foamed materials using two foaming agents: aluminum powder (powder, max. particle size 60 µm, purity 99.9%, Sigma-Aldrich, St. Louis, MO, USA) and hydrogen peroxide (perhydrol 36%, Biomus, Lublin, Poland). Polyglycolic acid (Sigma-Aldrich) and cellulose (Sigma-Aldrich) were added as stabilizers. To prepare the samples, the precursors were mixed with the alkaline activator (NaOH) for 5 min. Next, foaming agents and the stabilizer were added, mixed for 30 s, and left for foaming. The list of produced materials is shown in Table 2. The properties of the foams are related to the introduced foaming agent, but this is not the only factor affecting the quality of the materials. In the production of samples, the process of mixing the materials, the concentration of additives used, and the temperature of the geopolymerization process are also important factors. Blowing agents such as aluminum oxide or hydrogen peroxide, in combination with calcium oxides present in the matrix and binder, emit hydrogen, which gives porosity to the geopolymer structure. During this reaction, hydrated calcium silicate is released, while hydrogen, being lighter than air, escapes from the geopolymer mass, increasing the mold volume. After some time, the hydrogen completely escapes from the mass and is replaced with air, filling the spaces (pores) of the foamed geopolymer. Aluminum powder reacts with calcium oxides and calcium hydroxides. In turn, hydrogen peroxide (perhydrol) also reacts with calcium chlorides. Table 2 lists the composition of the manufactured materials.

In this part, two materials based on mine tailings have been selected. We rejected the material from ashes because the Bialystok has very poor results, and data about foaming fly ash class F (such as Skawina) are relatively common. There are not many publications that investigate the foaming process for mine tailing-based geopolymers. These particular mine tailings have been selected because of their availability and reasonable properties gained on the previous steps presented in this article.

Strength parameters were assessed for the obtained samples, which are presented in Table 7. The lowest densities were observed in samples foamed with hydrogen peroxide.

Conversely, the optimal properties for compressive strength were observed in samples foamed with aluminum powder.

Table 7. Results of the density and compressive strength tests.

Index	Solution	Foaming Agent	Stabilizer	Density, [kg/m^3]	Compressive Strength [MPa]
Wieczorek	NaOH 10 M	4.5 g Al	---	1292	14.4
		1.5 g Al	---	532	0.81
		30 mL Perhydrol 36%	20 mL polyglycol	529	0.62
		30 mL Perhydrol 36%	5 g cellulose	535	0.69
Staszic		4.5 g Al	---	861	4.3
		30 mL Perhydrol 36%	20 mL polyglycol	533	0.76
		30 mL Perhydrol 36%	5 g cellulose	501	0.81

There is a correlation between the obtained strength properties and the sample density. The compressive strength is increasing together with material density. Additionally, thermal conductivity tests were carried out for selected samples. The results for geopolymer foams based on mine wastes are presented in Table 8. The lowest thermal conductivity was found in the material based on mine wastes foamed with hydrogen peroxide, while materials foamed with aluminum powder were characterized by a relatively high thermal conductivity. Upon visual inspection of the obtained samples, the materials foamed with aluminum powder were characterized by lower porosity and lower volumetric compactness of porosity. The results of thermal conductivity confirm the conclusions put forward by other researchers that the greater the number of pores containing air, the lower the density of the material, and thus the lower the value of thermal conductivity [48].

Table 8. Summary of selected results for thermal conductivity testing.

No.	Index	Solution	Foaming Agent	Stabilizer	Thermal Conductivity [W/(m·K)]	Thermal Resistance [(m^2·K)/W]
1	Wieczorek	NaOH 10M	1.5 g Al		0.18522	0.1681
2	Wieczorek		30 mL Perhydrol 36%	5 g cellulose	0.11609	0.2593
3	Staszic		4.5 g Al		0.19403	0.1613
4	Staszic		30 mL Perhydrol 36%	5 g cellulose	0.15094	0.2084

4. Conclusions

The work aimed to select and prepare waste materials (mine waste, coal shale, fly ash, post-process waste) to be used as raw materials in the synthesis of inorganic polymers and subsequently the synthesis of waste-based geopolymers and the development of the foaming process. Hydrogen peroxide and aluminum powder were selected for the foaming process of geopolymer materials. The results indicated that geopolymer foams may successfully act as an insulation material. Research on geopolymer materials is a rapidly developing field of material science. Geopolymer materials are characterized by low thermal conductivities, good strength properties, and low densities.

- Mineralogical composition showed a slightly different composition of ashes and main tailings. In the case of main tailings, their composition was similar with different proportions of layered silicates such as quartz, kaolinite, muscovite, and illite.
- The chemical composition demonstrated that selected waste and ashes could be used as the base material to produce geopolymers, especially because of a sufficient amount of silica and alumina.

- The leachability and radioactivity tests for all analyzed materials were compliant with the applicable standards.
- Samples prepared using materials obtained from the mine have a density range of 1.69–1.75 g/cm^3. The density of samples made from industrial-waste ashes was slightly lower.
- The highest flexural strength was obtained for the Staszic sample (8.2 MPa). The remaining samples, comprised of coal shale and Skawina fly ash, exhibited similar flexural strength values of ~5.9 MPa. Samples from the waste incineration plant in Białystok had the lowest flexural strength.
- The Skawina sample with Skawina fly ash content exhibited the greatest compressive strength (58.8 MPa). The Staszic sample had a compressive strength nearly two times lower (31.4 MPa) than the Skawina sample. The Wieczorek sample had a slightly weaker compressive strength (27.9 MPa) than the Staszic sample. The most insufficient compressive strength was for the sample prepared using waste material from Białystok (7.2 MPa). The waste materials from Białystok do not have enough mechanical strength for building applications.
- Thermal conductivity tests demonstrated that the optimal foaming agent was hydrogen peroxide, which ensured thermal conductivity (0.11609 W/(m·K)). However, the foamed materials were characterized by a relatively low compressive strength (0.69 MPa). Therefore, future research will emphasize the selection of the most promising fillers, e.g., fibers and particles, and the manufacturing of a foamed composite material.

Author Contributions: Conceptualization, K.K. and B.F.; methodology, B.F., P.B. and M.Ł.; validation, B.K., M.M. and K.K.; formal analysis, M.Ł.; investigation, B.F. and P.B.; resources, M.Ł.; data curation, K.M. and M.M.; writing—original draft preparation, B.F. and P.B.; writing—review and editing, B.K., K.M. and M.Ł.; visualization, K.M.; supervision, K.K.; project administration, K.K.; funding acquisition, K.K. All authors have read and agreed to the published version of the manuscript.

Funding: 1. The presented works were carried out as part of the project: "Geopolymer foams with low thermal conductivity produced based on industrial waste as an innovative material for the circular economy", which is financed by the National Center for Research and Development under the LIDER X program. Grant No.: LIDER/31/0168/L-10/18/NCBR/2019. 2. The research was supported by the Polish National Agency for Academic Exchange under the complementary activities to those undertaken by the University as part of the University Alliance European entitled STARS EU academic cooperation with Ukrainian universities, grant no. BNI-UE-2023-8.

Institutional Review Board Statement: Not applicable.

Informed Consent Statement: Not applicable.

Data Availability Statement: The original contributions presented in this study are included in the article. Further inquiries can be directed to the corresponding author.

Conflicts of Interest: The authors declare no conflicts of interest. The funders had no role in the design of the study; in the collection, analyses, or interpretation of data; in the writing of the manuscript; or in the decision to publish the results.

References

1. Ma, C.-K.; Awang, A.Z.; Omar, W. Structural and Material Performance of Geopolymer Concrete: A Review. *Constr. Build. Mater.* **2018**, *186*, 90–102. [CrossRef]
2. Cong, P.; Cheng, Y. Advances in Geopolymer Materials: A Comprehensive Review. *J. Traffic Transp. Eng.* **2021**, *8*, 283–314. [CrossRef]
3. Furtos, G.; Prodan, D.; Sarosi, C.; Popa, D.; Moldovan, M.; Korniejenko, K. The Precursors Used for Developing Geopolymer Composites for Circular Economy—A Review. *Materials* **2024**, *17*, 1696. [CrossRef]
4. Gao, B.; Jang, S.; Son, H.; Lee, H.J.; Lee, H.J.; Yang, J.J.; Bae, C.-J. Study on Mechanical Properties of Kaolin-Based Geopolymer with Various Si/Al Ratio and Aging Time. *J. Korean Ceram. Soc.* **2020**, *57*, 709–715. [CrossRef]
5. Villa, C.; Pecina, E.T.; Torres, R.; Gómez, L. Geopolymer Synthesis Using Alkaline Activation of Natural Zeolite. *Constr. Build. Mater.* **2010**, *24*, 2084–2090. [CrossRef]

6. Ascensão, G.; Seabra, M.P.; Aguiar, J.B.; Labrincha, J.A. Red Mud-Based Geopolymers with Tailored Alkali Diffusion Properties and pH Buffering Ability. *J. Clean. Prod.* **2017**, *148*, 23–30. [CrossRef]
7. Feng, D.; Wang, J.; Wang, Y.; Xiao, X.; Hou, W.; Liang, S. Alkali-Activated Geopolymer Materials Prepared from Coal Gangue and Municipal Solid Waste Incineration Byproducts. *J. Build. Eng.* **2023**, *80*, 108074. [CrossRef]
8. Gadikota, C.R.; Chandra, D.S. Influence of Marine Environment on Durability and Microstructural Properties of Fly Ash-Based Geopolymer Concrete. *J. Build. Rehabil.* **2023**, *8*, 95. [CrossRef]
9. Zhang, Y.; Xiao, R.; Jiang, X.; Li, W.; Zhu, X.; Huang, B. Effect of Particle Size and Curing Temperature on Mechanical and Microstructural Properties of Waste Glass-Slag-Based and Waste Glass-Fly Ash-Based Geopolymers. *J. Clean. Prod.* **2020**, *273*, 122970. [CrossRef]
10. Jindal, B.B.; Sharma, R. The Effect of Nanomaterials on Properties of Geopolymers Derived from Industrial By-Products: A State-of-the-Art Review. *Constr. Build. Mater.* **2020**, *252*, 119028. [CrossRef]
11. Nawaz, M.; Heitor, A.; Sivakumar, M. Geopolymers in Construction-Recent Developments. *Constr. Build. Mater.* **2020**, *260*, 120472. [CrossRef]
12. Łach, M.; Kozub, B.; Bednarz, S.; Bąk, A.; Melnychuk, M.; Masłoń, A. The Influence of the Addition of Basalt Powder on the Properties of Foamed Geopolymers. *Materials* **2024**, *17*, 2336. [CrossRef] [PubMed]
13. Przybek, A.; Łach, M. Research on the Physical Properties of an Eco-Friendly Layered Geopolymer Composite. *Materials* **2024**, *17*, 4937. [CrossRef] [PubMed]
14. Hassan, A.; Arif, M.; Shariq, M. A Review of Properties and Behaviour of Reinforced Geopolymer Concrete Structural Elements-A Clean Technology Option for Sustainable Development. *J. Clean. Prod.* **2020**, *245*, 118762. [CrossRef]
15. Fan, X.; Zhu, J.; Gao, X. Sea/Coral Sand in Marine Engineered Geopolymer Composites: Engineering, Mechanical, and Microstructure Properties. *Int. J. Appl. Ceram. Technol.* **2024**, *22*, e14874. [CrossRef]
16. Gadikota, C.R.; Chandra, D.S. Influence of Slag Content on Mechanical, Durability and Microstructural Performance and Their Correlations in Fly Ash Geopolymers Exposed to Marine Environment. *Asian J. Civ. Eng.* **2024**, *25*, 2091–2108. [CrossRef]
17. Temuujin, J.; Minjigmaa, A.; Rickard, W.; Lee, M.; Williams, I.; Van Riessen, A. Fly Ash Based Geopolymer Thin Coatings on Metal Substrates and Its Thermal Evaluation. *J. Hazard. Mater.* **2010**, *180*, 748–752. [CrossRef] [PubMed]
18. Bąk, A.; Pławecka, K.; Łach, M. Comparison of Thermal Conductivity of Foamed Geopolymers Containing Phase Change Materials. *J. Phys. Conf. Ser.* **2023**, *2423*, 012003. [CrossRef]
19. Le, V.S.; Buczkowska, K.E.; Ercoli, R.; Pławecka, K.; Marian, N.M.; Louda, P. Influence of Incorporating Recycled Windshield Glass, PVB-Foil, and Rubber Granulates on the Properties of Geopolymer Composites and Concretes. *Polymers* **2023**, *15*, 2122. [CrossRef] [PubMed]
20. Feng, J.; Zhang, R.; Gong, L.; Li, Y.; Cao, W.; Cheng, X. Development of Porous Fly Ash-Based Geopolymer with Low Thermal Conductivity. *Mater. Des.* **2015**, *65*, 529–533. [CrossRef]
21. Ziejewska, C.; Grela, A.; Mierzwiński, D.; Hebda, M. Influence of Waste Glass Addition on the Fire Resistance, Microstructure and Mechanical Properties of Geopolymer Composites. *Materials* **2023**, *16*, 6011. [CrossRef] [PubMed]
22. Le, V.S.; Nguyen, V.V.; Sharko, A.; Ercoli, R.; Nguyen, T.X.; Tran, D.H.; Łoś, P.; Buczkowska, K.E.; Mitura, S.; Špirek, T.; et al. Fire Resistance of Geopolymer Foams Layered on Polystyrene Boards. *Polymers* **2022**, *14*, 1945. [CrossRef] [PubMed]
23. Yang, K.-H.; Lee, K.-H.; Song, J.-K.; Gong, M.-H. Properties and Sustainability of Alkali-Activated Slag Foamed Concrete. *J. Clean. Prod.* **2014**, *68*, 226–233. [CrossRef]
24. Zhang, Z.; Provis, J.L.; Reid, A.; Wang, H. Geopolymer Foam Concrete: An Emerging Material for Sustainable Construction. *Constr. Build. Mater.* **2014**, *56*, 113–127. [CrossRef]
25. Song, Y.; Xue, C.; Guo, W.; Bai, Y.; Shi, Y.; Zhao, Q. Foamed Geopolymer Insulation Materials: Research Progress on Insulation Performance and Durability. *J. Clean. Prod.* **2024**, *444*, 140991. [CrossRef]
26. Zhang, Z.; Provis, J.L.; Reid, A.; Wang, H. Mechanical, Thermal Insulation, Thermal Resistance and Acoustic Absorption Properties of Geopolymer Foam Concrete. *Cem. Concr. Compos.* **2015**, *62*, 97–105. [CrossRef]
27. Yatsenko, E.A.; Goltsman, B.M.; Izvarin, A.I.; Kurdashov, V.M.; Ryabova, A.V. Influence of Foaming Additives on the Porous Structure and Properties of Lightweight Geopolymers Based on Ash–Slag Waste. *Constr. Build. Mater.* **2024**, *443*, 137629. [CrossRef]
28. Jelle, B.P.; Gustavsen, A.; Baetens, R. The Path to the High Performance Thermal Building Insulation Materials and Solutions of Tomorrow. *J. Build. Phys.* **2010**, *34*, 99–123. [CrossRef]
29. Ducman, V.; Korat, L. Characterization of Geopolymer Fly-Ash Based Foams Obtained with the Addition of Al Powder or H_2O_2 as Foaming Agents. *Mater. Charact.* **2016**, *113*, 207–213. [CrossRef]
30. Kamseu, E.; NGouloure, Z.N.M.; Ali, B.N.; Zekeng, S.; Melo, U.C.; Rossignol, S.; Leonelli, C. Cumulative Pore Volume, Pore Size Distribution and Phases Percolation in Porous Inorganic Polymer Composites: Relation Microstructure and Effective Thermal Conductivity. *Energy Build.* **2015**, *88*, 45–56. [CrossRef]
31. Pławecka, K.; Bazan, P.; Lin, W.-T.; Korniejenko, K.; Sitarz, M.; Nykiel, M. Development of Geopolymers Based on Fly Ashes from Different Combustion Processes. *Polymers* **2022**, *14*, 1954. [CrossRef] [PubMed]
32. Öksüzer, N. The Effect of Calcination on Alkali-Activated Lightweight Geopolymers Produced with Volcanic Tuffs. *J. Aust. Ceram. Soc.* **2023**, *59*, 1053–1063. [CrossRef]

33. Ghani, U.; Hussain, S.; Noor-ul-Amin; Imtiaz, M.; Khan, S.A. Role of Calcination on Geopolymerization of Lateritic Clay by Alkali Treatment. *J. Saudi Chem. Soc.* **2021**, *25*, 101198. [CrossRef]
34. Provis, J.L.; Van Deventer, J.S.J. (Eds.) *Geopolymers: Structure, Processing, Properties and Industrial Applications*; Woodhead Publishing in materials; Woodhead: Oxford, UK; CRC Press: Boca Raton, FL, USA, 2009; ISBN 978-1-84569-449-4.
35. Rattanasak, U.; Pankhet, K.; Chindaprasirt, P. Effect of Chemical Admixtures on Properties of High-Calcium Fly Ash Geopolymer. *Int. J. Min. Met. Mater.* **2011**, *18*, 364–369. [CrossRef]
36. Chindaprasirt, P.; Chareerat, T.; Sirivivatnanon, V. Workability and Strength of Coarse High Calcium Fly Ash Geopolymer. *Cem. Concr. Compos.* **2007**, *29*, 224–229. [CrossRef]
37. Phoo-ngernkham, T.; Chindaprasirt, P.; Sata, V.; Pangdaeng, S.; Sinsiri, T. Properties of High Calcium Fly Ash Geopolymer Pastes with Portland Cement as an Additive. *Int. J. Min. Met. Mater.* **2013**, *20*, 214–220. [CrossRef]
38. Li, J.; Shan, Y.; Ni, P.; Cui, J.; Li, Y.; Zhou, J. Mechanics, Durability, and Microstructure Analysis of Marine Soil Stabilized by an Eco-Friendly Calcium Carbide Residue-Activated Coal Gangue Geopolymer. *Case Stud. Constr. Mater.* **2024**, *20*, e02687. [CrossRef]
39. Łach, M.; Korniejenko, K.; Walter, J.; Stefańska, A.; Mikuła, J. Decreasing of Leaching and Improvement of Geopolymer Properties by Addition of Aluminum Calcium Cements and Titanium Oxide. *Materials* **2020**, *13*, 495. [CrossRef]
40. Łach, M.; Korniejenko, K.; Komar, N.; Mikuła, J. A Study on the Physicochemical Properties of Different Post-Process Wastes from Thermal Processes. *IOP Conf. Ser. Mater. Sci. Eng.* **2019**, *660*, 012006. [CrossRef]
41. Arioz, E.; Arioz, Ö.; Koçkar, Ö.M. The Effect of Curing Conditions on the Properties of Geopolymer Samples. *Int. J. Chem. Eng. Appl.* **2013**, *4*, 423–426. [CrossRef]
42. Korniejenko, K.; Kejzlar, P.; Louda, P. The Influence of the Material Structure on the Mechanical Properties of Geopolymer Composites Reinforced with Short Fibers Obtained with Additive Technologies. *Int. J. Mol. Sci.* **2022**, *23*, 2023. [CrossRef] [PubMed]
43. Lee, W.K.W.; Van Deventer, J.S.J. Use of Infrared Spectroscopy to Study Geopolymerization of Heterogeneous Amorphous Aluminosilicates. *Langmuir* **2003**, *19*, 8726–8734. [CrossRef]
44. Wan, Q.; Rao, F.; Song, S.; García, R.E.; Estrella, R.M.; Patiño, C.L.; Zhang, Y. Geopolymerization Reaction, Microstructure and Simulation of Metakaolin-Based Geopolymers at Extended Si/Al Ratios. *Cem. Concr. Compos.* **2017**, *79*, 45–52. [CrossRef]
45. Autef, A.; Joussein, E.; Gasgnier, G.; Rossignol, S. Role of the Silica Source on the Geopolymerization Rate: A Thermal Analysis Study. *J. Non-Cryst. Solids* **2013**, *366*, 13–21. [CrossRef]
46. Zaharaki, D.; Komnitsas, K.; Perdikatsis, V. Use of Analytical Techniques for Identification of Inorganic Polymer Gel Composition. *J. Mater. Sci.* **2010**, *45*, 2715–2724. [CrossRef]
47. Martínez, L.M.T.; Kharissova, O.V.; Kharisov, B.I. (Eds.) *Handbook of Ecomaterials*; Springer International Publishing: Cham, Switzerland, 2019; ISBN 978-3-319-68254-9.
48. Łach, M.; Pławecka, K.; Bąk, A.; Lichocka, K.; Korniejenko, K.; Cheng, A.; Lin, W.-T. Determination of the Influence of Hydraulic Additives on the Foaming Process and Stability of the Produced Geopolymer Foams. *Materials* **2021**, *14*, 5090. [CrossRef] [PubMed]

materials

MDPI

Article

Thermodynamic and Kinetic Analysis of TiN Precipitation in Nickel-Based Superalloys During Solidification

Jie Zhang *, Chengwei Zeng, Haibin Zheng and Changlong Gu

College of Mechanical Engineering, Zhejiang University of Technology, Hangzhou 310014, China; 221122020320@zjut.edu.cn (C.Z.); 18858274591@163.com (H.Z.); 221123020460@zjut.edu.cn (C.G.)

* Correspondence: zhangjie0231@zjut.edu.cn; Tel.: +86-188-0018-5456

Abstract: Large TiN inclusions in nickel-based superalloys promote micropore formation, compromising the mechanical properties of the alloys. However, current research lacks a comprehensive coupled model that considers both solute element microsegregation and TiN precipitation specifically in nickel-based superalloys. This study investigated TiN precipitation during the solidification of IN718 alloys through a combined thermodynamic and kinetic approach. A modified Clyne–Kurz model was applied to account for multi-element microsegregation, enabling an integrated analysis of both microsegregation and precipitation processes. The results indicated that solute elements in the molten alloy segregated to varying degrees during solidification. At an initial nitrogen concentration of 25 ppm, TiN inclusions precipitated when the solid fraction reached 0.256, eventually resulting in a total TiN precipitation of 64 ppm by the end of solidification, while residual nitrogen in the liquid phase decreased to 1.3 ppm. Increasing the initial nitrogen concentration from 10 ppm to 40 ppm advanced the onset of TiN precipitation and raised the total amount from 16 ppm to 126 ppm. Further analysis indicated that cooling rates of 0.03 °C/s, 0.06 °C/s, and 0.18 °C/s did not significantly affect the final TiN accumulation.

Keywords: thermodynamics and kinetics; solidification; TiN precipitation; microsegregation; nickel-based superalloys

1. Introduction

The IN718 alloy is widely valued for its excellent high-temperature mechanical properties, corrosion resistance, and creep resistance, making it indispensable for high-performance applications in aerospace and other advanced industries [1]. However, to meet the stringent quality standards required in these fields, further refinement of the alloy is necessary. Aerospace components have stringent fatigue resistance requirements that are highly sensitive to the presence and level of inclusions within the alloy. Therefore, precise control over the size and quantity of inclusions is crucial for improving alloy quality.

Compared to oxide inclusions, TiN inclusions in nickel-based alloys are harder and more brittle, significantly affecting the alloy's fatigue life. Studies show that nitride inclusions around 6 μm in size have a comparable impact on fatigue life to oxide inclusions as large as 25 μm [2]. During alloy solidification, differences in the solubility of titanium and nitrogen between the solid and liquid phases lead to their redistribution at the solid–liquid interface, resulting in the enrichment of these elements. When the actual concentration of Ti and N in the liquid phase surpasses the equilibrium limit, primary nitrides (primarily TiN) begin to precipitate, often nucleating on oxides; subsequently, primary carbides may precipitate on the surface of these TiN particles through heterogeneous nucleation [3,4]. These TiN particles may aggregate into clusters or form complex carbonitrides with primary carbides. Under applied stress, microcracks can initiate at these larger inclusions, ultimately leading to fatigue failure of the alloy. Furthermore, blocky carbonitrides can obstruct interdendritic channels, limiting the flow and feedability of the molten alloy, which results in higher porosity within the alloy ingot and a notable reduction in overall alloy quality [5].

Over the past few decades, controlling TiN inclusions in IN718 alloys has been a primary research focus. Young and Mitchell [6] provided a comprehensive analysis of nitrogen additions and removals, highlighting that TiN primarily precipitates in interdendritic regions toward the end of solidification, where it contributes to micropore formation that diminishes the alloy's mechanical properties. Mitchell and Wang [7] investigated the solidification behavior and secondary phase precipitation in IN718. Their findings revealed that, as solidification progresses, NbC nucleates on pre-existing TiN particles. Depending on the C and N ratio in the alloy, these NbC particles may evolve into mixed (Nb + Ti) carbonitrides at later stages, which in turn alters the microstructure of the alloy. Sindhura et al. [8] further emphasized the importance of controlling nitrogen and titanium concentrations to limit undesirable TiN inclusions. Such inclusions cause microstructural issues like agglomeration and microporosity, ultimately weakening the alloy's mechanical performance. Leonardo et al. [3] introduced a method to minimize TiN precipitation through the VIM-VAR process, which effectively eliminates nitrides and produces TiN-free IN718 alloys. This technique helps prevent micropore formation and grain boundary weakening caused by TiN inclusions. Overall, in IN718 alloys, TiN particles not only act as heterogeneous nucleation sites that facilitate the formation of NbC particles but also cluster with other inclusions. These precipitates exhibit poor distribution uniformity within the structure and are difficult to improve through heat treatment or homogenization, ultimately affecting the alloy's microstructure and mechanical properties. The initial nitrogen concentration in the alloy significantly determines the timing and total amount of TiN precipitation. Therefore, controlling nitrogen content is crucial for minimizing TiN precipitation and enhancing the overall performance of the alloy.

Various microsegregation models have been developed to improve predictions of solute element redistribution between solid and liquid phases during alloy solidification. The Lever-Rule (L-R) model predicts solute distribution under the assumption of complete solute diffusion in both phases. In contrast, the Scheil model assumes complete diffusion in the liquid phase but none in the solid phase. However, more accurate predictions require accounting for the finite solute diffusion within the solid phase. The Brody–Fleming (B-F) model, introduced by Brody and Fleming [9], incorporates a back diffusion coefficient to address this need. Subsequent refinements to this coefficient were made by Clyne and Kurz [10], Ohnaka [11], Voller and Beckermann [12], and Won and Thomas [13]. Despite their advancements, these foundational models overlook the impact of inclusion precipitation on solute element microsegregation.

In fact, there is a strong interaction between the precipitation of inclusions and element microsegregation [14]. During solidification, solute element microsegregation enriches alloying elements in the residual liquid phase, promoting inclusion precipitation, which then depletes certain solutes from the system. This interaction highlights the importance of considering inclusion precipitation when predicting solute microsegregation. To address the issue of the interaction between inclusion precipitation and elemental microsegregation, several coupled models have been developed, integrating solute microsegregation and inclusion precipitation modules. For example, Yang et al. [15] used a coupled Scheil-based model to simulate solute segregation and TiN inclusion behavior during the directional solidification of the K418 alloy. Similarly, Que et al. [2] applied the Voller–Beckermann model (V-B model) alongside TiN precipitation and growth models to study TiN behavior during the continuous casting of Ti micro-alloyed steels. Liu and Gui et al. [16,17] employed the Clyne–Kurz model (C-K model) to predict TiN inclusions in steel, analyzing the effects of Ti and N content and cooling rates on TiN formation. Lee and Li et al. [18,19] employed the model based on Ohnaka's theory, incorporating the kinetics of inclusion precipitation and growth to predict the behavior of TiN inclusions during solidification. Similarly, Shu et al. [20] combined a modified Ohnaka-based model with a KWM model to predict inclusion size distributions during steel solidification. Won and Thomas [13] modified the Ohnaka parameters and, combined with the V-B model, extended the cell crystal growth model for the application to columnar crystals. Despite these advancements, most research has concentrated on inclusion behavior in steels, leaving a gap in studies on element microsegregation and inclusion precipitation during the solidification of nickel-based superalloys.

In this paper, the solidification process of the IN718 molten alloy during vacuum arc remelting was studied. A coupled model was developed and implemented using the modified C-K model proposed by Won and Thomas to integrate solute element microsegregation with a TiN precipitation prediction. Additionally, a thermodynamic analysis was used to evaluate the conditions for TiN inclusion formation. By using the coupling model, we evaluated the extent of element microsegregation and TiN precipitation levels. The effects of varying element concentrations and local cooling rates on TiN formation were explored. This study aims to provide a theoretical foundation for future research on the solidification behavior of IN718 alloys.

2. Thermodynamic and Kinetic Analysis of TiN Inclusion Formation in IN718

This section systematically examined the conditions for TiN inclusion precipitation in IN718 ingots during vacuum arc remelting (VAR), based on thermodynamic and kinetic models. A microsegregation model was applied to calculate the segregation levels of titanium and nitrogen in the residual liquid phase throughout solidification process. The coupled model predicted the amount of TiN precipitation at each stage, as well as the cumulative precipitation over the entire solidification process. Additionally, the impact of the microsegregation of various elements on the precipitation timing, size, and quantity of TiN inclusions was examined.

2.1. The Features of TiN Inclusions at Different Locations

Experiments were conducted on the ingot prepared by vacuum arc remelting (VAR), with metallographic samples taken from the edge, mid-radius, and center of the ingot. After grinding and polishing, the samples were examined under a metallographic microscope without corrosion treatment. Figure 1 shows the features of TiN inclusions at various locations within the ingot.

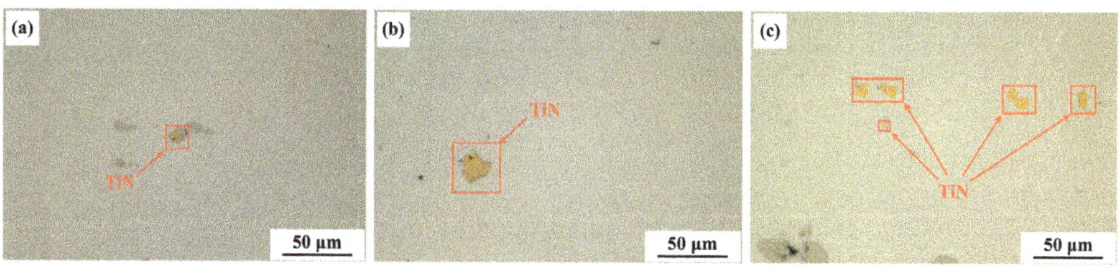

Figure 1. Features of TiN inclusions at various locations: (**a**) edge; (**b**) 1/2 radius; and (**c**) center.

These images reveal that TiN inclusions are primarily block-shaped and generally smaller than 25 μm. At the ingot edge, TiN inclusions are relatively small, around 8 μm. As the distance to the center decreases, both the size and quantity of TiN inclusions increase. The trend can be attributed to the slower cooling rates that lead to larger secondary dendrite arm spacing near the ingot center, which, in turn, facilitated the precipitation and growth of TiN.

In the IN718 alloy, TiN and NbC share similar crystal structures, making some degree of niobium and carbon incorporation in TiN particles unavoidable. To determine the elemental composition of TiN particles, we conducted scanning electron microscopy (SEM) observations on several representative TiN particles in the alloy samples. The SEM results, shown in Figure 2, indicated that Ti and N were the primary components in TiN particles, though small amounts of C and Nb were also present. For a more detailed elemental analysis, semi-quantitative point analyses were performed on the TiN particles. Figure 3 shows the locations and results of these analyses, revealing that Ti and N contents in the analyzed areas are approximately 38–48% and 46–49%, respectively, while Nb and C contents are only about 3–5% and 8–10%. The Ti-to-Nb ratio is approximately 13:1, and the N-to-C ratio is around 5:1. Thus, TiN particles are composed primarily of Ti and N, containing extremely low levels of Nb and C. Given the low Nb and C levels, their effects on TiN particles were neglected in this study, treating TiN as pure TiN for calculation purposes. The thermodynamic conditions for TiN formation are discussed in the following section.

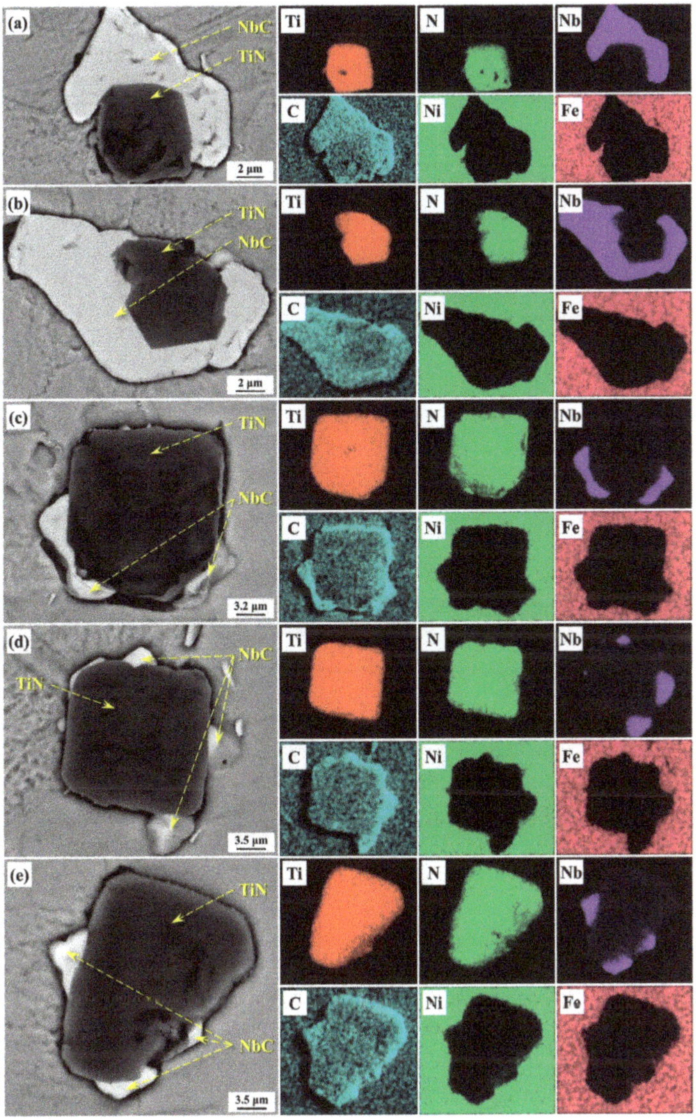

Figure 2. The SEM images and corresponding element mappings of typical nitride particles in the IN718 alloy: (**a**) nitride particle #1; (**b**) nitride particle #2; (**c**) nitride particle #3; (**d**) nitride particle #4; (**e**) nitride particle #5.

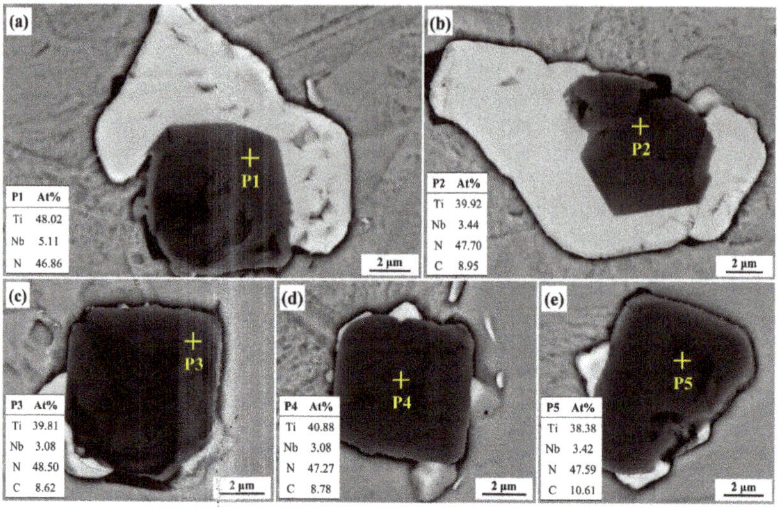

Figure 3. The results of point analyses for typical nitride particles in IN718 alloy: (**a**) nitride particle #1; (**b**) nitride particle #2; (**c**) nitride particle #3; (**d**) nitride particle #4; (**e**) nitride particle #5.

2.2. *Thermodynamic Analysis of TiN Precipitation Equilibrium*

The chemical composition of the IN718 ingot used in this study is presented in Table 1. A thermodynamic analysis was conducted to examine the precipitation process of TiN inclusions, with the relevant chemical reaction equation and precipitation conditions provided below [21].

$$[\mathrm{Ti}] + [\mathrm{N}] = \mathrm{TiN(s)}\ \Delta G^{\Theta} = -379000 + 149T \quad (1)$$

$$\Delta G \leq 0 \quad (2)$$

where ΔG^{Θ} is the standard Gibbs free energy, J/mol; ΔG is the Gibbs free energy of the reaction, J/mol; and T is the temperature, K.

Table 1. Chemical composition of the IN718 alloy (wt.%).

C	Si	Mn	P	S	Cr	Fe	Cu
0.018	0.05	0.01	0.003	0.0009	17.88	18.73	0.01
Mo	**Ti**	**Nb**	**Mg**	**Al**	**B**	**Co**	**Pb**
2.96	0.94	4.91	0.001	0.52	0.0032	0.01	0.0001
Bi	**Ca**	**Se**	**O**	**N**	**H**	**Ni**	
0.00005	0.001	0.0005	0.00041	0.0025	0.00003	Bal.	

According to the definition of Gibbs free energy, the equation for ΔG is given as follows:

$$\Delta G = \Delta G^{\Theta} + RT \ln \frac{a_{\mathrm{TiN}}}{f_{\mathrm{Ti}} \Omega [\mathrm{Ti}]_{\%} \cdot f_{\mathrm{N}} \Omega [\mathrm{N}]_{\%}} \quad (3)$$

where a_{TiN} denotes the activity of TiN in the molten alloy. According to Henry's law, f_{Ti} and f_{N} represent the activity coefficients of Ti and N in molten alloy, respectively, with 1% concentration as the standard; $\Omega[\mathrm{Ti}]_{\%}$ and $\Omega[\mathrm{N}]_{\%}$ represent the actual concentrations of Ti and N in the molten alloy, respectively; R is the ideal gas constant, 8.314 J·mol^{-1}·K^{-1}; for pure TiN, the activity is 1. Then, Equation (3) is rewritten as follows:

$$\begin{aligned}\Delta G &= \Delta G^{\Theta} - RT(\ln f_{\mathrm{Ti}} + \ln f_{\mathrm{N}} + \ln \Omega[\mathrm{Ti}]_{\%} \cdot \Omega[\mathrm{N}]_{\%}) \\ &= \Delta G^{\Theta} - \frac{RT}{\mathrm{lge}}(\mathrm{lg}f_{\mathrm{Ti}} + \mathrm{lg}f_{\mathrm{N}} + \mathrm{lg}\Omega[\mathrm{Ti}]_{\%} \cdot \Omega[\mathrm{N}]_{\%})\end{aligned} \quad (4)$$

The natural constant e ≈ 2.718; while $\mathrm{lg}f_{\mathrm{Ti}}$ and $\mathrm{lg}f_{\mathrm{N}}$ can be determined using the activity interaction equations:

$$\mathrm{lg}f_i = \sum\left(e_i^j \cdot [\mathrm{pct}j] + r_i^j \cdot [\mathrm{pct}j]^2\right) \quad (5)$$

where e_i^j and r_i^j represent the first-order and second-order activity interaction coefficients of solute element *j* with element *i* in the molten alloy, and [pct*j*] denotes the actual concentration of element *j*. The chemical composition of the IN718 alloy used in this study is presented in Table 1. Table 2 lists the first-order and second-order activity interaction coefficients used in this study's thermodynamic calculations.

Table 2. Activity interaction coefficient of components in molten nickel (1873 K).

	Element	C	Si	Mn	P	S	Cr	Cu	Mo	Ti	Nb
e_i^j	N	0.09	0.05	−0.05	0.05	0.01	−0.08	0.01	−0.04	−0.21	−0.07
	Ti	0	0	0	0	0	0.025	0	0	0.048	0
r_i^j	N	0	0	0	0	0	0	0	0	0	0
	Ti	0	0	0	0	0	0	0	0	−0.001	0
	Element	**Al**	**B**	**Co**	**O**	**N**	**Ta**	**Fe**	**W**	**V**	**Zr**
e_i^j	N	0	0.09	−0.01	0.05	0	−0.07	−0.01	−0.02	−0.13	−0.23
	Ti	0.0037	0	0	−1.8	−2.041	0	0	0	0	0
r_i^j	N	0	0	0	0	0	0	0	0	0	0
	Ti	0	0	0	0	0	0	0	0	0	0

From Equation (5), values for $\mathrm{lg}f_{\mathrm{Ti}} \approx 0.487$ and $\mathrm{lg}f_{\mathrm{N}} \approx -2.314$ are obtained. Subsequently, an expression for the concentration product of Ti and N is derived.

$$\mathrm{lg}(\Omega[\mathrm{Ti}]_{\%} \cdot \Omega[\mathrm{N}]_{\%}) = \frac{\mathrm{lge} \cdot (\Delta G^{\Theta} - \Delta G)}{RT} + 1.827 \quad (6)$$

When $\Delta G = 0$, TiN precipitates in the liquid phase, and the equilibrium concentration product for TiN precipitation can be determined.

$$\mathrm{lg}(\Omega[\mathrm{Ti}]_{\%}\Omega[\mathrm{N}]_{\%}) = \frac{\mathrm{lge} \cdot \Delta G^{\Theta}}{RT} + 1.827 \quad (7)$$

The calculation results of the JMatPro7.0 show that the solidus and liquidus temperatures of the IN718 alloy are 1360.15 K and 1640.95 K, respectively. By substituting T_L and T_S into Equation (7), the equilibrium concentration products of Ti and N required for TiN precipitation at both temperatures are obtained as follows:

$$\Omega[\mathrm{Ti}]_{\%}\Omega[\mathrm{N}]_{\%} = 0.0035(T_L = 1640.95\mathrm{K}) \quad (8)$$

$$\Omega[\mathrm{Ti}]_{\%}\Omega[\mathrm{N}]_{\%} = 0.000011(T_S = 1360.15\mathrm{K}) \quad (9)$$

A stability curve of TiN inclusion precipitation in the IN718 ingot was constructed from Equations (8) and (9), as shown in Figure 4. In this figure, the black and red lines represent the equilibrium curves of TiN precipitation of the molten alloy at the liquidus temperature and the solidus temperature, respectively, while point A denotes the actual concentration product of Ti and N. The data showed that the actual concentration product of Ti and N in IN718 exceeded the equilibrium concentration product at the solidus temperature but remained below it at the liquidus temperature, indicating that TiN inclusions precipitated during solidification. In fact, due to the differences in solubility of solute atoms between the solid and liquid phases, solute elements tend to accumulate at the solidification front. This elemental segregation behavior leads to an increase in the actual concentration product of Ti and N, ultimately causing TiN inclusions to precipitate earlier during the solidification process. Therefore, accurately evaluating the actual solute concentration at the solidification front is essential.

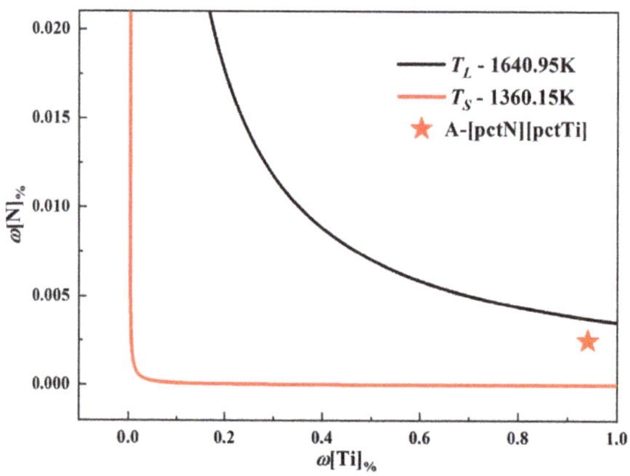

Figure 4. Stability curve of TiN inclusion precipitation.

2.3. *Microsegregation Models for Solute Elements*

2.3.1. Mathematical Models

The catastrophic influence of solute redistribution on microstructural homogeneity was realized decades ago for its ability to cause non-equilibrium phase, cracks, and other problems. Several models were developed to characterize the behaviors of these microsegregation elements. Based on the Lever Rule, the initial model assumes complete diffusion to the equilibrium of all solutes in both the liquid and the solid phase, which may be written as the following:

$$\frac{C_L}{C_0} = \frac{1}{[1-(1-k)f_S]} \tag{10}$$

where C_L is the concentration of the solute element concerned in the liquid part at the solidification front; C_0 is the initial concentration before solidifying; f_S is the solid fraction; and k is the equilibrium partition coefficient for the given element. The validation of this model is problematic due to the slow diffusion rate in the solid phase for large solute atoms.

The Scheil equation [22], the opposite limiting case to the L-R model, assumes complete diffusion in the liquid phase, no diffusion in the solid phase, and local equilibrium at the solid liquid interface. This nonequilibrium model is the following:

$$\frac{C_L}{C_0} = (1-f_S)^{(k-1)} \tag{11}$$

This model works well at the initial stage of solidification, while it turns out to be a 'disaster' as C_L approaches infinite at $f_S \rightarrow 1$. Hence, only when the cooling rates exceed 10 °C/s can this equation be applied.

Many simple models were proposed to predict microsegregation during steel solidification in view of finite nonzero diffusion in the solid phase. With fixed dendrite arm spacing taken into account, Brody and Fleming [9] proposed a general model. Ohnaka [11] and Clyne and Kurz [10] modified the back-diffusion parameters in this model. This equation leads to the following:

$$\frac{C_L}{C_0} = [1 - (1 - 2\Omega k) f_S]^{\frac{(k-1)}{(1-2\Omega k)}} \tag{12}$$

$$\Omega = \alpha\left[1 - \exp\left(-\frac{1}{\alpha}\right)\right] - \frac{1}{2}\exp\left(-\frac{1}{2\alpha}\right) \tag{13}$$

$$\alpha = \frac{4D_S t_f}{(\lambda_S/10000)^2} \tag{14}$$

where α is a Fourier number; D_S is the diffusion coefficient of solute in the solid phase; t_f is the local solidification time (LST); and λ_S is the secondary dendrite arm spacing (SDAS). Ω tends to α as $\alpha \to 0$, and the C-K Equation (12) tends to the Scheil Equation (11). Ω tends to $1/2$ as $\alpha \to \infty$, and the C-K Equation (12) tends to the 'L-R' Equation (10). The introduction of these parameters, i.e., D_S, t_f, and λ_S, covers all rates of solute diffusion in the solid phase, which means a broad range of applicability. In the present work, Equations (12), (14), (17) and (18) are regarded as the basis of calculations.

More advanced solutions and models [23–28] for microsegregation were developed for high accuracy and board adaptability at the expense of accurate solidification process parameters, etc. Considering the effect of coarsening on the dilution of the liquid phase, Voller and Beckermann [12] added an additional back-diffusion term to the Fourier number, as follows:

$$\alpha^+ = \alpha + \alpha^C \tag{15}$$

This model could match full coarsening model results as $\alpha^C = 0.1$, which extended the model from plate to columnar dendrites. Columnar crystals are the major form of dendritic growth during the VAR process; consequently, this model also has a certain basis for the solidification process in VAR. As a result, Equation (13) changes into the following:

$$\Omega = \alpha^+\left[1 - \exp\left(-\frac{1}{\alpha^+}\right)\right] - \frac{1}{2}\exp\left(-\frac{1}{2\alpha^+}\right) \tag{16}$$

Based on the C-K model, Won and Thomas accepted the proposals of Ohnaka and Voller and Beckermann, considering the influence of columnar dendrites and coarsening on the back diffusion coefficient:

$$\Omega = \alpha^+\left[1 - \exp\left(-\frac{1}{\alpha^+}\right)\right] - \frac{1}{2}\exp\left(-\frac{1}{2\alpha^+}\right) \tag{17}$$

$$\alpha^+ = 2(\alpha + \alpha^C) \tag{18}$$

2.3.2. Calculation of $T_{S\text{-}L}$ During Solidification of the IN718

Previous thermodynamic theoretical works showed that the temperature at the solidification front of the molten alloy ($T_{S\text{-}L}$) is a function of the liquidus temperature (T_L), solidus temperature (T_S), and solid-phase fraction (f_S) [29]. This relationship is expressed as follows:

$$T_{S\ \ L} = T_0 - \frac{T_0 - T_L}{1 - f_S \frac{T_L - T_S}{T_0 - T_S}} \tag{19}$$

where T_0 denotes the melting point of pure nickel, 1728.15 K; T_L and T_S are the liquidus and solidus temperatures, respectively (determined via JMatPro7.0 software), K; $T_{S\text{-}L}$ is the temperature of solidification front, K; and f_S represents the solid fraction.

Figure 5 illustrates the relationship between the temperature at the solidification front ($T_{S\text{-}L}$) and the solid fraction (f_S). The results showed that solidification of the alloy began at 1367.8 °C when the solid fraction was zero. As the temperature decreased, the solid fraction in the alloy gradually increased, with the solidification rate accelerating until full solidification was reached at 1087 °C, where the solid fraction was 1.

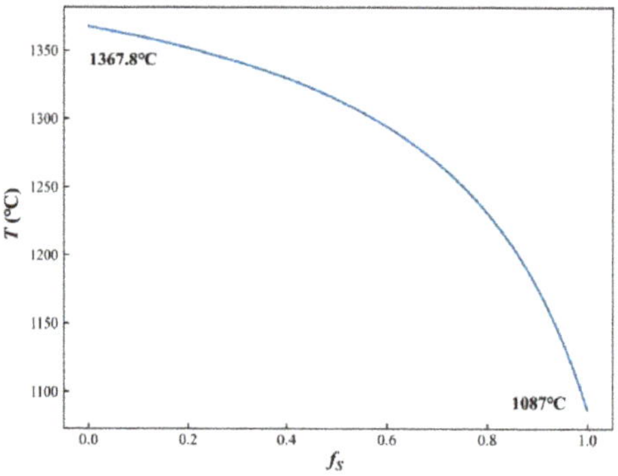

Figure 5. Curve of temperature at the solidification front as a function of solid fraction.

Table 3 provides data on equilibrium partition coefficients and the diffusivity of solute elements in γ iron, showing that diffusion coefficients of solute elements vary with temperature. This study incorporated the temperature dependence of diffusivity into the model for accurate analysis.

Table 3. Equilibrium partition coefficients (*ks*) and diffusion coefficients (D_S) of solute elements in γ phase [30].

Element	k	D_S (m^2/s)
C	0.3	0.000015 Exp (−143,511/*RT*)
S	0.035	0.00024 Exp (−223,426/*RT*)
Ti	0.7	0.000015 Exp (−250,956/*RT*)
Nb	0.39	0.000083 Exp (−266,479/*RT*)
Mg	0.02	0.0000055 Exp (−249,366/*RT*)
Al	0.59	0.00059 Exp (−241,417/*RT*)
O	0.03	0.000575 Exp (−168,615/*RT*)
N	0.48	0.000091 Exp (−168,490/*RT*)

2.3.3. Calculation of Cooling Rate

To proceed with the thermodynamic calculations, the value of C_L as f_S increases is essential. In Equation (12), C_0 and k are constants, indicating that Ω is a key parameter. The local solidification time is defined in Equation (14) as follows:

$$t_f = \frac{T_L - T_S}{C_R} \tag{20}$$

where T_L and T_S are the liquidus and solidus temperatures of the alloy (determined via JMatPro7.0 software), K; and C_R is the cooling rate, °C/min.

Previous experiments measured the secondary dendrite arm spacing of the IN718 alloy under various cooling rates and established a functional relationship between secondary dendrite arm spacing and the cooling rate [31]. However, errors were identified in the original fitted results, which prompted a re-fitting in this study, yielding the following:

$$\lambda_S = 183.72 \times {C_R}^{-\frac{1}{3}} - 13.749 \tag{21}$$

where λ_S represents the secondary dendrite arm spacing of the IN718 alloy, μm.

Based on the theoretical foundation of Equation (21), the local cooling rate within the alloy ingot could be estimated by measuring the secondary dendrite arm spacing. In this study, Equation (21) was incorporated into a program that calculates and outputs the cooling rate when the secondary dendrite arm spacing was provided. Additionally, the local solidification time (t_f) was determined by substituting the calculated cooling rate into Equation (20).

Due to variations in water flow rate and cooling conditions from the outer to the inner regions, the secondary dendrite arm spacing in the ingot's solidification structure ranged from 72 to 138 μm. Figure 6 displays dendrite morphologies corresponding to typical secondary dendrite arm spacing values. The figure shows that dendrites coarsen progressively as secondary dendrite arm spacing increases. To assess the impact of different cooling rates on microsegregation and TiN precipitation, three representative secondary dendrite arm spacing values (70 μm, 105 μm, and 140 μm) were selected for further analysis.

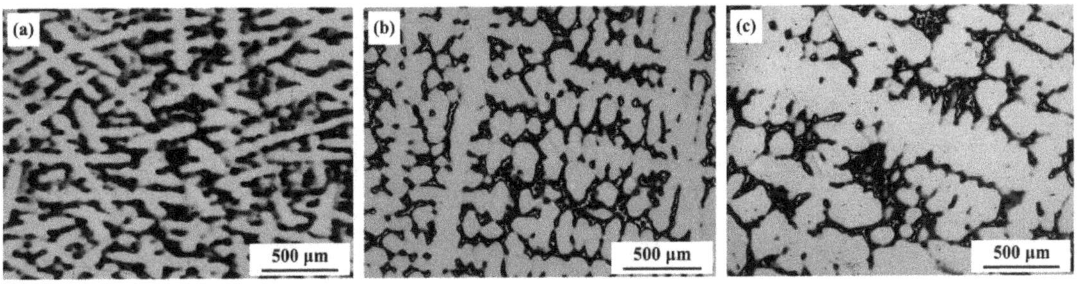

Figure 6. Dendrite morphology corresponding to different secondary dendrite arm spacing in solidification microstructure: (**a**) 72 μm; (**b**) 108 μm; and (**c**) 138 μm.

2.4. Calculation of TiN Precipitation

2.4.1. Microsegregation Calculations of Solute Elements

During solidification, microsegregation caused the concentrations of N and Ti in the residual liquid phase at the solidification front to increase as the solid fraction grew. TiN inclusions began to precipitate when the actual concentration product of Ti and N in the residual melt exceeded the equilibrium solubility product of TiN, resulting in the consumption of N and Ti in the liquid phase.

In a fully coupled model, the consumption of both N and Ti during TiN precipitation should be considered [32]. However, in this study, with an initial N concentration of only 0.0025%, it was assumed that all N was consumed during TiN precipitation, while only 0.0085% of Ti was used. Given that the Ti content in the IN718 alloy was approximately 0.94%, which significantly exceeded the amount that could be consumed by TiN precipitation, this model neglected Ti consumption and focused solely on N consumption.

Before TiN precipitation, the concentration of each alloying elements was calculated using Equation (22):

$$\frac{C_L}{C_0} = [1 - (1 - 2\Omega k) f_S]^{\frac{(k-1)}{(1-2\Omega k)}} \tag{22}$$

After TiN precipitation, the concentrations of Ti and other elements continued to be calculated using the same equation, but with adjustments to the N concentration in the residual liquid phase. The N concentration at step n was determined using Equation (23):

$$C_{L,n} = C_{L,\mathrm{eq},n-1} \frac{[1 - (1 - 2\Omega_n k_N) f_S]^{\frac{(k_N-1)}{(1-2\Omega_n k_N)}}}{[1 - (1 - 2\Omega_{n-1} k_N)(f_S - \Delta f_S)]^{\frac{(k_N-1)}{(1-2\Omega_{n-1} k_N)}}} \tag{23}$$

where $C_{L,n}$ is the N concentration in the calculation cell at step, wt.%; $C_{L,\mathrm{eq},n-1}$ is the equilibrium N concentration in the calculation cell at step $(n-1)$, wt.%; Ω_{n-1} and Ω_n

are the Ω parameter values for steps $n-1$ and n, respectively; and k_N is the equilibrium partition coefficient of N between the solid and liquid phases.

2.4.2. Effect of Microsegregation of Other Alloying Elements on TiN Precipitation

During solidification, microsegregation behavior is common among all alloying elements. The enrichment of these elements in the residual liquid phase alters the activity coefficients of N and Ti, thereby influencing TiN precipitation. Consequently, the program developed for this study accounted for the microsegregation effects of six key alloying elements: C, S, Nb, Mg, Al, and O. The software was equipped with a flexible interface, which enabled the seamless integration of additional equilibrium partition coefficients and diffusion coefficients for alloying elements as they became available.

2.5. The Unit and Process of Calculation

During solidification, primary dendrites initiate growth, with secondary dendrites forming symmetrically on either side of the primary dendrites due to the temperature gradient at the solidification front. Simultaneously, the solid–liquid interface on the secondary dendrites gradually advances toward the center of the interdendritic region. As shown in Figure 7, the computational unit of the coupling model used in this study started at the central axis of a secondary dendrite in the mushy zone and extended to half the length of the secondary dendrite arm spacing.

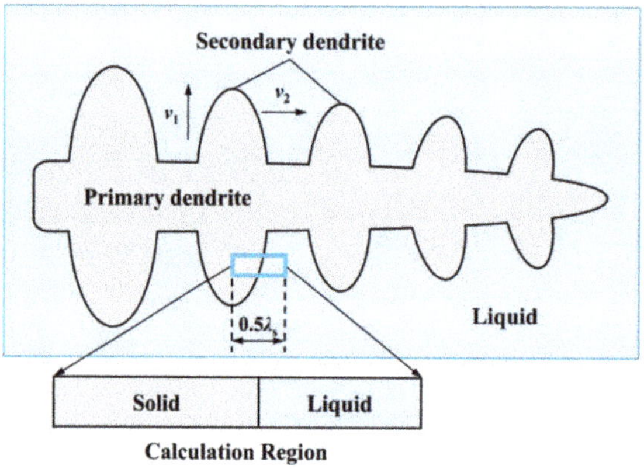

Figure 7. Schematic of the solidification calculation unit.

The solidification region, defined as half of the secondary dendrite arm spacing (λ_S), was divided into Q segments, each with a length of ΔX, as shown in Figure 8. The calculation step size was Δf_S. For each computational step, the solid fraction increased from f_S to $f_S + \Delta f_S$, advancing the solidification front by one unit and increased the number of solid phase cells from m to $m + 1$.

Before starting the calculations used the coupled model, the following assumptions were made:

(1) Mass transfer in the calculation region was governed solely by diffusion. Solute diffusion was limited in the solid phase but remained uniform in the liquid phase.
(2) Dendrite growth followed a parabolic shape, as shown in Figure 7.
(3) The solid–liquid interface was treated as a planar surface.
(4) At the solidification front, the reaction between Ti and N in the liquid phase reached equilibrium, satisfying the relationship in Equation (4).
(5) The calculated unit represented half of the secondary dendrite arm spacing, as shown in Figure 7.
(6) The precipitation of other types of inclusions during solidification was disregarded.

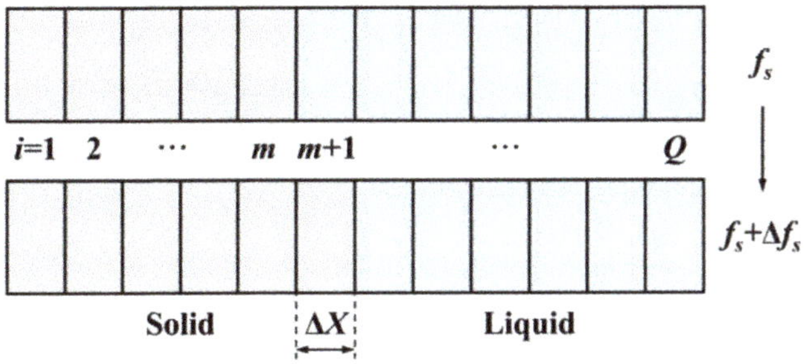

Figure 8. Schematic of the division of the calculation region.

The overall calculation process of the coupled model was depicted in Figure 9. In each iteration step, f_S was incremented by Δf_S (with $f_S = 0$ for the initial step). First, the concentrations of Ti and N in the molten alloy were calculated using the modified C-K model, followed by an evaluation using Equation (2) to determine whether TiN inclusions precipitated. If no precipitation occurred, the process moved to the next step. If TiN precipitation was detected, the microsegregation module was coupled with the TiN precipitation module to compute the concentrations of Ti and N in the residual liquid phase. Finally, the TiN precipitation amount and its cumulative value were calculated for each step. The process continued until $f_S = 1$, marking the end of the calculation.

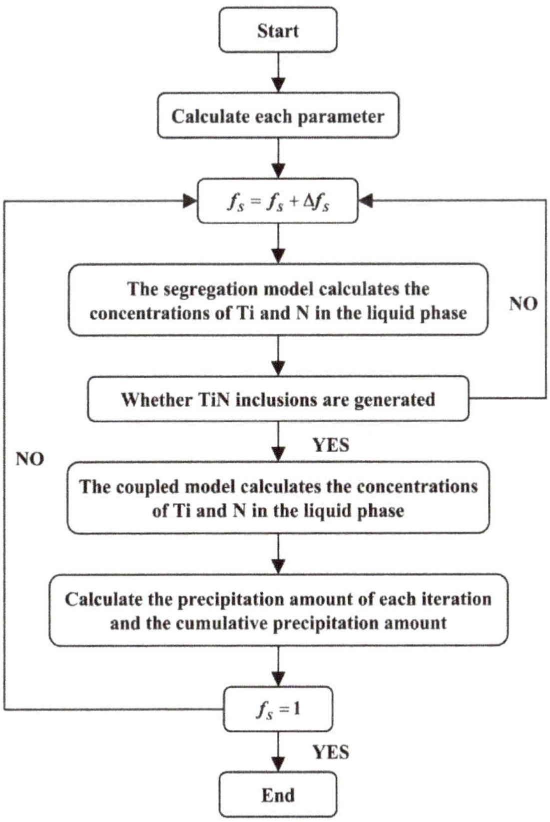

Figure 9. The main calculation process of the model.

3. Results and Discussion

3.1. Discussion of Fundamental Results

3.1.1. Thermodynamic Calculation Results

Using Equations (1)–(3), the ΔG for the precipitation reaction of TiN inclusions at different solid fractions was calculated to determine whether TiN precipitated. Using a secondary dendrite arm spacing of 105 μm as an example, Figure 10 shows the variation in the Gibbs free energy (ΔG) during solidification. The graph indicated that, as solidification progressed, the solid fraction (f_S) increased and ΔG decreased. For $0 \leq f_S < 0.256$, $\Delta G > 0$, indicating that no TiN precipitation occurred in the residual liquid phase. At $f_S = 0.256$, $\Delta G = 0$, marked the critical point for TiN precipitation. Substituting f_S into Equation (19) yielded a critical temperature for TiN precipitation at the solidification front of 1619.81 °C. For $0.256 < f_S \leq 1$, $\Delta G < 0$, indicating the onset of TiN precipitation, which continued to increase as f_S rose. The maximum amount of TiN precipitation was reached when solidification was complete ($f_S = 1$). Reducing the initial concentrations of N and Ti in the molten alloy effectively lowered the free energy of the TiN precipitation reaction, delaying the threshold for TiN formation. This adjustment also decreased the total amount of TiN precipitated at the end of solidification and reduced the potential for TiN inclusion agglomeration.

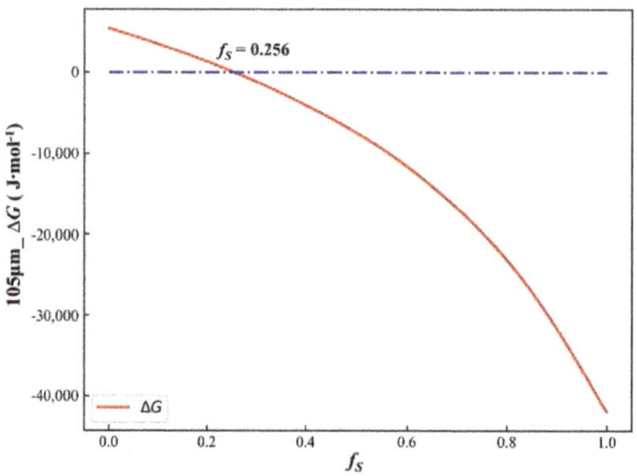

Figure 10. The ΔG change curve of the TiN precipitation reaction during solidification.

3.1.2. Calculation Results of the Coupled Model

By accounting for the consumption of N due to TiN precipitation (through the coupling of the microsegregation module with the TiN precipitation module), the variation in N concentration in the residual liquid phase under coupled conditions was obtained. Using a secondary dendrite arm spacing of 105 μm as an example, the computational results are shown in Figure 11. The results indicated that, without considering the impact of TiN precipitation, the N concentration in the residual liquid phase gradually increased during solidification, reached 52.2 ppm by the end of the process. When the effects of both microsegregation and TiN precipitation on the N concentration were included, the results were as follows: for $f_S < 0.256$, the outputs from the coupled model and the microsegregation model alone were identical. At $f_S = 0.256$, TiN precipitation began in the residual liquid phase. For $f_S > 0.256$, as solidification progressed, the N concentration in the residual liquid phase decreased, and the extent of N microsegregation diminished. By the end of solidification, the N concentration dropped to just 1.3 ppm, indicating that nearly all N precipitated as primary nitride or carbonitride during the solidification process.

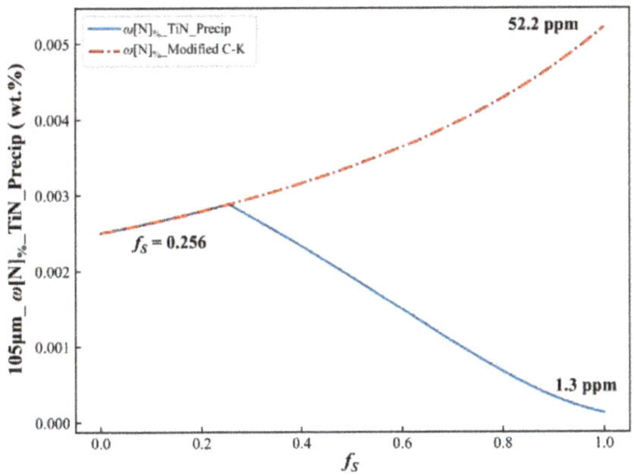

Figure 11. N concentration curves in the residual liquid phase were predicted using the modified C-K model and the coupled model, respectively.

3.1.3. Microsegregation of Other Alloying Elements

The influence of microsegregation of major alloying elements on TiN precipitation was examined. Using a secondary dendrite arm spacing of 105 μm as an example, the microsegregation of six key elements—C, O, Mg, Nb, S, and Al—was predicted using both the modified C-K model and the Scheil model. The results are shown in Figure 12. The microsegregation data for Ti and N were excluded from this comparison, as their behavior was influenced by TiN precipitation.

Figure 12 shows the concentration change curves for six major alloying elements calculated by the Scheil model and the modified C-K model. These curves reveal that, regardless of whether the modified C-K model or the Scheil model is used, the concentration of each alloying element in the residual liquid phase increases as solidification progresses, with a marked acceleration in the later stages. During the initial stages of solidification, the concentrations calculated by the Scheil model closely align with those obtained using the modified C-K model. However, in the later stages, the Scheil model predicts much higher concentrations than the modified C-K model, which does not align with physical expectations. Table 4 compares the microsegregation ratios of these six elements as calculated by the modified C-K model. The data indicated that the order of microsegregation ratios during solidification was Mg > S > O > Nb > C > Al. Notably, the microsegregation of Mg and S became particularly pronounced as solidification advanced, highlighting their significant impact on TiN precipitation.

Table 4. The results of microsegregation of six major alloying elements were calculated using the modified C-K model.

	Initial Concentration C_0 (wt.%)	Terminal Concentration C_L (wt.%)	Concentration Gradient ΔC (wt.%)	Microsegregation Ratio $\Delta C/C_0$
C	0.018	0.060	0.042	2.3
S	0.0009	0.0308	0.0299	33.2
Nb	4.91	20.66	15.75	3.2
Mg	0.001	0.146	0.145	145.0
Al	0.52	0.96	0.44	0.8
O	0.00041	0.01368	0.01327	32.4

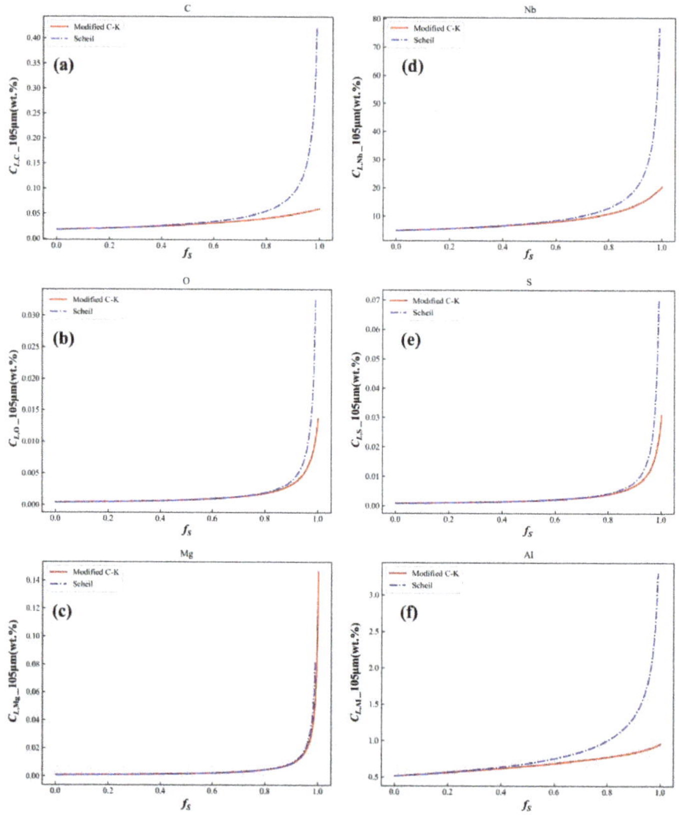

Figure 12. Microsegregation of the six major alloying elements calculated by the Scheil model and the modified C-K model: (**a**) C; (**b**) O; (**c**) Mg; (**d**) Nb; (**e**) S; (**f**) Al.

3.1.4. Effect of Different Models

The microsegregation behavior of Ti and N was analyzed using the L-R model, Scheil model, modified C-K model, and the coupled model, with a secondary dendrite arm spacing of 105 μm. For accuracy, the results for $f_S > 0.99$ were excluded due to the Scheil model's limitations at the final stage of solidification. Figure 13 presents the concentration variation curves for Ti and N.

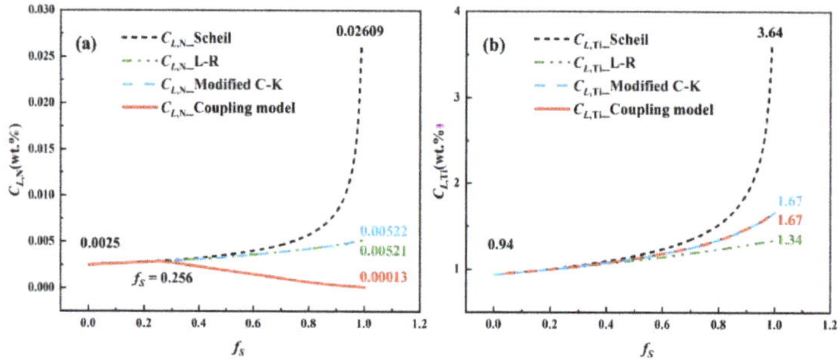

Figure 13. Curves of Ti and N concentrations predicted by different microsegregation models: (**a**) N; and (**b**) Ti.

Figure 13a,b show that, at the start of solidification, the initial concentrations of N and Ti were 0.0025 wt.% (25 ppm) and 0.94 wt.%, respectively. Before TiN precipitation ($f_S < 0.256$), the concentration results for Ti and N in the residual liquid phase were nearly identical across all models.

However, after TiN precipitation began, the behavior of N diverged. The L-R, Scheil, and modified C-K models predicted increasing N concentrations as solidification progressed, whereas the coupled model showed a decreasing trend due to N consumption by TiN precipitation. The highest N concentration at the end of solidification was 0.02609 wt.% (260.9 ppm), as predicted by the Scheil model. The L-R model and modified C-K model provided similar results, at 0.00521 wt.% (52.1 ppm) and 0.00522 wt.% (52.2 ppm), respectively, between the Scheil and coupled model. The coupled model yielded the lowest N concentration, 0.00013 wt.% (1.3 ppm), reflecting the significant N consumption associated with TiN precipitation. The modified C-K model's results closely resembled those of the L-R model due to the N element diffusing faster in the solid phase.

For Ti, all four models predicted an increasing concentration as solidification advanced. The Scheil model forecasted the highest Ti concentration at 3.64 wt.% at the end of solidification. The coupled model and modified C-K model produced similar results of 1.67 wt.%, intermediate between the predictions of the Scheil and L-R models, because the effect of TiN precipitation on Ti concentration was not considered. The L-R model predicted the lowest Ti concentration, at 1.34 wt.%.

3.1.5. Effects of Coupled Model

Using a secondary dendrite arm spacing of 105 μm and a step size of 0.001 as an example, the TiN precipitation amount at each step and the total accumulated precipitation were calculated. The variation curves for stepwise precipitation and accumulated precipitation as functions of the solid fraction are shown in Figure 14. Figure 14a indicates that, when $f_S < 0.256$, the concentration product of Ti and N in the residual liquid phase is below the equilibrium concentration product, resulting in no TiN precipitation. Once $f_S \geq 0.256$, TiN begins to precipitate in the residual liquid phase. As the solid fraction increases, the amount of TiN precipitated at each step gradually decreases until it reaches zero at the end of solidification. Figure 14b illustrates that, when $f_S < 0.256$, there is no accumulation of TiN in the residual liquid phase. When $f_S \geq 0.256$, the accumulated amount of TiN increases progressively with the solid fraction. However, the rate of accumulation slows as solidification continues, eventually reaching zero at the end of solidification. By the completion of solidification, the total accumulated TiN precipitation reaches 64 ppm.

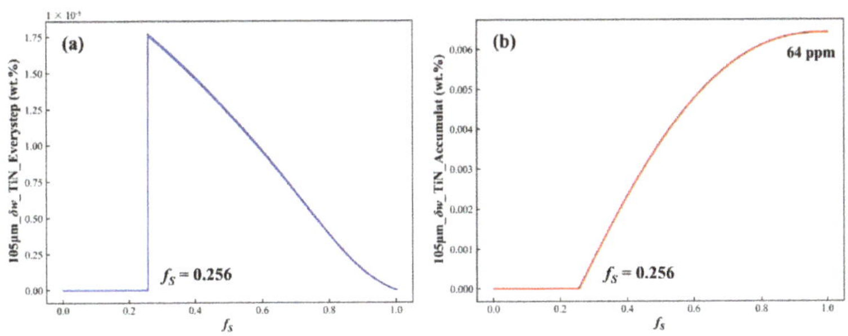

Figure 14. (**a**) TiN precipitation at each step; (**b**) accumulated TiN precipitation during solidification.

3.2. Effect of Initial Concentration of N on TiN Precipitation

Using a secondary dendrite arm spacing of 105 μm as an example, the accumulated precipitation amounts of TiN in the residual liquid phase were calculated for initial N concentrations of 10, 25, and 40 ppm. The variation curves of accumulated TiN precipitation as a function of the solid fraction are shown in Figure 15. The results indicated that, as the initial N concentration increased, TiN began to precipitate earlier, and the accumulated precipitation at the end of solidification rose significantly. At an initial N concentration of 10 ppm, TiN started to precipitate at f_S = 0.606. As solidification progressed, the precipitation gradually increased, reaching an accumulated amount of 16 ppm by the end of solidification. When the initial N concentration was raised to 25 ppm, TiN began to precipitate at f_S = 0.256, and the accumulated precipitation at the end of solidification increased notably to 64 ppm. At an initial N concentration of 40 ppm, TiN precipitation started before solidification began, and the final accumulated precipitation reached 126 ppm. The results indicated that higher initial nitrogen concentrations led to earlier TiN precipitation and a greater total amount of TiN formed. The increased number of TiN particles provided abundant heterogeneous nucleation sites for NbC precipitation. During the later stages of solidification, these TiN and NbC particles further evolved into mixed (Nb + Ti) carbonitrides. These precipitates exhibited poor distribution uniformity within the structure and were difficult to adjust effectively through subsequent heat treatment or homogenization, ultimately becoming part of the solidified structure and negatively impacting the mechanical properties of the alloy [7]. Therefore, controlling the nitrogen concentration in the alloy was crucial for optimizing the microstructure and enhancing the mechanical properties of the alloy.

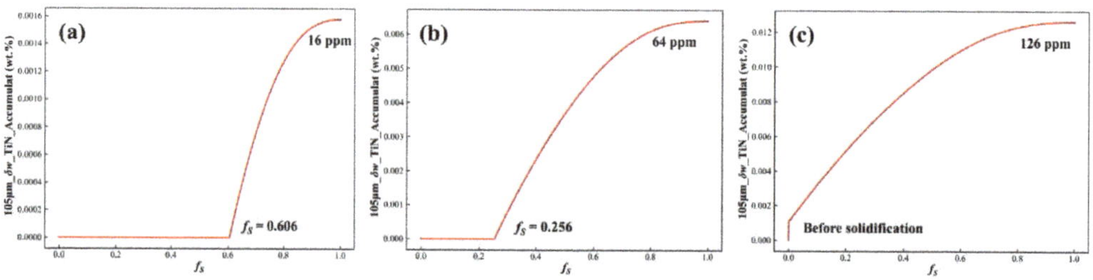

Figure 15. Curves of TiN accumulated precipitation with solid fraction under different initial N concentrations: (**a**) 10 ppm; (**b**) 25 ppm; and (**c**) 40 ppm.

3.3. *Effect of Cooling Rate on TiN Precipitation*

The curve of the secondary dendrite arm spacing against the cooling rate was established based on the re-fitting Equation (21). The local cooling rates corresponding to secondary dendrite arm spacings of 70 μm, 105 μm, and 140 μm were predicted. The results are shown in Figure 16. This figure shows that the secondary dendrite arm spacing gradually decreases as the cooling rate increases.

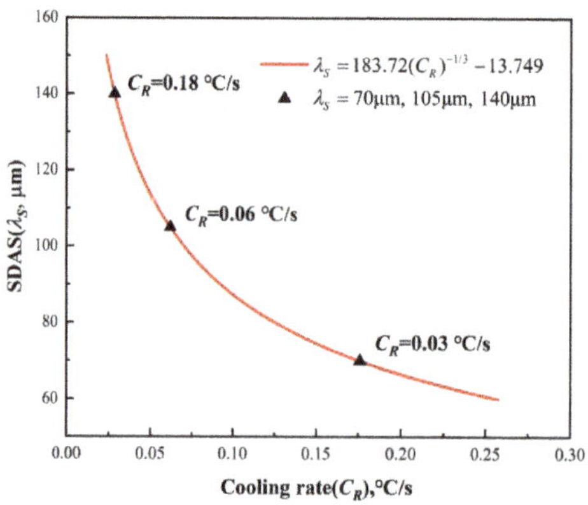

Figure 16. Curve of secondary dendrite arm spacing at different cooling rates.

For secondary dendrite arm spacing values of 70 μm, 105 μm, and 140 μm, the corresponding local cooling rates were 0.18 °C/s, 0.06 °C/s, and 0.03 °C/s, respectively. Using the coupled model, the N concentration in the residual liquid phase and the accumulated TiN precipitation at these cooling rates were estimated, with the results shown in Figure 17. This figure revealed that increasing the cooling rate from 0.03 °C/s to 0.18 °C/s had few effects on the N concentration and accumulated TiN precipitation. When the impact of TiN precipitation on the N concentration was ignored, the N concentration at the end of solidification remained 52 ppm for all three cooling rates. When TiN precipitation was considered, the final N concentration dropped to 1.3 ppm for all rates, with a total of 64 ppm of TiN precipitation throughout the solidification process. These findings suggested that increasing the cooling rate had little effect on solute microsegregation or TiN precipitation within dendritic cells. However, higher cooling rates effectively reduced secondary dendrite arm spacing, thereby decreasing the size of microsegregation units (half of the secondary dendrite arm spacing), which significantly influenced solute microsegregation and TiN precipitation.

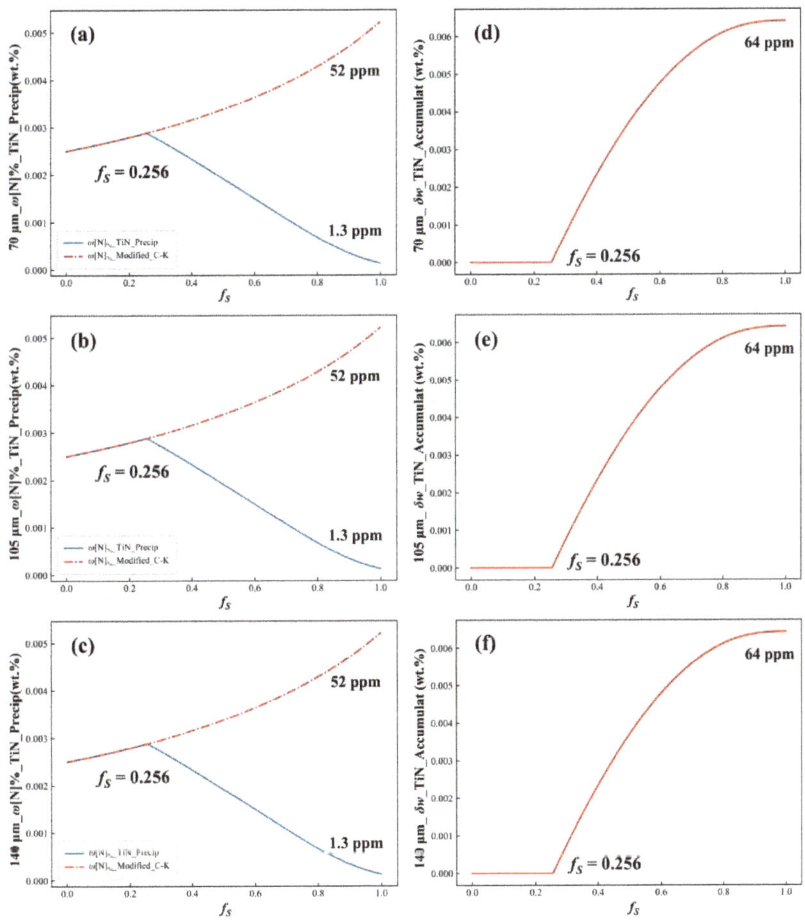

Figure 17. The concentration of N and TiN accumulated precipitation in the residual liquid phase during solidification at different cooling rates: (**a**,**d**) C_R = 0.18 °C/s; (**b**,**e**) C_R = 0.06 °C/s; (**c**,**f**) C_R = 0.03 °C/s.

4. Conclusions

Large TiN inclusions in nickel-based superalloys promote micropore formation, compromising the mechanical properties of the alloys. However, current research lacks a comprehensive coupled model that considers both solute element microsegregation and TiN precipitation specifically in nickel-based superalloys. This study integrated microsegregation and TiN precipitation calculations to conduct a comprehensive thermodynamic analysis of element microsegregation and TiN precipitation during the solidification of IN718 alloys. The effects of the coupled model, multi-element microsegregation, initial N concentration, and the cooling rate on TiN precipitation were evaluated. The main conclusions are as follows:

1. The microsegregation ratios of solute elements (excluding Ti and N) in the IN718 molten alloy followed the order: Mg > S > O > Nb > C > Al. Niobium showed the most significant increase in concentration during solidification (approximately 15.75 wt.%), due to its high initial concentration. This finding highlights the importance of considering multi-element microsegregation in thermodynamic calculations of TiN precipitation in the IN718 alloy.
2. Without coupling the TiN precipitation module with the microsegregation module, the N concentration in the residual liquid phase rose to 52.2 ppm by the end of solidification. When the modules were coupled, the N concentration decreased to 1.3 ppm after TiN precipitation, demonstrating a marked difference between the two scenarios.
3. Thermodynamic analysis indicated that TiN inclusions began to precipitate when the solid fraction exceeded 0.256. As solidification progressed, the amount of TiN precipitation increased and reached 64 ppm by the end of solidification.
4. With an increase in the initial N concentration from 10 ppm to 40 ppm, the solid fraction at which TiN precipitation began shifted from 0.606 to a point before the onset of solidification, and the accumulated TiN precipitation during solidification increased from 16 ppm to 126 ppm.
5. The variation in N concentration and accumulated TiN precipitation was predicted for cooling rates of 0.03 °C/s, 0.06 °C/s, and 0.18 °C/s. The results were nearly identical, suggesting that changes in cooling rate had little effect on TiN precipitation.

Author Contributions: Conceptualization, J.Z.; Methodology, J.Z. and C.Z.; Formal analysis, C.Z. and H.Z.; Investigation, J.Z. and C.Z.; Data curation, C.Z.; Writing—original draft, J.Z. and C.Z.; Writing—review & editing, J.Z., H.Z. and C.G.; Visualization, C.G.; Supervision, J.Z. All authors have read and agreed to the published version of the manuscript.

Funding: This research received no external funding.

Institutional Review Board Statement: Not applicable.

Informed Consent Statement: Not applicable.

Data Availability Statement: The original contributions presented in the study are included in the article, further inquiries can be directed to the corresponding author.

Conflicts of Interest: The authors declare no potential conflicts of interest with respect to the research, authorship, and/or publication of this article.

References

1. Reed, R.C. *The Superalloys: Fundamentals and Applications*; Cambridge University Press: New York, NY, USA, 2006; p. 18, ISBN 0-521-85904-2.
2. Que, G.; Song, S.; Nyembwe, A.; Xue, Z.; Qi, J.; Deng, Z. TiN Inclusions Distribution in the Composite Cross Section of Titanium Micro-Alloyed Steel Slab: Thermo-Chemical Modeling and Automatic Statistic Analysis. *J. Mater. Res. Technol.* **2023**, *26*, 5383–5392. [CrossRef]
3. Leonardo, I.M.; Da Hora, C.S.; Dos Reis Silva, M.B.; Sernik, K. Production of Nitride-Free 718 by the VIM-VAR Processing Route. In Proceedings of the 9th International Symposium on Superalloy 718 & Derivatives: Energy, Aerospace, and Industrial

Applications, Champion, PA, USA, 13 May 2018; Ott, E., Liu, X., Andersson, J., Bi, Z., Bockenstedt, K., Dempster, I., Groh, J., Heck, K., Jablonski, P., Kaplan, M., et al., Eds.; Springer International Publishing: Cham, Switzerland, 2018; pp. 303–315.
4. Yang, L.; Webler, B.A.; Cheng, G. Precipitation Behavior of Titanium Nitride on a Primary Inclusion Particle during Solidification of Bearing Steel. *J. Iron Steel Res. Int.* **2017**, *24*, 685–690. [CrossRef]
5. Yuan, C.; Guo, J.T.; Zhou, L.Z. Effect of Nitrogen on Microstructure and Properties of a Cast Cobalt-Base Superalloy. *Adv. Mater. Res.* **2011**, *278*, 472–478. [CrossRef]
6. Young, E.C.; Mitchell, A. Some Aspects of Nitrogen Addition and Removal during Special Melting and Processing of Iron and Nickel Base-Alloys. *High Temp. Mater. Process.* **2001**, *20*, 79–102. [CrossRef]
7. Mitchell, A.; Wang, T. Solidification and Precipitation in IN718. In Proceedings of the 5th International Symposium on Superalloys 718, 625, 706 and Various Derivatives, Pittsburgh, PA, USA, 17–20 June 2001; Loria, E.A., Ed.; TMS: Pittsburgh, PA, USA, 2001; pp. 81–90.
8. Sindhura, D.; Sravya, M.V.; Murthy, G.V.S. Comprehensive Microstructural Evaluation of Precipitation in Inconel 718. *Metallogr. Microstruct. Anal.* **2019**, *8*, 233–240. [CrossRef]
9. Brody, H.D.; Flemings, M.C. Solute redistribution in dendritic solidification. *Trans. TMS-AIME* **1966**, *236*, 615–624.
10. Clyne, T.W.; Kurz, W. Solute Redistribution during Solidification with Rapid Solid State Diffusion. *Metall. Trans. A* **1981**, *12*, 965–971. [CrossRef]
11. Ohnaka, I. Mathematical Analysis of Solute Redistribution during Solidification with Diffusion in Solid Phase. *Trans. Iron Steel Inst. Jpn.* **1986**, *26*, 1045–1051. [CrossRef]
12. Voller, V.R.; Beckermann, C. A Unified Model of Microsegregation and Coarsening. *Metall. Mater. Trans. A* **1999**, *30*, 2183–2189. [CrossRef]
13. Won, Y.-M.; Thomas, B.G. Simple Model of Microsegregation during Solidification of Steels. *Metall. Mater. Trans. A* **2001**, *32*, 1755–1767. [CrossRef]
14. Gui, L.; Long, M.; Huang, Y.; Chen, D.; Chen, H.; Duan, H.; Yu, S. Effects of Inclusion Precipitation, Partition Coefficient, and Phase Transition on Microsegregation for High-Sulfur Steel Solidification. *Metall. Mater. Trans. B* **2018**, *49*, 3280–3292. [CrossRef]
15. Yang, F.; Cao, J.; Shi, L.; Yu, J.; Deng, K.; Ren, Z. Influence of Microsegregation on the TiN Inclusions Formation Behavior in a K418 Superalloy during the Continuous Unidirectional Solidification Process. *ISIJ Int.* **2023**, *63*, 436–447. [CrossRef]
16. Liu, T.; Chen, D.; Long, M.; Liu, P.; Duan, H.; Gui, L.; Fan, H.; Chen, H. Control of Coarse Precipitates of Titanium Nitride in High-Strength Low-Alloy Steel. *Met. Sci. Heat Treat.* **2020**, *61*, 534–542. [CrossRef]
17. Gui, L.; Long, M.; Zhang, H.; Chen, D.; Liu, S.; Wang, Q.; Duan, H. Study on the Precipitation and Coarsening of TiN Inclusions in Ti-Microalloyed Steel by a Modified Coupling Model. *J. Mater. Res. Technol.* **2020**, *9*, 5499–5514. [CrossRef]
18. Lee, M.H.; Park, J.H. Synergistic Effect of Nitrogen and Refractory Material on TiN Formation and Equiaxed Grain Structure of Ferritic Stainless Steel. *Metall. Mater. Trans. B* **2018**, *49*, 877–893. [CrossRef]
19. Li, B.; Shi, X.; Guo, H.; Guo, J. Study on Precipitation and Growth of TiN in GCr15 Bearing Steel during Solidification. *Materials* **2019**, *12*, 1463. [CrossRef]
20. Shu, Q.; Visuri, V.-V.; Alatarvas, T.; Fabritius, T. Model for Inclusion Precipitation Kinetics During Solidification of Steel Applications in MnS and TiN Inclusions. *Metall. Mater. Trans. B* **2020**, *51*, 2905–2916. [CrossRef]
21. Pak, J.-J.; Jeong, Y.-S.; Tae, S.-J.; Kim, D.-S.; Lee, Y.-Y. Thermodynamics of Titanium and Nitrogen in an Fe-Ni Melt. *Metall. Mater. Trans. B* **2005**, *36*, 489–493. [CrossRef]
22. Kurz, W.; Fisher, D.J. *Fundamentals of Solidification*, 3rd ed.; Trans Tech Publications Ltd.: Zurich, Switzerland, 1992; p. 280, ISBN 0-87849-522-3.
23. Kobayashi, S. Solute Redistribution during Solidification with Diffusion in Solid Phase: A Theoretical Analysis. *J. Cryst. Growth* **1988**, *88*, 87–96. [CrossRef]
24. Matsumiya, T.; Kajioka, H.; Mizoguchi, S.; Ueshima, Y.; Esaka, H. Mathematical Analysis of Segregations in Continuously-Cast Slabs. *Trans. Iron Steel Inst. Jpn.* **1984**, *24*, 873–882. [CrossRef]
25. Battle, T.P.; Pehlke, R.D. Mathematical Modeling of Microsegregation in Binary Metallic Alloys. *Metall. Trans. B* **1990**, *21*, 357–375. [CrossRef]
26. Wang, C.Y.; Beckermann, C. A Unified Solute Diffusion Model for Columnar and Equiaxed Dendritic Alloy Solidification. *Mater. Sci. Eng. A* **1993**, *171*, 199–211. [CrossRef]
27. Miettinen, J. Mathematical Simulation of Interdendritic Solidification of Low-Alloyed and Stainless Steels. *Metall. Trans. A* **1992**, *23*, 1155–1170. [CrossRef]
28. Miettinen, J. Calculation of Solidification-Related Thermophysical Properties for Steels. *Metall. Mater. Trans. B* **1997**, *28*, 281–297. [CrossRef]
29. Ma, Z.; Janke, D. Characteristics of Oxide Precipitation and Growth during Solidification of Deoxidized Steel. *ISIJ Int.* **1998**, *38*, 46–52. [CrossRef]
30. Choudhary, S.K.; Ghosh, A. Mathematical Model for Prediction of Composition of Inclusions Formed during Solidification of Liquid Steel. *ISIJ Int.* **2009**, *49*, 1819–1827. [CrossRef]
31. Song, H.; Miao, Z.; Shan, A.; Xu, W.; Lu, J. Application of Confocal Scanning Laser Microscope in Studying Solidification Behavior of Alloy 718. In *Superalloy 718 and Derivatives*; Ott, E.A., Groh, J.R., Banik, A., Dempster, I., Gabb, T.P., Helmink, R., Liu, X., Mitchell, A., Sjöberg, G.P., Wusatowska-Sarnek, A., Eds.; Wiley: Beijing, China, 2010; pp. 168–180, ISBN 978-1-118-49522-3.

32. Wang, P.; Li, C.; Wang, L.; Zhang, J.; Xue, Z. Thermodynamic Analysis of TiN Precipitation in SWRH92A High Carbon Tire Cord Steel Under the Influence of Solute Micro-Segregations During Solidification. *Metall. Mater. Trans. B* **2021**, *52*, 2056–2071. [CrossRef]

materials

MDPI

Article

Design of Shape Forming Elements for Architected Composites via Bayesian Optimization and Genetic Algorithms: A Concept Evaluation

David O. Kazmer [1,*], Rebecca H. Olanrewaju [1], David C. Elbert [2] and Thao D. Nguyen [2]

[1] Department of Plastics Engineering, University of Massachusetts Lowell, Lowell, MA 01854, USA; rebecca_olanrewaju@student.uml.edu

[2] Department of Mechanical Engineering, Johns Hopkins University, Baltimore, MD 21218, USA; elbert@jhu.edu (D.C.E.); vicky.nguyen@jhu.edu (T.D.N.)

* Correspondence: david_kazmer@uml.edu; Tel.: +1-978-934-2962

Abstract: This article presents the first use of shape forming elements (SFEs) to produce architected composites from multiple materials in an extrusion process. Each SFE contains a matrix of flow channels connecting input and output ports, where materials are routed between corresponding ports. The mathematical operations of rotation and shifting are described, and design automation is explored using Bayesian optimization and genetic algorithms to select fifty or more parameters for minimizing two objective functions. The first objective aims to match a target cross-section by minimizing the pixel-by-pixel error, which is weighted with the structural similarity index (SSIM). The second objective seeks to maximize information content by minimizing the SSIM relative to a white image. Satisfactory designs are achieved with better objective function values observed in rectangular rather than square flow channels. Validation extrusion of modeling clay demonstrates that while SFEs impose complex material transformations, they do not achieve the material distributions predicted by the digital model. Using the SSIM for results comparison, initial stages yielded SSIM values near 0.8 between design and simulation, indicating a good initial match. However, the control of material processing tended to decline with successive SFE processing with the SSIM of the extruded output dropping to 0.023 relative to the design intent. Flow simulations more closely replicated the observed structures with SSIM values around 0.4 but also failed to predict the intended cross-sections. The evaluation highlights the need for advanced modeling techniques to enhance the predictive accuracy and functionality of SFEs for biomedical, energy storage, and structural applications.

Keywords: architected composites; additive manufacturing; artificial intelligence; optimization

1. Introduction

Recent advances in modeling, simulation, and manufacturing are accelerating the discovery of new materials while also reducing the time and cost of bringing new products to market. The materials genome initiative [1,2] reinforces these trends by seeking to accelerate materials discovery to enable materials-by-design wherein material and ultimate product properties are predicted from first principles; the end goal is to benefit humanity by creating new materials that support novel applications while reducing cost and environmental impact. Architected and cellular material composites offer promising avenues for creating efficient structures with previously impossible properties. Examples include lightweight structures and "white-space" materials [3] with gradient properties [4], negative Poisson's ratio [5], and other tunable properties [6]. While nature provides abundant examples of architected materials [7], a review of the literature suggests that most of the recent work relies on additive manufacturing methods to achieve complex material placement at the mesoscale and microscale. Kladovasilakis et al. [8] provide a recent review with a classification of architected materials based on the geometry, e.g., stochastic, homogeneous, and pseudo-periodic architectures. There are at least two issues with the use of additive manufacturing (AM) as the production route for architected materials. First, while layer-based AM processes such as Continuous Liquid Interface Production (CLIP) [9] and Computed Axial Lithography (CAL) [10] are substantially faster than extrusion-based and laser-based processes, they remain extremely slow and energy intensive relative to net-shape manufacturing processes [11,12]. Second, products made by AM processes have significant failure modes related to finite layer thickness [13], fatigue [14], and quality assurance [15]. Together, these issues preclude the adoption of architected materials in mass production with the most prevalent applications being lattice-based structural designs with mesoscale features [16].

This article provides the first investigation of the design and use of shape forming elements as a method for the mass production of architected material composites. The goal is to provide a structured methodology for designing architected composites that supports efficient and scalable production. The approach is a two-dimensional extension of one-dimensional layer multiplying elements [17] used in polymer processing to create multi-layered sheets [18,19], tubes [20,21], and simple hierarchical structures [22]. As described, a shape forming element (SFE) converts an input cross-section to an output cross-section with differing geometry through the use of multiple flow channels that deform, rotate, and shift the local distributions of materials; multiple SFE stages are then used to achieve higher levels of complexity of architected composites with high production rates when implemented in processes such as extrusion and injection molding. The article then describes implementations of design automation using Bayesian optimization and generic algorithms. Validation of the SFE designs using flow simulation and rapid prototyping is described, and significant implementation issues with the SFE concept related to optimization of the architected composites and methods of manufacture are identified. Accordingly, suggested modeling extensions are discussed to enable broad and practical applications in various industries, including biomedical and aerospace.

2. Materials and Methods

2.1. *Shape Forming Elements*

Consider a cross-section composed of two materials processed through sets of flow channels provided within two shape forming elements (SFEs) as shown in Figure 1. There are at least five operations that each SFE can provide: (1) cutting, (2) rotation, (3) shifting, (4) combining, and (5) reshaping. The first four operations are the focus of this article; reshaping SFEs are in concurrent development but in need of further research with respect to design automation. Other SFE functions such as flipping, copying, and deleting are easy to model in theory but difficult to realize in practice without transient actions implemented within the manufacturing process.

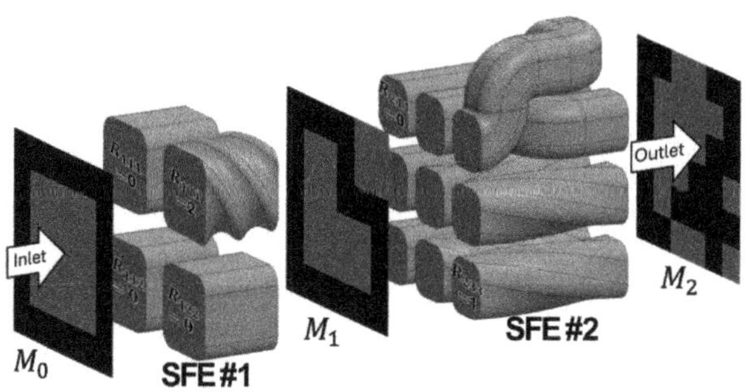

Figure 1. Flow channel designs implemented in a first SFE with the rotation matrix of Equation (3) and a second SFE with a rotation matrix of Equation (6) also with a shift matrix of Equation (8). The material traverses from left to right resulting in the three material cross-sections M_0, M_1, and M_2 corresponding to Equations (1), (5) and (9).

The input section of Figure 1, M_0, is modeled as two materials in a 6 × 6 matrix:

$$M_0 = \begin{bmatrix} 1 & 1 & 1 & 1 & 1 & 1 \\ 1 & 0 & 0 & 0 & 0 & 1 \\ 1 & 0 & 0 & 0 & 0 & 1 \\ 1 & 0 & 0 & 0 & 0 & 1 \\ 1 & 0 & 0 & 0 & 0 & 1 \\ 1 & 1 & 1 & 1 & 1 & 1 \end{bmatrix} \tag{1}$$

The use of the {0,1} material representation allows for straightforward computation of the relative material concentrations. For the defined M_0, material 1 is located at 20 of 36 overall locations, so the concentration of material 1 is 0.55. While a 6 × 6 matrix is used for illustration purposes, larger matrices are later analyzed for providing improved resolution across multiple shape forming operations. Multiple materials and their blends can be modeled by extension.

A rotation matrix for each i-th SFE, R_{ijk}, is defined for each jk-th port representing a partition of the input material matrix M_i. The number and locations of the ports define the number of cuts and partitions, P_{jk}, of the input material matrix. For example, suppose that a first SFE has a 2 × 2 set of ports, as shown in Figure 1. Then, M_0 can be represented as a set of P_{jk} as

$$M_0 = \begin{bmatrix} P_{11} = \begin{bmatrix} 1 & 1 & 1 \\ 1 & 0 & 0 \\ 1 & 0 & 0 \end{bmatrix} & P_{12} = \begin{bmatrix} 1 & 1 & 1 \\ 0 & 0 & 1 \\ 0 & 0 & 1 \end{bmatrix} \\ P_{21} = \begin{bmatrix} 1 & 0 & 0 \\ 1 & 0 & 0 \\ 1 & 1 & 1 \end{bmatrix} & P_{22} = \begin{bmatrix} 0 & 0 & 1 \\ 0 & 0 & 1 \\ 1 & 1 & 1 \end{bmatrix} \end{bmatrix} \tag{2}$$

The material processed through each jk-th port can be rotated relative to the center-line axis of the swept section in its flow direction. For the example of Figure 1 in which $R_{121} = 2$, a rotation matrix in which only the top right port is rotated 180° in the first SFE is:

$$R_{1jk} = \frac{\pi}{2}\begin{bmatrix} 0 & 2 \\ 0 & 0 \end{bmatrix}, \tag{3}$$

where in the output $M_{i,rot}$ from the rotation function, *rot*(), for the i-th SFE is:

$$M_{i,rot} = rot(M_{i-1}, R_i) = \left[\left[rot90\left(P_{ijk}, R_{ijk}\right)\right]\right]. \tag{4}$$

Here, the *rot*90() operation is the built-in Matlab (Mathworks, Waltham, MA, USA) function for rotating a matrix by 90° counterclockwise R_{jk} times, allowing for the straightforward manipulation of matrix orientations. An interactive design function *Design_SFE*() is provided in the Supplementary Materials along with other routines for optimization as later described. Appendix C provides a synopsis of these and other Matlab built-in and developed functions used in the performance of the research; all developed functions are available as described in the Supplementary Materials Section.

The result of Equations (2)–(4) is the first output material, M_1:

$$M_1 = \begin{bmatrix} P_{111,rot} = \begin{bmatrix} 1 & 1 & 1 \\ 1 & 0 & 0 \\ 1 & 0 & 0 \end{bmatrix} & P_{112,rot} - \begin{bmatrix} 1 & 0 & 0 \\ 1 & 0 & 0 \\ 1 & 1 & 1 \end{bmatrix} \\ P_{121,rot} = \begin{bmatrix} 1 & 0 & 0 \\ 1 & 0 & 0 \\ 1 & 1 & 1 \end{bmatrix} & P_{122,rot} = \begin{bmatrix} 0 & 0 & 1 \\ 0 & 0 & 1 \\ 1 & 1 & 1 \end{bmatrix} \end{bmatrix} \tag{5}$$

Successive rotation and shifting operations may be defined and applied. For example, suppose that material M_1 is fed into a second SFE with a set of 3×3 ports and a rotation matrix defined as:

$$R_2 = \frac{\pi}{2}\begin{bmatrix} 0 & 0 & 0 \\ 0 & 3 & 1 \\ 0 & 2 & 1 \end{bmatrix} \quad (6)$$

Then, $M_2 = rot(M_{1,\text{rot}}, R_2)$ would be

$$M_2 = \begin{bmatrix} rot90\left(\begin{bmatrix} 1 & 1 \\ 1 & 0 \end{bmatrix}, 0\right) & rot90\left(\begin{bmatrix} 1 & 1 \\ 0 & 1 \end{bmatrix}, 0\right) & rot90\left(\begin{bmatrix} 0 & 0 \\ 0 & 0 \end{bmatrix}, 0\right) \\ rot90\left(\begin{bmatrix} 1 & 0 \\ 1 & 0 \end{bmatrix}, 0\right) & rot90\left(\begin{bmatrix} 0 & 1 \\ 0 & 0 \end{bmatrix}, 3\right) & rot90\left(\begin{bmatrix} 1 & 1 \\ 0 & 1 \end{bmatrix}, 1\right) \\ rot90\left(\begin{bmatrix} 1 & 0 \\ 1 & 1 \end{bmatrix}, 0\right) & rot90\left(\begin{bmatrix} 0 & 0 \\ 1 & 1 \end{bmatrix}, 2\right) & rot90\left(\begin{bmatrix} 0 & 1 \\ 1 & 1 \end{bmatrix}, 1\right) \end{bmatrix} = \begin{bmatrix} 1 & 1 & 1 & 1 & 0 & 0 \\ 1 & 0 & 0 & 1 & 0 & 0 \\ 1 & 0 & 0 & 0 & 1 & 1 \\ 1 & 0 & 0 & 1 & 1 & 0 \\ 1 & 0 & 1 & 1 & 1 & 1 \\ 1 & 1 & 0 & 0 & 1 & 0 \end{bmatrix} \quad (7)$$

Each SFE can also shift the material flows from the input port locations to different output port locations using defined shift matrices, $S_{X,ijk}$ and $S_{Y,ijk}$, in which integer elements represent the relative movement between the *jk*-th ports located along the respective *X* and *Y* axes in the *i*-th SFE. For example, the second SFE of Figure 1 swaps the top center and top right port outputs by defining $S_{X,2jk}$ as:

$$S_{X,2jk} = \begin{bmatrix} 0 & 1 & -1 \\ 0 & 0 & 0 \\ 0 & 0 & 0 \end{bmatrix} \quad (8)$$

Applying the *shift*() function to the shift matrix of Equation (8) with the material distribution of Equation (7) yields:

$$M_{2,shift} = shift\left(M_{2,rot}, S_{X,2jk}\right) = \begin{bmatrix} 1 & 1 & 0 & 0 & 1 & 1 \\ 1 & 0 & 0 & 0 & 0 & 1 \\ 1 & 0 & 0 & 0 & 1 & 1 \\ 1 & 0 & 0 & 1 & 1 & 0 \\ 1 & 0 & 1 & 1 & 1 & 1 \\ 1 & 1 & 0 & 0 & 1 & 0 \end{bmatrix} \quad (9)$$

Figure 1 depicts the flow channels of the first and second SFE examples along with the modeled architected composite cross-sections wherein the lighter and darker sections, respectively, represent materials 0 and 1.

2.2. Design Automation

A salient feature of the shape forming element method is that many designs can be achieved across multiple SFE stages. However, achieving a target architecture by manual design is literally puzzling, akin to a tiling puzzle [23], with non-obvious solutions. A significant issue is that the rotation of the material partitions can be hard to visualize across SFE stages to achieve desired material architectures. Furthermore, the number of combinations that may be considered is staggering. The number of combinations is on the order of d^p where p represents the number of design parameters and d represents the number of degrees of freedom for each design parameter (typically four considering only port rotation but much greater when shifting and other transforms are considered). The later design examples have a design space comprising around 4^{55} or 1.27×10^{30} combinations. Thus, design automation was implemented using Bayesian optimization and genetic algorithms.

Arróyave and McDowell [24] provide a review of systems approaches for materials design in which they suggest that "the most efficient methods for materials discovery are based on variants of Bayesian optimization (BO)". According to Brochu [25], "Bayesian optimization employs the Bayesian technique of setting a prior over the objective function and combining it with evidence to get a posterior function. This permits a utility-based selection of the next observation to make on the objective function, which must take into account both exploration (sampling from areas of high uncertainty) and exploitation (sampling areas likely to offer improvement over the current best observation)". Thus, Bayesian optimization allows for the exploration of a vast design space more efficiently than traditional methods, which is particularly advantageous in the context of complex material architectures where intuitive design approaches might fail.

Striving for intelligent design through design automation, integer optimization was first implemented using the Matlab function *bayesopt*() to design a series of SFEs to minimize an objective function G according to the flowchart of Figure 2. The algorithm iteratively performs successive levels of design optimization, starting with the initial level (iLevel = 0). In this initial level, the algorithm loops through multiple input cross-sections (later demonstrated) and evaluates the objective function for each and selecting the best. The Bayesian optimization then progresses for each SFE, allowing the algorithm to refine and optimize the number and specific rotation of the flow channels to minimize the objective value. After each iteration, the selected design and resulting output section are adopted for subsequent analysis until the number of SFE stages is met (iLevel = m) or there is no improvement in the objective function (ΔG = 0).

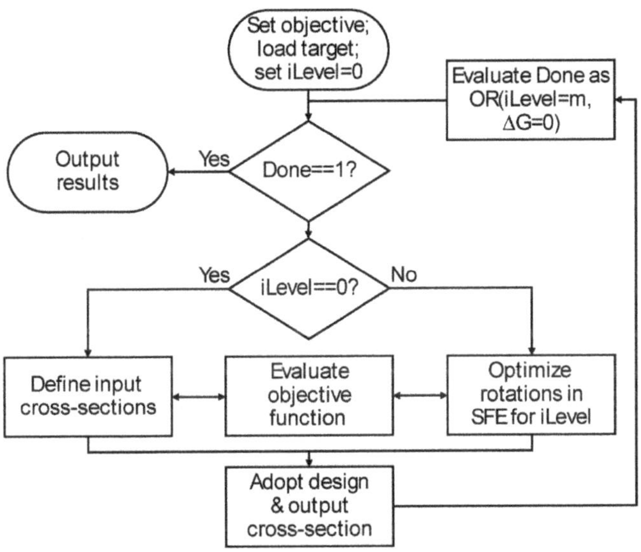

Figure 2. Implemented algorithm for shape function element (SFE) design using sequential Bayesian optimization.

To assist the evolution of the architected cross-section across sequential SFE and Bayesian optimization stages, a first objective function G_1 was developed so that the output material matches a target cross-section design represented as a binary image, I_{Target}. Mathematically, this first objective is achieved by minimizing G_1, which is defined as:

$$G_1 = w \cdot MSSE(I_{Target}, I_{Output}) + (1-w)(1-SSIM)(I_{Target}, I_{Output}) \tag{10}$$

Here, $MSSE$ represents the mean sum squared error representing a pixel-by-pixel comparison of the target image, I_{Target}, and the output image, I_{Output}, computed as:

$$MSSE(I_{Target}, I_{Output}) = \sum (I_{Target} - I_{Output})^2 / N^2, \tag{11}$$

where the two images are binary (as in the representation of Equation (1) and scaled to size N; this definition is similar to the Matlab function *immse*(). Meanwhile, $SSIM$ represents the structural similarity index for measuring image quality as defined by Wang et al. [26]. Since $SSIM$ approaches a value of 1 with perfect correlation, the value of $(1 - SSIM)$ is calculated as the error norm in the objective function for minimization purposes. The purpose of the scalar weighting value, w, is to balance between the trade-off between local pixel matching driven by MSSE and overall structural similarity driven by $SSIM$. A value of 0.5 was typically used as the weighting value with similar execution time and results for non-extreme values, e.g., $w \in (0.1, 0.9)$.

As later characterized, the testing and validation of Bayesian optimization using the sequential algorithm of Figure 2 frequently resulted in poor results. The reason was that the optimal design requires the selection of input cross-section and early SFE stages that are far from optimal as measured by the objective function. In other words, the sequential Bayesian optimization falls prey to the same limitation of human sequential design thinking in which the architected design only comes together in the latter SFE stages based on unobvious intermediate cross-sections. As such, a different integer optimization was implemented according to the flowchart of Figure 3 in which all design parameters are concurrently optimized. The vector of design parameters is defined as:

$$x = \left[i_{Input}, \{(m_1, n_1), \ldots, (m_N, n_N)\}, \left\{\left[R_{1jk}\right], \ldots, \left[R_{Njk}\right]\right\}\right], \tag{12}$$

where R_{ijk} is the rotation matrix for a given stage i per Equation (3), m_i and n_i are the number of SFE ports in the x and y directions, and i_{Input} is an integer representing the input material cross-section. This implementation allows for non-square fluid ports, although square ports are also implemented with the assumption that $m_i = n_i$ and a corresponding reduction in the length of the design vector x.

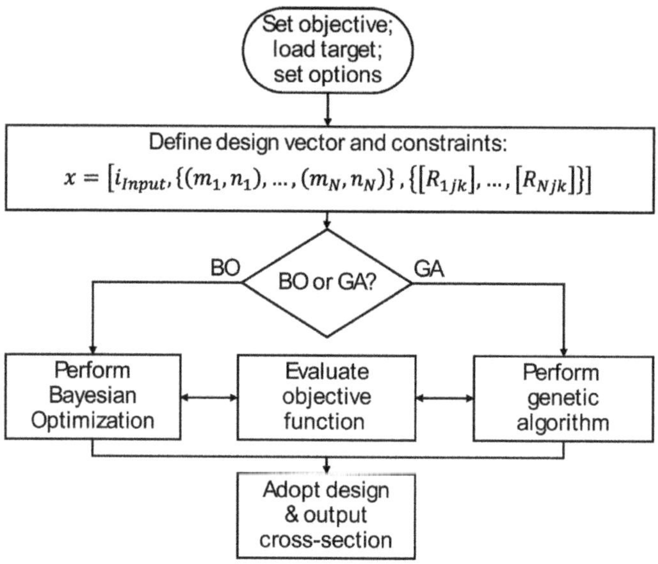

Figure 3. Implemented algorithm for shape function element (SFE) design using concurrent Bayesian optimization or genetic algorithms.

Given this problem structure, it was straightforward to also implement a solution based on genetic algorithms (GAs) as shown in the flowchart of Figure 3. This implementation calls Matlab's *ga*() function that mimics the process of natural selection to iteratively improve a population of candidate solutions [27]. The GA's convergence is guided by constrained optimization as described by Conn, Gould, and Toint in their works on globally convergent augmented Lagrangian algorithms [28]. The resulting GA method is generally well adapted to handling large systems with complex constraints. Subsequent testing showed that GA tended to outperform BO, which tended to converge to local, suboptimal solutions.

The first objective G_1 per Equation (10) was developed for SFE design automation to match a target image. For demonstration purposes, a second objective, G_2, is defined as:

$$G_2 = SSIM(I_{White}, I_{Output}) \tag{13}$$

Here, I_{White} represents a plain white image such that the intent is to maximize the difference between the output and a uniform material. In other words, the minimization of G_2 should drive the SFEs to produce a highly complex, architected material like a T-square or other fractal [29] as related to the concepts of negentropy [30] and Gibbs' free energy [31]. While the objectives G_1 and G_2 are used as examples, other objective functions are needed to derive architectures to maximize material and product performance.

2.3. Physical Prototyping

SFEs were also physically embodied and tested for demonstration purposes. A co-extrusion setup as shown in Figure 4 was implemented with dual injection cylinders having a bore diameter of 22 mm and a depth of 40 mm. The plungers were designed with a diametral clearance of 0.1 mm and a series of four 0.5 mm grooves to provide a dynamic seal that minimizes leakage under pressure. SFEs were designed with input and output boundaries as an 11 by 11 mm square. This section size allows for a 4 by 4 grid of flow channels each having a port geometry of 2 by 2 mm with a separating wall thickness of 1 mm. Prior work with these processes (e.g., [32]) as well as design guidelines from AM service providers indicated that the hydraulic port diameter should be greater than 2 mm with a length/diameter ratio greater than 10 to avoid clogging. In addition, a minimum wall thickness of 0.8 mm is needed between the passageways within an SFE to provide sufficient structural integrity during additive manufacturing and end use.

For the purposes of this article, the extruder and SFEs were produced in PA12 by Autotiv (Salem, NH, USA) through a selective laser sintering process having a resolution of approximately 60 microns. The implemented design for the G_1 objective is shown in Figure 4. Three-dimensional (3D) computer aided design (CAD) models for this and the G_2 designs are provided as described in the Supplemental Information. The stacked assembly was bolted together after which the bores for the inner and outer materials were, respectively, filled with 11.4 cm^3 of blue and yellow modeling clay (Amazon part no. B000KI7XA2 and B0015ZV6CK); previous studies have successfully utilized these materials for similar flow validation purposes, demonstrating its relevance in characterizing non-Newtonian flow in extrusion processes [33,34]. The assembly was then placed in an Arbor press that extruded the clay with a force of 200 kgf on the pistons with an extrusion time of 6 s. These imposed conditions correspond to a material pressure of 2.6 MPa and a volumetric flow rate of 3.8 cm^3/s that were adopted for flow simulation.

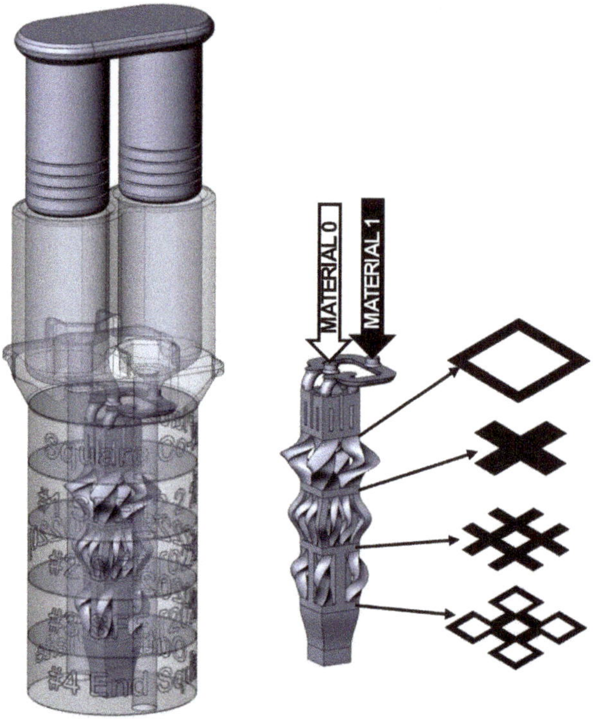

Figure 4. Three-dimensional models and cross-sections of the prototyped experimental design for a G_1 objective.

2.4. Flow Simulation

To understand the material distribution resulting from each stage of SFEs, transient multi-phase flow simulation including moving interfaces was performed with Comsol 4.6 (Stockholm, Sweden) using the level set method. The level set method [35] represents moving material interfaces using a fixed mesh by modeling the isocontour of the materials as a smooth step function between a first material represented as 0 in one domain and a second material represented as 1 in the second domain. The viscosity η for each material was modeled as isothermal but non-Newtonian with the Williamson equation [36]:

$$\eta(\dot{\gamma}) = \frac{\eta_0}{1 + (\lambda\dot{\gamma})^{1-n}} \tag{14}$$

In this model, η_0 represents the Newtonian limit of the viscosity [Pa.s] at a zero-shear rate $\dot{\gamma}$ [s^{-1}] and reference temperature, n is the power-law index, and λ is a characteristic relaxation time [s]. The material coefficients correspond to industrial recycled polyolefins of current interest [37]. Appendix B provides further details including the model coefficients, simulation conditions, and other details to access the CAD developed from the SFE design optimization; additional results are provided in Appendix B to investigate the roles of mesh refinement and viscosity modeling on development of the material cross-sections.

3. Results

3.1. Manually Created Designs

A Matlab script, *Design_SFE*(), available as described in the Supplemental Information, was authored to interactively create architected designs for systems up to four shape forming layers. One relatively simple design, referred to as "boxes", is shown in Figure 5 in which the first, second, and third rotation matrices, respectively, provide sets of 2 by 2, 4 by 4, and 3 by 3 ports. For demonstration purposes, selected ports are rotated by 180 degrees to convert a co-extruded square cross-section to a "plus" shape, then a "hash" shape, and finally to the "boxes" shape. The starting section M_0 is represented by an integer, i_{input}, that maps the two starting materials to (1) a square cross-section, (2) an inverted square cross-section, (3) a bi-layer system, and (4) a tri-layer system.

$$i_{Input} = 1$$

$$R_1 = \begin{bmatrix} 2 & 2 \\ 2 & 2 \end{bmatrix}$$

$$R_2 = \begin{bmatrix} 0 & 2 & 2 & 0 \\ 2 & 2 & 2 & 2 \\ 2 & 2 & 2 & 2 \\ 0 & 2 & 2 & 0 \end{bmatrix}$$

$$R_3 = \begin{bmatrix} 2 & 0 & 2 \\ 0 & 0 & 0 \\ 2 & 0 & 2 \end{bmatrix}$$

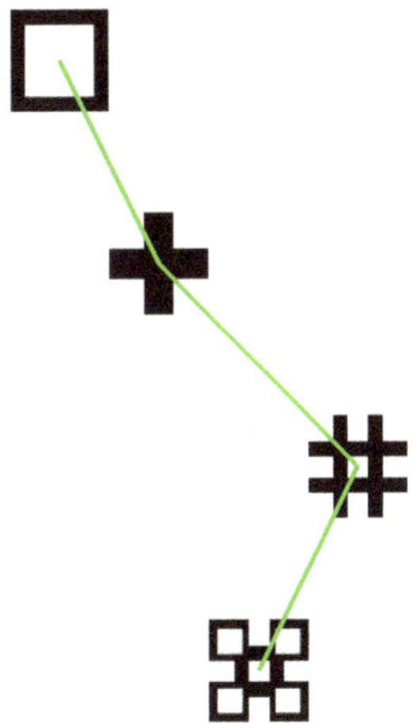

Figure 5. An interactively created "boxes" design specifying the input cross-section, rotation matrices having 2, 4, and 3 ports per side, and the resulting composite cross-sections.

3.2. Automated Design Solutions

Even when restricting the ports to a square geometry, the number of possible architected designs increases very quickly with the number of SFEs and number of ports in each SFE. A Matlab script, *SFE_Opt*(), was authored to implement design automation as described with respect to Figure 3 and Equation (12). To test the methods, the first objective G_1 was minimized to match the output of the automated designs with the interactively designed cross-section of Figure 5. The design vector, x, was developed assuming square ports so that its length of 52 elements included one material input type, three integers corresponding to the number of ports per SFE side, and three sets of 16 integers representing up to 16 port rotations per SFE. The developed script was coded and tested to be fully parallelized and also scalable with respect to the number of SFEs and ports. The material sections at each SFE level were each represented by a set of 220 by 220 pixels having values of 0 or 1; this size of 220 was selected as the product of 3, 4, 4, and 5 to provide reasonable image resolution with minimal computing resources.

Figure 6 provides example results for the Bayesian optimization (BO, at left) and genetic algorithms (GA, at right); the outputs are the same though the two algorithms produce different rotation matrices for the flow channels in the SFEs. The Supplemental Information provides the corresponding convergence plots for each of the solutions. Both the BO and GA methods were able to reproduce the design of Figure 5. However, the number of iterations and CPU time varied drastically. The BO called the SFE processing function only 300 times but consumed significant compute time updating its Gaussian process models and optimizing its acquisition function to determine the next point to evaluate. As a result, the BO method typically required on the order of 1200 s to converge. By comparison, the GA method performs approximately 100 times the number of function evaluations than the BO to support GA's generative iteration. However, given that the image matrix operations are relatively trivial and speedy, the GA method was much faster, typically requiring 20 s to converge.

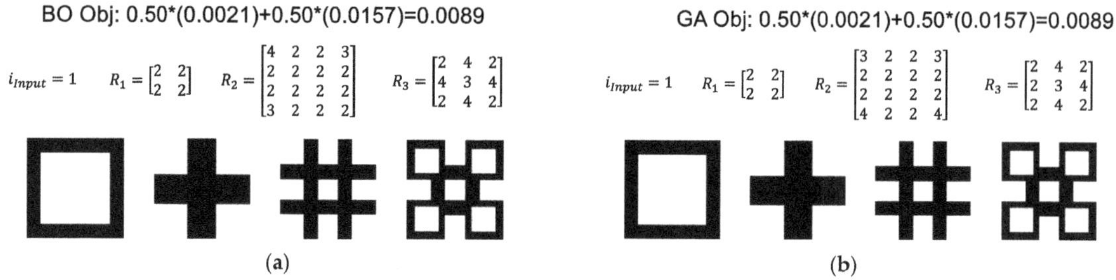

Figure 6. Solutions automatically generated to minimize G_1 to match the target "boxes" image through (**a**) Bayesian optimization and (**b**) genetic algorithms. Corresponding objective function values are respectively indicated.

Turning to the G_2 objective to minimize the structural similarity index with a white square, the same script *SFE_Opt*() was implemented with the results of Figure 7 for (top) square and (bottom) non-square ports. In the latter case, the non-square ports were modeled by augmenting the design vector, x, to allow the number of ports in the horizontal and vertical directions to vary such that the number of design parameters increased from 52 to 55. The modeling of non-square input and output ports was performed by mapping the partition of each input material using Matlab's meshgrid() and then interpolating the cross-section to the output shape using Matlab's interp2() function. This interpolation can result in non-binary material values (e.g., 0.5) depending on the size and number of the rotations as well as the size of the modeled material matrix.

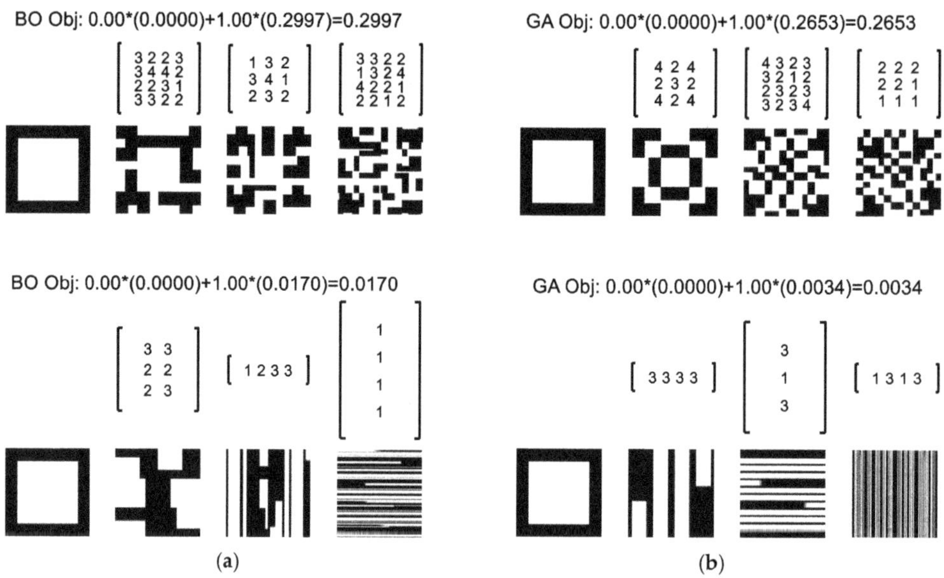

Figure 7. Solutions automatically generated to minimize G_2 (structural similarity index) with (**a**) Bayesian optimization and (**b**) genetic algorithms using (**top**) square and (**bottom**) rectangular ports.

Figure 7 illustrates a black and white pictorial representation of the successive architected material cross-sections resulting from the SFE design matrices provided above each section. The primary conclusion with respect to Figure 7 is that the relaxation of the square port assumption provides a significant improvement in the objective function; this result is not dissimilar to the finding of Li, Dul, and Kim [38] that material performance improved when a larger number of material microstructures is allowed in optimization. The secondary conclusion is that the GA method again outperformed the BO method for both the square and non-square problems.

To further investigate the use of the BO method, the hierarchical implementation of Figure 2 was also implemented to minimize the G_2 objective assuming square ports. In this implementation, each of the input types is evaluated relative to the SSIM. The lowest SSIM is then adopted for further processing with the SFEs with a varying number of ports (1×1, 2×2, 3×3, and 4×4). Compared to the concurrent implementation of Figure 3, the hierarchical method greatly reduces the size of the design vector, x, to exactly the number of parameters needed for each optimization stage. The results are shown in Figure 8 wherein the different rows correspond to different SFE stages, the number of ports in each SFE stage is presented from 1×1 at left to 4×4 at right, and the green lines correspond to the selected solution at each step in the hierarchy. The results show that the co-extruded square and inverted square have the same G_2 relative to a white target image; this result is expected given their similar structure. The bi-layer and tri-layer have higher G_2 values so the first input type is accepted. The optimization of the first SFE tests 1 by 1, 2 by 2, 3 by 3, and 4 by 4 sets of fluid ports as depicted with the columns going from left to right. The 3 by 3 is found to provide the lowest G_2 value and so is adopted. The analysis then proceeds to the second and third SFE stages that respectively result in the selection of a 4 by 4 and then 3 by 3 stage.

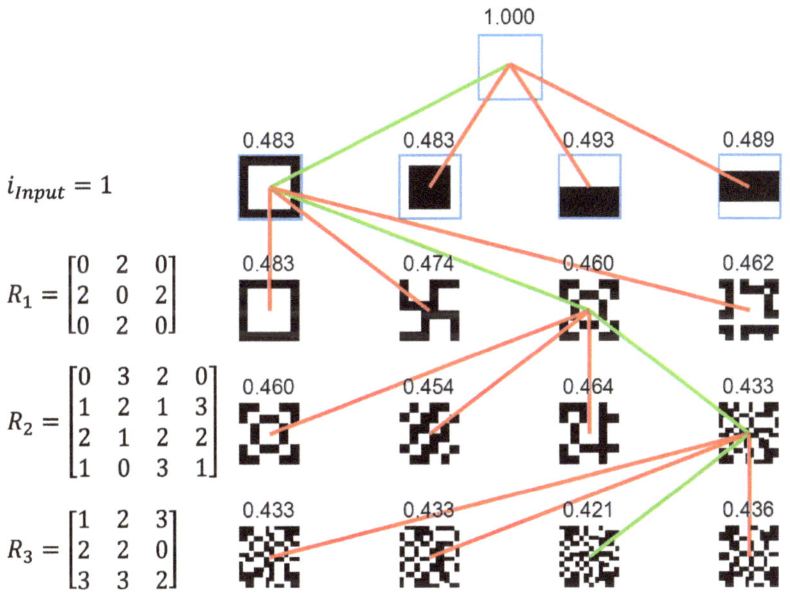

Figure 8. Hierarchical Bayesian optimization of the G_2 objective showing each stage of the optimization with accepted (green) design and inferior (red) solutions.

In Figure 8, each SFE stage provides a reduction in the G_2 value. However, the hierarchical adoption results in a suboptimal solution relative to the concurrent selection of all design parameters, x, yielding the results at top in Figure 7 (BO: $G_2 = 0.300$ and GA: $G_2 = 0.265$). As previously discussed, the performance of the hierarchical solution is limited since the optimal value at an upstream stage does not guarantee a globally optimal result given the drastic changes in the architecture's topology associated with downstream partitions and rotations. It may be that a closed-form (algebraic) solution or an inverse solution methodology (designing from the output back to the input) exists that can efficiently lead to globally optimal solutions.

3.3. Validation

Figure 9 provides the validation results for the G_1 "boxes" objective including the design intent on the left, sections from physical prototyping in the center, and cut plots from multi-phase flow simulation on the right. The top row of Figure 9 shows that the initial condition is reasonably achieved with the physical prototype though there is some rounding as well as variation in the wall thickness of the outer material around the perimeter. By comparison, the flow simulation predicts a more octagonal shape with minor rounding of the material interfaces.

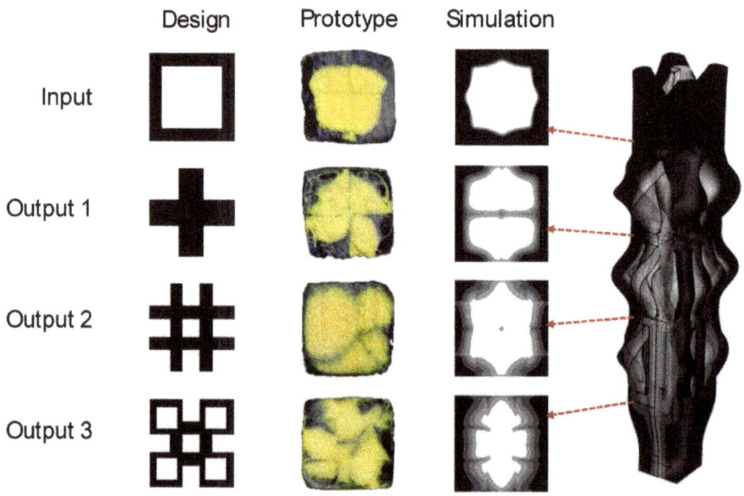

Figure 9. Validation results for the G_1 objective including design intent on the **left**, physical prototyping in the **center**, and multi-phase flow simulation on the **right**.

The second row corresponds to the output of the first SFE in which each flow channel in a 2 × 2 grid is rotated by 90 degrees. The physical prototype shows inconsistent results in that the flows of the top two (upper) quadrants are not significantly altered from their input state, while the bottom two (lower) quadrants clearly demonstrate some rotation in the material distribution. However, the bottom two quadrants undergo only about 45° rather than the desired 90° of rotation. Meanwhile, the flow simulation predicts a minor distribution between the center and the edges, resulting in an hourglass shape. The lack of accuracy in the flow simulation is troubling, since it precludes model-based optimization. Regardless, both the prototype and simulation results suggest that more deformation is needed to achieve the desired output; this increased deformation could be achieved by making the flow channels longer with further rotation.

The third and fourth rows of Figure 9 provide the prototyped and simulated results for the second and third SFEs, respectively. It is clear that control of the material distribution is lacking. In the prototyped material, the proportion of flow from the top-left quadrant is significantly greater than the other quadrants. This result is surprising since the flow channels in the SFEs are similarly sized and fed by fully symmetric feed channels. Regardless, the prototyped results clearly show the incursion of the outer material into the core with multiple material interfaces. By comparison, the flow simulation underpredicts the amount of flow deformation and the extent of the material interfaces observed in the physical system. Regarding gradation, the prototyped material showed little gradation between the two materials while the flow simulation suggests significant gradation as the flow progressed.

Figure 10 provides the validation results for the G_2 objective to maximize the information content of the material architecture (minimize structural similarity index to a white square) including the design intent on the left, sections from physical prototyping in the center, and cut plots from multi-phase flow simulation on the right. The nature of the results in Figure 10 is similar to those discussed for the results in Figure 9. The prototyped results again clearly show the incursion of the outer material into the core with multiple material interfaces. By comparison, the flow simulation again underpredicts the amount of flow deformation and the extent of the material interfaces observed in the physical system.

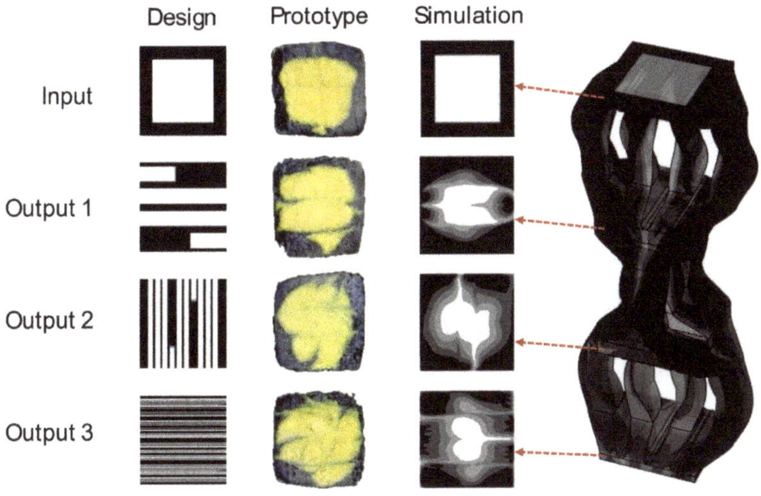

Figure 10. Validation results for the G_2 objective including design intent on the **left**, physical prototyping in the **center**, and multi-phase flow simulation on the **right**.

3.4. Structural Similarity Analysis of Resultant Sections

The developed cross-sections across different stages of the prototyped and simulated manufacturing process are compared to the intended designs using the structural similarity index (SSIM). This methodology involves preprocessing images to a uniform size of 300 × 300 pixels and converting them into grayscale to ensure consistency in comparison. The SSIM is then calculated between the images categorized into three paired comparisons: Design to Prototype (D to P), Design to Simulation (D to S), and Prototype to Simulation (P to S).

The results from this analysis for the images of Figure 9 are provided in Table 1. The SSIM values for the inlet ($i = 0$) are around 0.8 across all comparisons, indicating a reasonable translation of the design into both prototype and simulated forms. However, in the subsequent SFE stages ($i = 1$ through 3), there is a notable decline in the SSIM values. After the first SFE ($i = 1$), the SSIM values drop significantly to 0.1924 (D to P) and 0.1950 (D to S), though there is a moderate similarity between the prototype and simulation (0.4893). The similarity values are fairly stable after the second SFE stage with a slight decrease in similarity between the prototype and simulation (0.4364) compared to the first SFE stage. After the third SFE, there is a slight improvement in all SSIM values though more similarity between the prototype and simulation than with the intended design.

Table 1. Structural similarity indices for results of Figure 9 by image comparison.

SFE Stage	Design to Prototype	Design to Simulation	Prototype to Simulation
0 (inlet)	0.7547	0.7949	0.7875
1	0.1924	0.1950	0.4893
2	0.2002	0.1965	0.4364
3 (outlet)	0.3175	0.3456	0.4418

The structural similarity indices were also calculated for the images of Figure 10 in which the objective was to minimize the SSIM relative to white space. The results are provided in Table 2. As the material progresses through subsequent SFE stages, a marked decline is again observed in the SSIM values (0.2602, 0.3944, 0.3813 in the second set; 0.1873, 0.1958, 0.4513 in the third; and 0.0287, 0.0230, 0.3902 in the fourth), underscoring a growing discrepancy between the expected and actual results.

Table 2. Structural similarity indices for results of Figure 10 by image comparison.

SFE Stage	Design to Prototype	Design to Simulation	Prototype to Simulation
0 (inlet)	0.7547	0.7949	0.7875
1	0.2602	0.3944	0.3813
2	0.1873	0.1958	0.4513
3 (outlet)	0.0287	0.0230	0.3902

The consistent trend in both sets of results suggests a fundamental issue with the material processing stages not adhering to the design assumptions, leading to a decrease in predictive accuracy as the material undergoes further processing. The consistent reduction in similarity across these stages highlights the challenges in maintaining design integrity during material transformation processes. It underscores the need for refining our modeling and simulation approaches to better account for the complexities and variabilities inherent in material processing, aiming to enhance the fidelity and reliability of the outcomes at each stage of development as subsequently discussed.

4. Discussion

4.1. Optimization of Material Architecture

The use of SFEs can enable new material architectures with enhanced properties. This article described examples with objectives for (1) matching a pre-defined cross-section and (2) maximizing the negentropy relative to a uniform material. However, none of these examples claimed to be optimal, and a framework for establishing optimal architecture on an application-specific basis needs to be developed. There is significant related research such as the handling of objectives and constraints in material architectures [39,40], consideration of interfacial properties [41,42], leveraging gradient-based property distributions [43,44], utility and cost of material architectures [45,46], nonlinear material properties as a function of their processing history [47–49], gradation of the materials as a function of processing [50,51], and sustainability concerns [45,47], among others. Theoretically, a fundamental goal is to prove the optimality of an architected material composite for a given application. More practically, the goal is to ensure acceptable robust performance, i.e., fitness for use given uncertainty in material properties and end-use requirements [52,53]. Finite element analysis of the architected composites can be employed to provide model-based topology optimization of the stresses and structural performance.

4.2. Coding and Decomposition

Both the BO and GA optimizations were implemented with the design vector, x, of Equation (13) for square and rectangular flow ports without a shifting function. In an efficient implementation, the number of design parameters in this vector is equal to the number of ports in each SFE stage such that an SFE with a 2 by 2 set of ports would have four parameters, while an SFE with a 4 by 4 set of ports would have 16 parameters. In the current implementation, however, the design vector comprised a fixed number of parameters to accommodate the maximal number of ports. In other words, the rotation matrices for all the SFEs are sized to (the maximum number of ports per side)^2, e.g., 4^2 or 5^2.

This coding has the effect of not only increasing the dimensionality of the optimization problem but also confounding the trialed changes to the dummy design parameters alongside those of the truly significant design parameters. Consider the G_1 target matching example of Figure 6. If the design parameters representing the input cross-section and number of ports per SFE stage are explicitly set as $x_{1:4} \rightarrow \{1, 2, 4, 3\}$ and removed from the optimization problem, then the number of design parameters in the matrix x is reduced from 1 (for the type of input material cross-section) plus 3 (for the number of ports per side for each SFE) plus 3×4^2 (for three SFEs each having up to 4 by 4 ports to rotate), which equals 52 to $2^2 + 3^2 + 4^2 = 29$. This change results in a $4^{(52-29)}$ or 7.04×10^{13} reduction in the number of combinations to be explored. While the GA was robust with respect to converging to an acceptable solution with reasonable compute time, both implementations would benefit through the use of improved coding to provide the minimal number of design parameters and a more efficient design search. The number of design parameters could also be reduced through the implementation of logic supporting symmetry conditions.

Such improvements would facilitate the application of SFEs to a higher number of stages, greater number of ports per stage, shifting of ports, and other relaxations made to the current set of assumptions. More generally, however, is the need for architecture decomposition into a minimal number of stages with a minimal number of ports while also relaxing the assumptions of 1:1 mapping between the input and output ports as well as the narrow use of square or rectangular ports. While these assumptions were made to demonstrate the SFE concept and methodology, much more capable systems are envisioned with further modeling and AI advances. Badini et al. [54] provide some AI-inspired approaches that are closely related to the architectures that may be realized with SFEs.

4.3. Shape Forming Element Design

A square grid of input and output ports for each SFE was assumed for the initial exploration of the SFE concept. While this square port assumption was relaxed in the second example investigating the G_2 objective (minimal SSIM) with rectangular ports, the broader objective of the shape forming elements is to operate with more general transforms mapping the input to output for each SFE to create complex material architectures. The input and output ports can be designed with other shapes (triangular, circular, and non-convex), non-bijective connections (two inlets to one outlet, one inlet to two outlets, etc.), and intermediate flow channel operations to provide custom transforms between the input and output sections. Figure 11 provides some example transforms, including their preliminary flow channel designs and predicted material distributions.

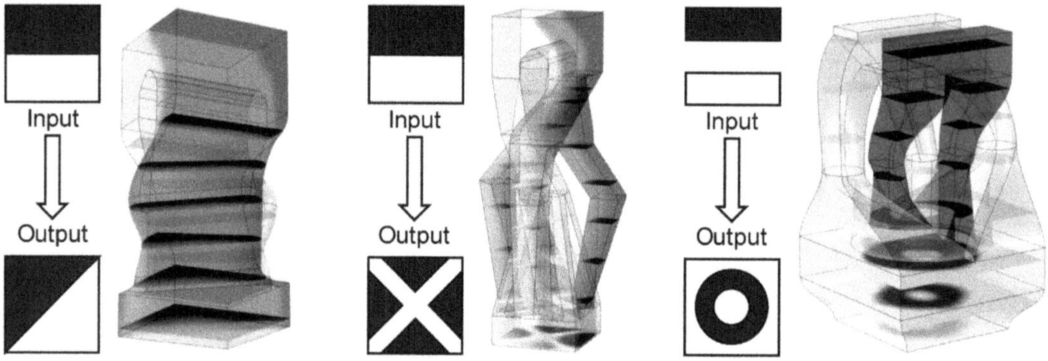

Figure 11. Examples of custom input-output transforms beyond square and rectangular mappings.

The design of such custom transforms can be accomplished manually, but the work is non-trivial and tedious. The triangular transform (on the left in Figure 11) was first attempted with a simple loft connecting the rectangular input and triangular output. The resulting material distribution in the output section was highly circular and concentrated at the corners with the shortest flow path. The material distribution with the diagonal transform design was improved by implementing a first loft to an intermediate slot to distribute the flows across the hypotenuse of the triangle and then a second loft to the triangular output. The "X" design (middle of Figure 11) is the third design iteration with a routing of four black triangular legs around the white core material. The design still needs additional flow balancing to achieve a more uniform output material distribution. The "bullseye" design at right was the easiest to design and provided the desired output with three nested flow layers (innermost white circle, intermediate black annulus, and outer white annulus).

More generally, design automation of the flow geometry to create these material transforms is a critical extension that can be enabled by modeling and artificial intelligence including consideration manufacturing constraints. For a given input–output transform, the routing, cross-section shapes, and associated design parameters can be determined using the methods similar to the described BO or GA tied to flow simulation for performance prediction; each of the flow simulations shown in Figure 11 required approximately 300 s of compute time and could be readily integrated into a design system. The processing performance measures in the objective function should include at least the output material distribution and its uniformity of flow rates. Constraints related to the pressure drop, minimum flow channel dimensions, and minimum wall stock between flow channels would ensure manufacturability and processability. The minimum wall stock is critical to the creation of physically robust SFEs that can withstand the stresses imparted by the flow material during manufacturing. In practice, no failure has yet been observed in polymer or metal prototypes with a 1 mm minimum wall thickness during validation experiments. In the future, an application programming interface (API) to 3D CAD could be used to directly build the SFEs with final expert approval.

5. Conclusions

The concept of shape forming elements (SFEs) was described for the first time, aiming to provide a framework for the design of architected composites for industrial applications. The enablement of high-throughput manufacturing techniques for architected composites presents significant opportunities to enhance material properties through innovative material and process design strategies. Specifically, shape forming elements (SFEs) provide a systematic approach to the intelligent manufacturing of architected materials with high production rates, low cost, and a low carbon footprint.

The method was demonstrated for square and rectangular ports with a 1:1 mapping of inputs to outputs. Design automation by Bayesian optimization and genetic algorithms provided reasonable solutions for multi-objective optimization with a design vector on the order of 50 parameters. The provided results shows that the shape forming elements enable a wide array of architectures driven by the orientation and shifting of the constitutive materials within each SFE stage. Relaxation of the assumptions in the current implementation will lead to the more precise control needed to achieve more complex and capable composite architectures. While the current work demonstrated rotation, shifting, and limited shaping between rectangular ports, a more complex reshaping of input and output sections is straightforward to model but needs more advanced integration with higher level systems optimization to achieve improved material architectures. Other SFE functions such as copying and deleting would be desirable and can be computationally modeled but are difficult to practically implement in simple physical systems.

The use of shape forming elements could represent a significant step forward in the design of new material systems that combine light weight with high performance, which are increasingly demanded in industries ranging from aerospace to biomedical engineering. However, the results clearly show the need for further research to identify the optimal architected designs and methods for their manufacture.

Supplementary Materials: The following supporting information can be downloaded at https://www.mdpi.com/article/10.3390/ma17215339/s1. (1) The Matlab script Design_SFE.m to manually design shape forming elements; see the header for directions. (2) The Matlab script SFE_Opt.m to automate SFE design using Bayesian optimization or genetic algorithms providing reproduction of the presented results. There are many options and functions implemented therein to explore beyond what is described in the paper. (3) The Matlab script Compare_Results.m and accompanying images for quantitative analysis of the results. (4) The STEP files corresponding to the developed CAD supporting the G_1 and G_2 solutions, respectively, corresponding to the "Boxes" and low structural similarity index solutions that may be used for prototyping and simulation purposes. All information is provided as is without warranty or support. If you do use these, please reference this article.

Author Contributions: Conceptualization, D.C.E., D.O.K. and T.D.N.; methodology, D.O.K., T.D.N. and R.H.O.; software, D.O.K.; validation, D.O.K. and R.H.O.; formal analysis, D.O.K.; investigation, D.O.K.; resources, D.O.K. and T.D.N.; data curation, D.C.E.; writing—original draft preparation, D.O.K.; writing—review and editing, D.O.K., T.D.N. and R.H.O.; visualization, D.O.K.; supervision, D.O.K. and T.D.N.; project administration, T.D.N.; funding acquisition, D.O.K. and T.D.N. All authors have read and agreed to the published version of the manuscript.

Funding: Portions of this work were supported by U.S. Federal Award #FA9550-22-1-0222 "Design and Processing of High-Performance Architected Blends" and National Science Foundation Award #2118808 "DMREF/Collaborative Research: Integrated Material Design and Processing—Application to Recycled Plastics". Any opinions, findings, and conclusions or recommendations expressed in this material are those of the author(s) and do not necessarily reflect the views of the NSF. Any opinions, findings, and conclusions or recommendations expressed in this research are those of the author(s) and do not necessarily reflect the views of the sponsors.

Data Availability Statement: The original contributions presented in the study are included in the article/Supplementary Materials, further inquiries can be directed to the corresponding author.

Conflicts of Interest: The authors declare no conflicts of interest.

Appendix A. Optimization Details

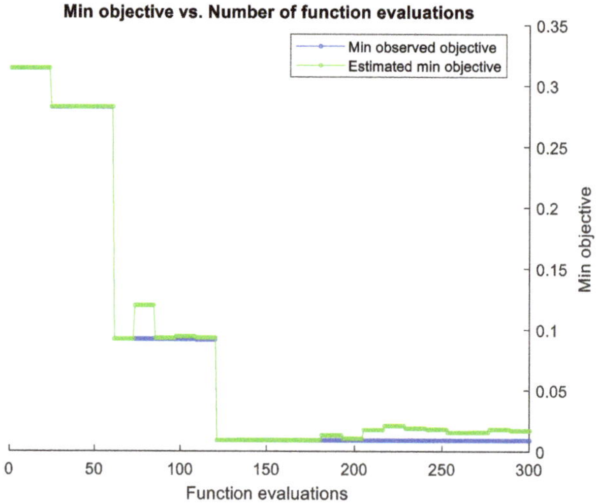

Figure A1. Convergence plot for converged design solution for G1 objective with Bayesian optimization (compute time: 1201 s).

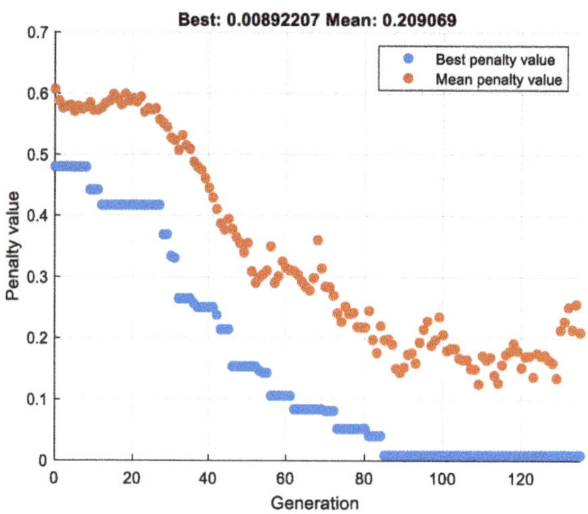

Figure A2. Convergence plot for converged design solution for G1 objective with genetic algorithm (compute time: 21.32 s).

Appendix B. Flow Simulation Details

Transient, non-Newtonian, multi-phase flow simulations using the level set method were performed with Comsol 4.6 using a segregated solver that iteratively solves the fluid flow and level set (material concentration) fields. Materials 0 and 1 (properties indicated in Table A1) were specified at the top inlet ports with a uniform velocity corresponding to a volumetric flow rate of 5 cm^3/s; an atmospheric pressure boundary condition was specified at the outlet. Figure A3 provides the quarter-model design, simulation, and concentration results for a quarter-symmetry model (with symmetric conditions specified for both the flow fields and material levels on the right and front planes). This first model contained

1,465,781 elements and required a solution time of 4080 s, physical memory of 8.82 GB, and virtual memory of 10.74 GB. All simulations were conducted on an AMD Ryzen 9 7900X CPU operating at 4.70 GHz with 24 logical processors.

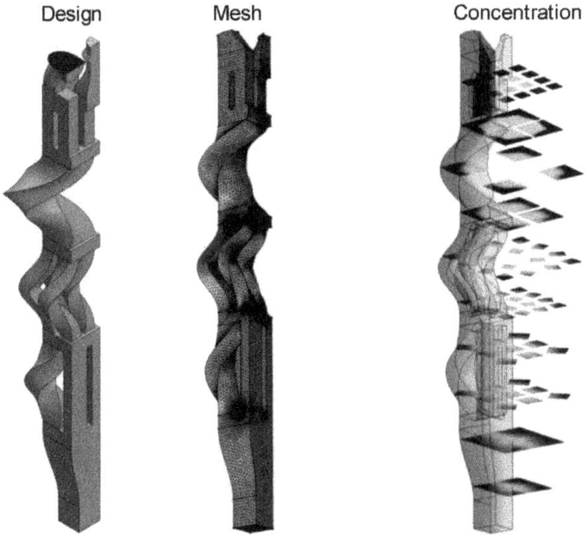

Figure A3. Design, mesh, and concentration results from the Comsol level-set simulation.

To investigate the role of mesh dependence on the development of the developed material flow fields, a higher fidelity mesh was created by changing the Comsol mesh setting from "Fine" to "Finer." This mesh refinement resulted in an increase in the number of tetrahedral elements to 3,703,025. The simulated processing time was also increased from 20 to 30 s to ensure a more fully developed flow field. These changes resulted in an increase in the run time to 30,195 s as well as an increase in the physical and virtual memory to 16.62 and 20.07 GB, respectively. The material concentration results for the different levels of mesh refinement are provided in Figure A4. The results suggest that the material distribution is more refined with additional mesh refinement but still far from predicting the material behavior observed via extrusion with the physical prototype.

Similar mesh refinement was also conducted for the second example of Figure 10 in which the maximum edge length of the tetrahedral elements was reduced from 0.2 to 0.1 mm. This resulted in a mesh having 6,787,935 that required 52,926 s with physical and virtual memory of 28.21 and 33.78 GB, respectively. Furthermore, to evaluate the possibility that the extruded materials are not shear thinning as modeled with the described Williamson equation, materials 0 and 1 were both modeled as Newtonian fluids with respective viscosities of 162 and 582 Pa.s. The results are shown in Figure A5 and suggest a similar refinement of the flow behavior during material processing with no additional discernible effect of the change in the viscosity model relative to the impact of mesh refinement previously demonstrated in Figure A4.

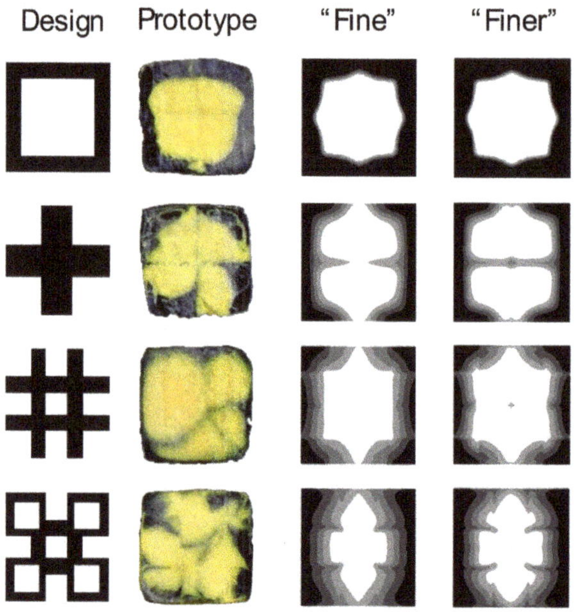

Figure A4. Design, prototype results, and "fine" and "finer" flow simulation results for the example of Figure 9.

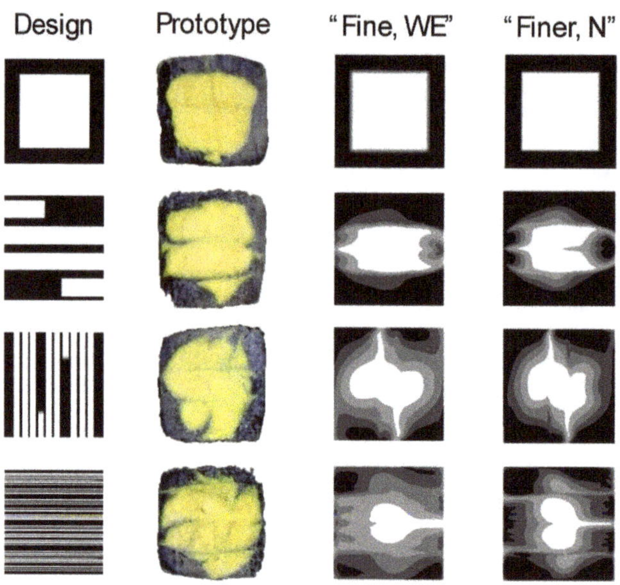

Figure A5. Design, prototype results, and "Fine, WE" and "Finer, N" flow simulation results for the example of Figure 10 wherein "WE" represent the non-Newtonian fluid model with the Williamson equation while "N" represents the Newtonian viscosity model.

The solver convergence is shown in Figure A6 for the "Finer, N" condition with all three modeled SFEs and the die filled with material 0 initially at rest. The perturbations at iterations 100 and 150 correspond to the movement of the level set interface between materials 0 and 1 traveling through the latter SFE flow channels. The pressure history as a function of time is shown in Figure A7. The early fluctuations are associated with the travel of the level set interface through the flow channels in the SFEs with the propagation of material 1 that has higher viscosity (582 Pa·s) than the material 0 (162 Pa·s). The pressure plateau after 15 s suggests that the material interfaces depicted in Figure 10 are fully developed as simulated.

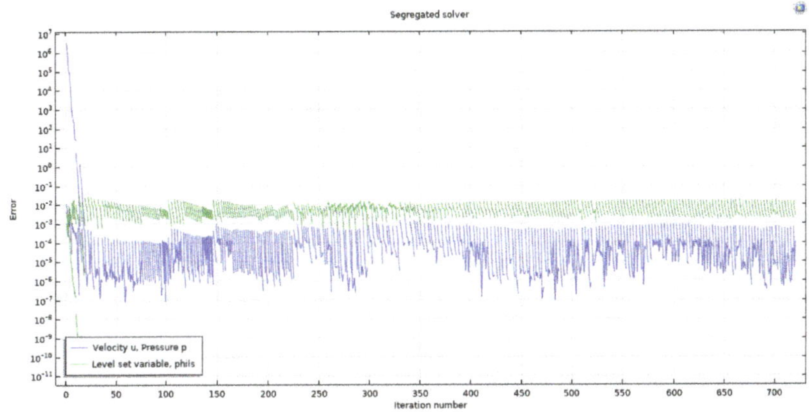

Figure A6. Convergence plot for Comsol flow simulation of SFE design example #2 (Figure 10).

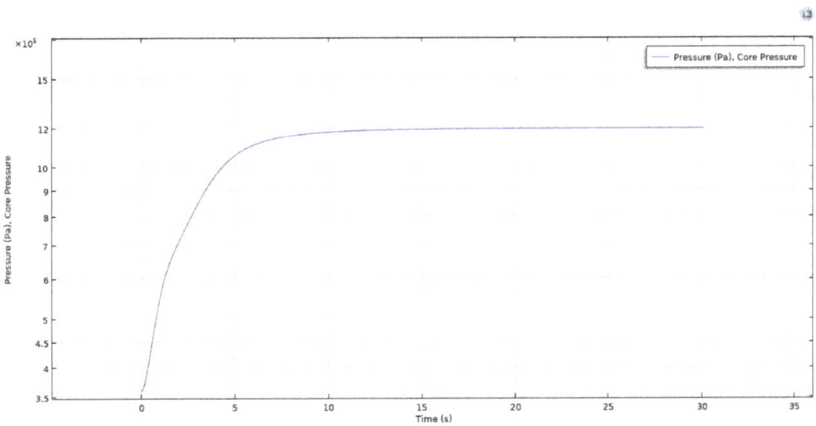

Figure A7. Pressure history plot for Comsol flow simulation of SFE design example #2 (Figure 10).

Table A1. Flow simulation material data for materials 0 and 1.

Material Property	Material 0	Material 1
MFR (g/10 min, 2.16 kg)	8.63 ± 0.05	5.67 ± 0.12
Melt density, kg/m^3	850	850
Willamson, zero-shear viscosity, η_0(230 °C) [Pa]	1127	1935
Willamson, power-law index, n	0.249	0.283
Willamson, characteristic relaxation time, λ [s]	0.0232	0.0437

Appendix C. Description of Matlab Functions

Bayesopt: The bayesopt function in Matlab is a statistical technique for optimizing nonlinear objective functions that are expensive to evaluate. It operates by constructing a probabilistic model of the objective function and iteratively selecting the most promising data points to evaluate based on current estimates. The function integrates prior belief about the objective function and updates the model as new data are gathered, balancing exploration of the design space with exploitation of known good regions. This approach is particularly advantageous for optimizing complex material architectures where the direct evaluation of every design configuration is computationally prohibitive.

Compare_Results: The Compare_Results.m Matlab script was written for this article to compare the images from various stages of material processing experiments, specifically for Design (D), Prototype (P), and Simulation (S). This script automates the loading, processing, and comparison of images corresponding to these stages using the structural similarity index (SSIM) to quantitatively assess their similarity. The script can handle both grayscale and black/white images with a Boolean option (bGray) to toggle between these modes based on the analysis needs; each image is resized to a standard dimension (300×300 pixels) to ensure consistency in comparison.

Design_SFE: The Design_SFE.m function is a Matlab script authored in this research that is designed to create and visualize shape forming elements (SFEs) with configurable square port rotations for material distribution in architected composites. The function includes a graphical user interface (GUI) that allows users to interactively set the number of ports per stage and adjust the rotation of each port using keyboard inputs. The script initializes a figure with controls to navigate through stages and modify port configurations, supporting the design of complex flow channels in multi-layered structures. This tool is built to aid in the visualization and manual adjustment of SFE configurations, enabling a hands-on approach to designing optimized material architectures for various manufacturing processes.

Ga: The ga function in Matlab implements genetic algorithms (GAs), which is a class of optimization techniques inspired by natural selection that are effective for solving complex optimization problems with large solution spaces. This function works by generating a population of candidate solutions, which are then subjected to operations such as selection, crossover, and mutation to evolve the population toward better solutions over successive generations. Each candidate is evaluated based on a fitness function, and the best-performing individuals are more likely to contribute to the next generation. Genetic algorithms are particularly useful for exploring high-dimensional design spaces and finding global optima in problems where traditional gradient-based optimization methods might fail due to nonlinearity or discontinuities.

Immse: The immse function in Matlab is an acronym for IMage Mean Squared Error, which is used to compute the mean squared error between two images. This function is typically employed to quantify the difference between a reference image and a test image, providing a measure of the distortion or error introduced during image processing or transmission. The mean squared error calculated by immse reflects the average of the squares of the pixel intensity differences between corresponding pixels of the two images. It is a critical tool in image analysis, and it is particularly useful in optimization routines for image processing where reducing the error metric is a common objective, such as in the manuscript's context for validating composite architectures.

Interp2: The interp2 function in Matlab performs two-dimensional interpolation on matrices, which is used for estimating intermediate values between discrete data points in a grid. This function can accept various methods of interpolation such as linear (used in this research), cubic, or spline, allowing users to choose the one that best suits their application needs. In this application, it supports the creation of smooth representations of the material concentrations in matrices of varied sizes and aspect ratios as discussed in the context of transforming material cross-sections in the manuscript.

Meshgrid: The meshgrid function in Matlab is used to create two 2D arrays containing the coordinates of points on a grid, where one array holds the x-coordinates and the other holds the y-coordinates. It is a method for mapping functions over a grid, enabling vectorized operations on matrices that represent points across a meshed domain. This function is used to support the interpolation of material distributions when mapping material concentrations in the partition of the output from one SFE to an input partition of the next SFE and they have different pixel sizes.

Rot90: The rot90 function in Matlab rotates a matrix or an array by 90 degrees counterclockwise. This function is used to manipulate image data or reorient matrices to align with different coordinate systems or perspectives. A second argument specifies the number of times the matrix should be rotated by 90 degrees, enabling flexibility in achieving the desired orientation. For example, the function call rot90(A,2) would rotate the matrix A by 180 degrees.

SFE_Opt: The SFE_Opt.m function is a Matlab script authored for this research that optimizes shape forming elements (SFEs) through integer-valued Bayesian optimization (BO) and genetic algorithms (GAs). This script is designed to optimize the design of SFEs used in creating architected material composites by adjusting parameters such as port shapes, sizes, and rotations to achieve specific performance objectives. Users can configure the optimization process through Boolean options to choose between using Bayesian optimization or genetic algorithms, specify the shape of ports (square or non-square), and decide on varying input types which directly affect the degrees of freedom in the optimization problem. Key options in the script allow users to penalize the number of ports, append results to an archive for later review, and display cross-sectional views and objective values during optimization runs. The script also accommodates non-rectangular ports and includes placeholders for future functionalities like solving from the bottom up or sequentially through stages. Different objectives functions are also implemented including placeholders and trial functions related to mechanical properties.

SSIM: The ssim function in Matlab calculates the structural similarity index (SSIM) between two images, which is a method for measuring the similarity between two images. SSIM is a perception-based model that considers image degradation as a perceived change in structural information, incorporating aspects of luminance, contrast, and structure. The SSIM function returns a decimal value between 0 and 1 that, respectively, indicates no and perfect similarity.

References

1. White, A. The materials genome initiative: One year on. *MRS Bull.* **2012**, *37*, 715–716. [CrossRef]
2. De Pablo, J.; Jackson, N.E.; Webb, M.; Chen, L.-Q.; Moore, J.; Morgan, D.; Pollock, T.; Schlom, D.; Toberer, E.; Analytis, J. Advancing and Accelerating Materials Innovation: New Frontiers for the Materials Genome Initiative. 2019. Available online: https://www.nist.gov/publications/advancing-and-accelerating-materials-innovation-new-frontiers-materials-genome (accessed on 10 August 2024).
3. Schaedler, T.A.; Carter, W.B. Architected cellular materials. *Annu. Rev. Mater. Res.* **2016**, *46*, 187–210. [CrossRef]
4. Seharing, A.; Azman, A.H.; Abdullah, S. A review on integration of lightweight gradient lattice structures in additive manufacturing parts. *Adv. Mech. Eng.* **2020**, *12*, 1687814020916951. [CrossRef]
5. Li, T.; Chen, Y.; Hu, X.; Li, Y.; Wang, L. Exploiting negative Poisson's ratio to design 3D-printed composites with enhanced mechanical properties. *Mater. Des.* **2018**, *142*, 247–258. [CrossRef]
6. Sinha, P.; Mukhopadhyay, T. Programmable multi-physical mechanics of mechanical metamaterials. *Mater. Sci. Eng. R Rep.* **2023**, *155*, 100745. [CrossRef]
7. Lakes, R. Materials with structural hierarchy. *Nature* **1993**, *361*, 511–515. [CrossRef]
8. Kladovasilakis, N.; Tsongas, K.; Karalekas, D.; Tzetzis, D. Architected materials for additive manufacturing: A comprehensive review. *Materials* **2022**, *15*, 5919. [CrossRef] [PubMed]
9. Lee, B.J.; Hsiao, K.; Lipkowitz, G.; Samuelsen, T.; Tate, L.; DeSimone, J.M. Characterization of a 30 μm pixel size CLIP-based 3D printer and its enhancement through dynamic printing optimization. *Addit. Manuf.* **2022**, *55*, 102800. [CrossRef]
10. Toombs, J.T.; Luitz, M.; Cook, C.C.; Jenne, S.; Li, C.C.; Rapp, B.E.; Kotz-Helmer, F.; Taylor, H.K. Volumetric additive manufacturing of silica glass with microscale computed axial lithography. *Science* **2022**, *376*, 308–312. [CrossRef] [PubMed]
11. Kazmer, D.; Peterson, A.M.; Masato, D.; Colon, A.R.; Krantz, J. Strategic cost and sustainability analyses of injection molding and material extrusion additive manufacturing. *Polym. Eng. Sci.* **2023**, *63*, 943–958. [CrossRef]

12. Kokare, S.; Oliveira, J.; Godina, R. Life cycle assessment of additive manufacturing processes: A review. *J. Manuf. Syst.* **2023**, *68*, 536–559. [CrossRef]
13. Lai, Y.C.; Yang, C.C.; Levon, J.A.; Chu, T.M.G.; Morton, D.; Lin, W.S. The effects of additive manufacturing technologies and finish line designs on the trueness and dimensional stability of 3D-printed dies. *J. Prosthodont.* **2023**, *32*, 519–526. [CrossRef] [PubMed]
14. Shanmugam, V.; Das, O.; Babu, K.; Marimuthu, U.; Veerasimman, A.; Johnson, D.J.; Neisiany, R.E.; Hedenqvist, M.S.; Ramakrishna, S.; Berto, F. Fatigue behaviour of FDM-3D printed polymers, polymeric composites and architected cellular materials. *Int. J. Fatigue* **2021**, *143*, 106007. [CrossRef]
15. Blakey-Milner, B.; Gradl, P.; Snedden, G.; Brooks, M.; Pitot, J.; Lopez, E.; Leary, M.; Berto, F.; Du Plessis, A. Metal additive manufacturing in aerospace: A review. *Mater. Des.* **2021**, *209*, 110008. [CrossRef]
16. Benedetti, M.; Du Plessis, A.; Ritchie, R.; Dallago, M.; Razavi, N.; Berto, F. Architected cellular materials: A review on their mechanical properties towards fatigue-tolerant design and fabrication. *Mater. Sci. Eng. R Rep.* **2021**, *144*, 100606. [CrossRef]
17. Schrenk, W.; Alfrey, T., Jr. Coextruded multilayer polymer films and sheets. In *Polymer Blends*; Academic Press: Cambridge, MA, USA, 1978; pp. 129–165. [CrossRef]
18. Li, Z.; Olah, A.; Baer, E. Micro-and nano-layered processing of new polymeric systems. *Prog. Polym. Sci.* **2020**, *102*, 101210. [CrossRef]
19. Li, M.; Liu, W.; Zhang, F.; Zhang, X.; Omer, A.A.A.; Zhang, Z.; Liu, Y.; Zhao, S. Polymer multilayer film with excellent UV-resistance & high transmittance and its application for glass-free photovoltaic modules. *Sol. Energy Mater. Sol. Cells* **2021**, *229*, 111103.
20. Yu, G.; Ji, Y.; Qin, J.; Hong, W.; Li, C.; Zhang, G.; Wu, H.; Guo, S. Producing microlayer pipes and tubes through multiplication coextrusion and unique annular die: Simulation and experiment. *Ind. Eng. Chem. Res.* **2021**, *60*, 18408–18420. [CrossRef]
21. Schneider, T.; Danda, C.; Ling, G.; Colton, M.F.; McCauley, K.M.; Maia, J. Microlayer and nanolayer tubing and piping via layer multiplication coextrusion. I. Validation. *J. Appl. Polym. Sci.* **2020**, *137*, 48683. [CrossRef]
22. Neerincx, P.E.; Hofmann, M.; Gorodetskyi, O.; Feldman, K.; Vermant, J.; Meijer, H.E. One-step creation of hierarchical fractal structures. *Polym. Eng. Sci.* **2021**, *61*, 1257–1269. [CrossRef]
23. UEHARA, R. Computational complexity of puzzles and related topics. *Interdiscip. Inf. Sci.* **2023**, *29*, 119–140. [CrossRef]
24. Arróyave, R.; McDowell, D.L. Systems approaches to materials design: Past, present, and future. *Annu. Rev. Mater. Res.* **2019**, *49*, 103–126. [CrossRef]
25. Brochu, E.; Cora, V.M.; De Freitas, N. A tutorial on Bayesian optimization of expensive cost functions, with application to active user modeling and hierarchical reinforcement learning. *arXiv* **2010**, arXiv:1012.2599.
26. Wang, Z.; Bovik, A.C.; Sheikh, H.R.; Simoncelli, E.P. Image quality assessment: From error visibility to structural similarity. *IEEE Trans. Image Process.* **2004**, *13*, 600–612. [CrossRef]
27. Goldberg, D.E. *Genetic Algorithms in Search, Optimization and Machine Learning*; Addison-Wesley: Boston, MA, USA, 1989.
28. Conn, A.; Gould, N.; Toint, P. A globally convergent Lagrangian barrier algorithm for optimization with general inequality constraints and simple bounds. *Math. Comput.* **1997**, *66*, 261–288. [CrossRef]
29. Murodillaevich, N.F.; Amanbayevna, A.S.; Mirzayevich, N.O.; Sharifovich, N.I. Methods and algorithms of geometric modeling of fractal structured objects. In Proceedings of the 2021 International Conference on Information Science and Communications Technologies (ICISCT), Tashkent, Uzbekistan, 3–5 November 2021; pp. 1–5.
30. Brillouin, L. The negentropy principle of information. *J. Appl. Phys.* **1953**, *24*, 1152–1163. [CrossRef]
31. Gibbs, J.W. A method of geometrical representation of the thermodynamic properties by means of surfaces. *Trans. Conn. Acad. Arts Sci.* **1873**, 382–404.
32. Lahaie, R.G.; Hansen, C.J.; Kazmer, D.O. Development of Fused Deposition Modeling of Multiple Materials (FD3M) Through Dynamic Coaxial Extrusion. *3D Print. Addit. Manuf.* **2023**, *11*, 485–495. [CrossRef]
33. Marnot, A.; Dobbs, A.; Brettmann, B. Material extrusion additive manufacturing of dense pastes consisting of macroscopic particles. *MRS Commun.* **2022**, *12*, 483–494. [CrossRef]
34. Rojas, C.J.; Portilla, C.; Özkan, L. Model-based feedback control of filament geometry in extrusion-based additive manufacturing. *IFAC-PapersOnLine* **2024**, *58*, 403–408. [CrossRef]
35. Osher, S.; Fedkiw, R.P. Level set methods: An overview and some recent results. *J. Comput. Phys.* **2001**, *169*, 463–502. [CrossRef]
36. Williamson, R.V. The flow of pseudoplastic materials. *Ind. Eng. Chem.* **1929**, *21*, 1108–1111. [CrossRef]
37. Kazmer, D.O.; Nzeh, S.O.; Shen, B.; Elbert, D.C.; Nagarajan, R.; Sobkowicz-Kline, M.; Nguyen, T.D. Characterization, processing, and modeling of industrial recycled polyolefins. *Polym. Eng. Sci.* **2024**, *64*, 4801–4815. [CrossRef]
38. Li, L.; Du, Z.; Kim, H.A. Design of architected materials for thermoelastic macrostructures using level set method. *JOM* **2020**, *72*, 1734–1744. [CrossRef]
39. Li, H.; Luo, Z.; Zhang, N.; Gao, L.; Brown, T. Integrated design of cellular composites using a level-set topology optimization method. *Comput. Methods Appl. Mech. Eng.* **2016**, *309*, 453–475. [CrossRef]
40. Gao, J.; Cao, X.; Xiao, M.; Yang, Z.; Zhou, X.; Li, Y.; Gao, L.; Yan, W.; Rabczuk, T.; Mai, Y.-W. Rational designs of mechanical metamaterials: Formulations, architectures, tessellations and prospects. *Mater. Sci. Eng. R Rep.* **2023**, *156*, 100755. [CrossRef]
41. Huang, S.; Fu, Q.; Yan, L.; Kasal, B. Characterization of interfacial properties between fibre and polymer matrix in composite materials—A critical review. *J. Mater. Res. Technol.* **2021**, *13*, 1441–1484. [CrossRef]

42. Firooz, S.; Steinmann, P.; Javili, A. Homogenization of composites with extended general interfaces: Comprehensive review and unified modeling. *Appl. Mech. Rev.* **2021**, *73*, 040802. [CrossRef]
43. Surendranath, H.; Bruck, H.A.; Gowrisankaran, S. Enhancing the optimization of material distributions in composite structures using gradient architectures. *Int. J. Solids Struct.* **2003**, *40*, 2999–3020. [CrossRef]
44. Taheri, A.H.; Hassani, B.; Moghaddam, N. Thermo-elastic optimization of material distribution of functionally graded structures by an isogeometrical approach. *Int. J. Solids Struct.* **2014**, *51*, 416–429. [CrossRef]
45. Arnold, S.M.; Piekenbrock, M.; Ricks, T.M.; Stuckner, J. Multiscale analysis of composites using surrogate modeling and information optimal designs. In Proceedings of the AIAA Scitech 2020 Forum, Orlando, FL, USA, 6–10 January 2020; p. 1863.
46. Huang, J.; Liew, J.; Ademiloye, A.; Liew, K.M. Artificial intelligence in materials modeling and design. *Arch. Comput. Methods Eng.* **2021**, *28*, 3399–3413. [CrossRef]
47. Nassar, M.M.; Alzebdeh, K.I.; Pervez, T.; Al-Hinai, N.; Munam, A. Progress and challenges in sustainability, compatibility, and production of eco-composites: A state-of-art review. *J. Appl. Polym. Sci.* **2021**, *138*, 51284. [CrossRef]
48. De France, K.; Zeng, Z.; Wu, T.; Nyström, G. Functional materials from nanocellulose: Utilizing structure–property relationships in bottom-up fabrication. *Adv. Mater.* **2021**, *33*, 2000657. [CrossRef] [PubMed]
49. Yuan, S.; Li, S.; Zhu, J.; Tang, Y. Additive manufacturing of polymeric composites from material processing to structural design. *Compos. Part B Eng.* **2021**, *219*, 108903. [CrossRef]
50. Casalotti, A.; D'Annibale, F.; Rosi, G. Optimization of an architected composite with tailored graded properties. *Z. für Angew. Math. und Phys.* **2024**, *75*, 126. [CrossRef]
51. Bhat, C.; Prajapati, M.J.; Kumar, A.; Jeng, J.-Y. Additive Manufacturing-Enabled Advanced Design and Process Strategies for Multi-Functional Lattice Structures. *Materials* **2024**, *17*, 3398. [CrossRef]
52. Nguyen, V.-K.; Pham, H.-T.; Pham, H.-H.; Dang, Q.-K.; Minh, P.S. Reliability-Based Multi-Objective Optimization Design of a Compliant Feed Drive Mechanism for Micromachining. *Appl. Sci.* **2023**, *13*, 4968. [CrossRef]
53. Du, X.; Chen, W. Efficient uncertainty analysis methods for multidisciplinary robust design. *AIAA J.* **2002**, *40*, 545–552. [CrossRef]
54. Badini, S.; Regondi, S.; Pugliese, R. Unleashing the power of artificial intelligence in materials design. *Materials* **2023**, *16*, 5927. [CrossRef]

materials

MDPI

Article

Effects of Different Cooling Treatments on Heated Granite: Insights from the Physical and Mechanical Characteristics

Qinming Liang [1], Gun Huang [1,*], Jinyong Huang [1], Jie Zheng [2,*], Yueshun Wang [1] and Qiang Cheng [1]

[1] School of Resources and Safety Engineering, Chongqing University, Chongqing 400044, China
[2] School of Geology and Mining Engineering, Xinjiang University, Urumqi 830046, China
* Correspondence: hg023@cqu.edu.cn (G.H.); jieuan@163.com (J.Z.)

Abstract: The exploration of Hot Dry Rock (HDR) geothermal energy is essential to fulfill the energy demands of the increasing population. Investigating the physical and mechanical properties of heated rock under different cooling methods has significant implications for the exploitation of HDR. In this study, ultrasonic testing, uniaxial strength compression experiments, Brazilian splitting tests, nuclear magnetic resonance (NMR), and scanning electron microscope (SEM) were conducted on heated granite after different cooling methods, including cooling in air, cooling in water, cooling in liquid nitrogen, and cycle cooling in liquid nitrogen. The results demonstrated that the density, P-wave velocity (V_{P}), uniaxial compressive strength (UCS), tensile strength (σ_{t}), and elastic modulus (E) of heated granite tend to decrease as the cooling rate increases. Notably, heated granite subjected to cyclic liquid nitrogen cooling exhibits a more pronounced decline in physical and mechanical properties and a higher degree of damage. Furthermore, the cooling treatments also lead to an increase in rock pore size and porosity. At a faster cooling rate, the fracture surfaces of the granite transition from smooth to rough, suggesting enhanced fracture propagation and complexity. These findings provide critical theoretical insights into optimizing stimulation performance strategies for HDR exploitation.

Keywords: rock mechanical properties; cooling treatments; heated granite; experimental research

1. Introduction

With the increasing global demand for energy, the development and utilization of sustainable energy resources have emerged as a critical global issue [1–3]. Geothermal energy, recognized as a clean, efficient, and renewable resource, offers considerable potential for development [4–6]. In particular, Hot Dry Rock (HDR) geothermal systems, known for their high temperatures and substantial energy storage deep within Earth's crust, represent a promising geothermal resource [7,8]. It is estimated that the geothermal energy content of HDR is 800 times that of hydrothermal resources and 300 times that of the total energy from fossil fuels [9]. Effective exploitation of HDR resources can not only help alleviate the energy crisis but also reduce environmental pollution and greenhouse gas emissions, offering significant environmental and social benefits. The creation of sufficiently interconnected channels within the reservoir is crucial for commercial-scale heat extraction from HDR geothermal systems [10]. During the geothermal extraction process, the introduction of low-temperature fluids, such as water or liquid nitrogen, into the rock strata cools the high-temperature HDR reservoirs and generates thermal stress, inevitably damaging the reservoirs. This reduces the fracturing stress during hydraulic fracturing and enhances reservoir permeability [11], which significantly impacts the extraction efficiency of HDR [12]. However, different cooling rates have varying impacts on the physical and mechanical properties of high-temperature rocks, thereby affecting the extent of damage. Therefore, a thorough understanding of the effects of cooling rates on the physical and mechanical properties of rocks is particularly crucial for the production efficiency of geothermal engineering.

HDR reservoirs are characterized by high temperatures, which significantly alter the properties of rocks [13–16]. The effects of heating on rock properties have been well-documented, with changes in physical and mechanical properties primarily due to the evaporation of pore water [17–19], decomposition of mineral components [20–23], and transformations in the crystal structures within mineral grains [24–26]. The contact between the high-temperature rocks of HDR reservoirs and external cold media during extraction induces thermal stress due to cooling, thus affecting the rock properties. Water cooling, a widely used method, has been extensively studied in recent years. Wu et al. [27] found that heated granite subjected to water cooling during Brazilian split tests exhibited severe thermal damage, resulting in lower tensile strength (σ_t) compared to samples that were naturally cooling. Xue [28] explored the impact of water cooling on rock crack formation and morphology through the thermal/hydro/mechanical coupling model and hydraulic fracturing experiments on heated granite. The results indicated that water cooling significantly reduces σ_t and elastic modulus (E) of rocks, especially under high temperatures, promoting crack formation and expansion, thus leading to a more complex fracture network. Zhu et al. [29] observed that the density and mechanical strength of granite decreased with increasing cycles of cyclic water cooling, with notable reductions in uniaxial compressive strength (UCS) and E after 30 cycles. Scanning electron microscopy (SEM) results suggested that the significant increase in microcracks was the primary cause of the deterioration in the physical and mechanical properties of granite. Further studies by Cui et al. [30] showed that increasing the number of heating/cooling cycles gradually decreased the P-wave velocity (V_p) of granite, with a reduction in mass and an increase in volume, attributed to structural changes caused by the thermal expansion and contraction of mineral particles due to moisture evaporation under high temperatures. Moreover, Wang et al. [31] investigated the effects of different water cooling durations under varying temperatures on the fracture performance of granite. The results indicated that the influence of water cooling on fracture performance was slightly less than that of temperature. At the same temperature, the fracture toughness first decreased, then increased, and finally, decreased again as the water cooling duration extended. Fu et al. [32] analyzed the impact of water cooling treatment on the mechanical properties and damage behavior of clayey sandstone, indicating that E, cohesion, and internal friction angle of clayey sandstone significantly decreased, with the extent of damage closely related to the temperature and number of water cooling cycles. These findings collectively highlight the profound impact of water cooling on the structural integrity and mechanical properties of rocks, providing robust support for the extraction of HDR.

When the environmental temperature of reservoir rocks changes abruptly, thermal stress is generated within the rocks, determined by the temperature difference between the rock and the cooling medium. Therefore, some pioneering studies have attempted to use cooling media at lower temperatures to enhance the extraction efficiency of HDR [33]. Liquid nitrogen (LN_2), with a temperature as low as -196 °C under normal atmospheric conditions, creates a substantial temperature difference when introduced into HDR reservoirs, thereby altering the physical and mechanical properties and pore structures of rocks [34]. Su et al. [35] used LN_2 to cool heated marble, observing a significant increase in porosity, while l V_p, UCS, and E showed a decreasing trend, primarily due to thermal damage incurred during rapid cooling. Similarly, Ding et al. [36] found that liquid nitrogen cooling significantly increased the total porosity of heated limestone, especially after high-temperature treatment. The increase in internal porosity of the rocks reduces their mechanical properties. Additionally, Zhou et al. [37] discovered that LN_2 cooling significantly reduces the σ_t of rocks, with the degree of reduction varying at different temperatures. Before reaching a temperature threshold, the effect of liquid nitrogen cooling was found to be superior to that of water cooling; beyond this threshold, its effectiveness was less than that of water cooling. However, when temperatures exceeded 200 °C, liquid nitrogen cooling resulted in granite exhibiting more shear or mixed-mode fractures during failure, increasing the roughness of the fracture surface. Xi et al. [38] conducted Hopkinson bar tests on rocks after LN_2 cooling and found that the fragments had a higher fractal dimension, indicating that the rapid cooling by LN_2 caused a swift release of internal stresses, resulting in numerous small and complex fragments and varying degrees of damage within the rocks. Since V_p can effectively reflect the damage in rocks, Sun et al. [39] observed attenuation in V_p, amplitude, and main frequency with each LN_2 cooling cycle, where wave velocity attenuation could reach up to 85.01%. These studies clearly demonstrated that LN_2 cooling treatment has profound effects on the physical and mechanical behavior of various rocks. It is evident that not only the choice of cooling medium but also the rate at which cooling is applied plays a pivotal role in determining the extent and nature of rock damage in HDR systems. This understanding is essential for optimizing extraction processes and improving the overall efficiency of geothermal energy systems.

Existing research has extensively investigated the impacts of heating/cooling cycles on the physical and mechanical properties of rocks; some related research is summarized in Table 1. Furthermore, some researchers have observed significant differences in the dynamic mechanical properties of sandstone after heating under various cooling treatments (natural cooling, water cooling, and LN_2 cooling) [40]. This indicates that the rate of cooling can significantly affect the physical and mechanical properties of rocks. It can be inferred that due to the cooling shock from cooling media during the extraction process of Hot Dry Rock (HDR) reservoirs, different cooling rates can also significantly alter the physical and mechanical properties of the reservoir rocks. However, current research on the effects of different cooling rates on heated granite, a common matrix rock in HDR geothermal reservoirs, remains relatively scarce. Although some studies have explored the impacts of natural cooling, water cooling, and liquid nitrogen cooling on the physical and mechanical properties of heated rock, systematic studies on cyclic liquid nitrogen cooling are still limited. In particular, research comparing the effects of cyclic liquid nitrogen cooling with other cooling methods, such as natural cooling, water cooling, and single liquid nitrogen cooling, on the physical and mechanical properties of heated granite is still not comprehensive. Therefore, this study aims to investigate the deteriorative effects of different cooling rates (cooling in air, cooling in water, cooling in LN_2, and cyclic cooling in LN_2) on the physical and mechanical properties of heated granite. We designed a series of physical property tests and mechanical experiments on granite heated to different temperatures and subsequently subjected to various cooling treatments. The study first analyzed changes in granite density, V_p, UCS, σ_t, and E after different treatments. Then, the microstructural features of the samples were examined using NMR non-destructive testing and SEM. Finally, the damage mechanisms of heated granite treated with different cooling methods were analyzed. This research can provide theoretical insights for the extraction of geothermal resources.

Table 1. Related indicators in the published literature.

	Granite	Preparation Method	Physical and Mechanical Characteristics
Ref. [27]	Changsha gray granite	Slow heating followed by water or air cooling	Tensile strength
Ref. [29]	Gray and fine-grained granite	Cycle heating and water-cooling	Volume, mass, density, UCS, and E
Ref. [31]	Granite from Hubei province, China	Water cooling after heated	Fracture toughness
Ref. [37]	Granite from Rizhao, China	Liquid nitrogen cooling and water cooling	Tensile strength

2. Materials and Methods

Granite, as a typical rock type in Hot Dry Rock, was selected as the experimental material in our study. This study utilized granite samples sourced from Rizhao, Shandong Province, China, for subsequent experiments. Samples were selected from large rock masses that exhibited no visible cracks or signs of weathering. In accordance with the International Society for Rock Mechanics (ISRM), a rock-cutting machine was used to cut the rocks into standard specimens of 50 mm diameter × 100 mm height and disc specimens of 50 mm diameter × 25 mm height. All specimens were polished on both ends to ensure end flatness of less than 0.02 mm, as shown in Figure 1a. The 50 × 100 mm^2 specimens were used for wave velocity tests and uniaxial compressive strength tests, while the 50 × 25 mm^2 specimens were used for Brazilian split tests to determine tensile strength. MTS 815 Flex Test GT rock mechanics system produced in the United States and AG-250kN IS Electronic Precision Material Testing Machines produced in Japan were employed for the uniaxial compressive strength experiment and Brazilian split tests, respectively, as illustrated in Figure 1b,c. For MTS 815, the maximum axial load reaches 2800 kN, the maximum peripheral pressure is 80 MPa, the maximum pore water pressure is 80 MPa, and the temperature is up to 200 °C. The maximum axial load of the AG-250kN IS is 250 kN, the tester stiffness is 15 GN/m, and the range of the loading rate is 0.0005–1000 mm/s [41,42].

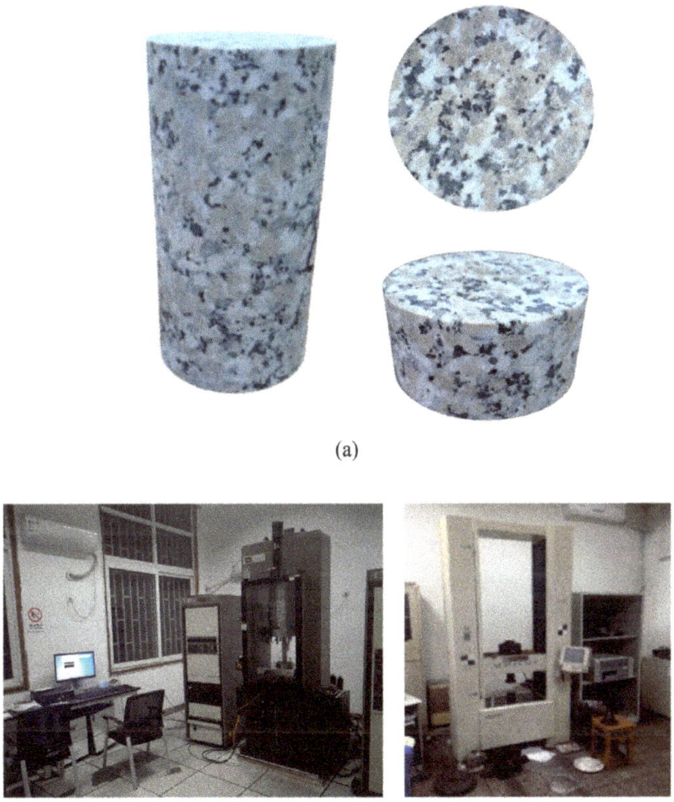

(a) (b) (c)

Figure 1. Samples and equipment. (**a**) Specimens for uniaxial compression strength experiment and Brazilian split tests. (**b**) MTS815 experimental apparatus. (**c**) AG-250kN IS Electronic Precision Material Testing Machine.

After processing, based on existing research, a heating rate of 2 °C/min was selected to heat the temperature to the target value to avoid thermal shock [5]. The target values in this study are 200 °C and 300 °C. Once the target temperature was reached, the temperature was maintained for 10 h to ensure uniform internal heating of the samples. After heating, the samples were subjected to four types of cooling treatments: cooling in air (CA), cooling in water (CW), cooling in liquid nitrogen (CL), and c cycle cooling in liquid nitrogen (CCL) for 5 cycles, as detailed in Table 2. After the cooling treatment, the granite specimens were dried again in an air-drying oven, and then physical and mechanical tests were performed. We first tested the density and wave speed of the granite standard pieces before and after the heating/cooling treatment. Before conducting mechanical experiments, we tested the pore structure of the heated granite samples after cooling treatments. Then, uniaxial compressive strength tests and tensile strength tests were conducted after the heating/cooling treatments. After the mechanical experiments, samples were sliced from the fracture surfaces for SEM testing to obtain the microstructural damage characteristics under different cooling methods.

Table 2. Summary of different heated temperature and cooling methods of granite.

Number	Heated Temperature	Cooling Methods
CA-200	200	CA
CA-300	300	CA
CW-200	200	CW
CW-300	300	CW
CL-200	200	CL
CL-300	300	CL
CCL-200	200	CCL
CCL-300	300	CCL

3. Results

3.1. Physical Properties

3.1.1. Density

Changes in the mass and volume of granite can reflect alterations in its internal structure to some extent. High temperatures and rapid cooling can cause changes in the internal pore structure of granite, such as the merging of pores or the formation of new pores, which also affect the rock's density. During the heating and rapid cooling process, the detachment of boundary particles in granite may occur, reducing the mass of the granite and thereby decreasing its density. Therefore, this paper first calculates the mass loss rate, volumetric expansion rate, and density of granite after treatment with different cooling methods. The formulas used for these calculations are as follows:

$$\Delta m = \frac{m_1 - m_0}{m_0} \times 100\% \quad (1)$$

$$\Delta V = \frac{V_1 - V_0}{V_0} \times 100\% \quad (2)$$

$$\rho = \frac{m_1}{V_1} \quad (3)$$

where m_0 and m_1 represent the mass of the granite sample before and after the heating/cooling treatment, respectively, g; V_0 and V_1 represent the volume of the granite sample before and after the heating/cooling treatment, respectively, cm^3.

At temperatures of 200 °C and 300 °C, both the mass loss rate and the volumetric expansion rate of granite show significant changes with varying cooling methods, as illustrated in Figure 2. The CCL treatment results in higher mass loss rates and volumetric expansions compared to the other three methods. The reasons for these trends can be attributed to two main factors. On the one hand, moisture evaporation plays a role, as granite contains certain amounts of bound and free water. During the heating process, a considerable amount of free water and some bound water evaporate in gaseous form, which is a key factor in the mass loss of granite. On the other hand, the detachment of surface particles is due to differences in the coefficients of thermal expansion among the various mineral components of granite. Heating causes uneven expansion within the rock's internal minerals, followed by uneven contraction during subsequent cooling. This uneven thermal expansion and contraction weaken the binding ability between rock particles, potentially leading to the detachment of surface particles. With increased cooling rates, granite undergoes more intense uneven contraction, potentially leading to greater surface particle detachment and, thus, larger mass loss.

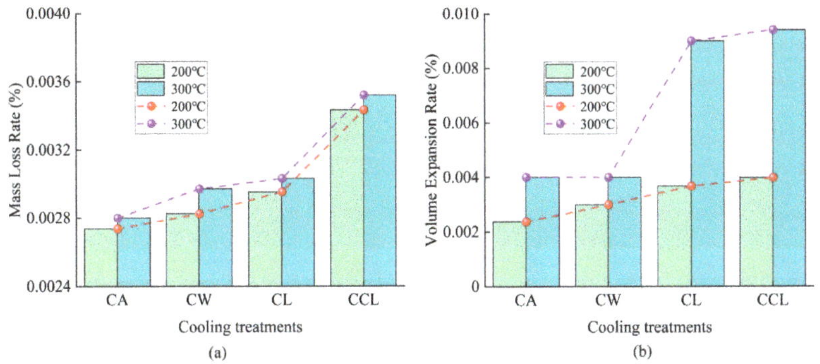

Figure 2. Changes in mass loss rate and volume expansion rate of granite under different heating temperatures and cooling treatments. (**a**) Results of the temperature at 200 °C. (**b**) Results of the temperature at 300 °C.

Changes in the density of granite following different cooling treatments are shown in Figure 3. At 200 °C, the density of granite remains virtually unchanged after CA and CW treatments, whereas it changes significantly after CL and CCL treatments. At 300 °C, the density of granite changes after CW, CL, and CCL treatments, with no significant change following CA treatments. This indicates that the greater the temperature difference during cooling treatment, the more likely it is that the density of granite will significantly decrease after treatment. The mass loss rates, volumetric expansion rates, and density of high-temperature granite all change after undergoing different cooling treatments, suggesting that the internal structure of the rock may be affected, which could further influence its wave speed characteristics. Therefore, we further analyze the changes in the V_p of granite before and after cooling treatment.

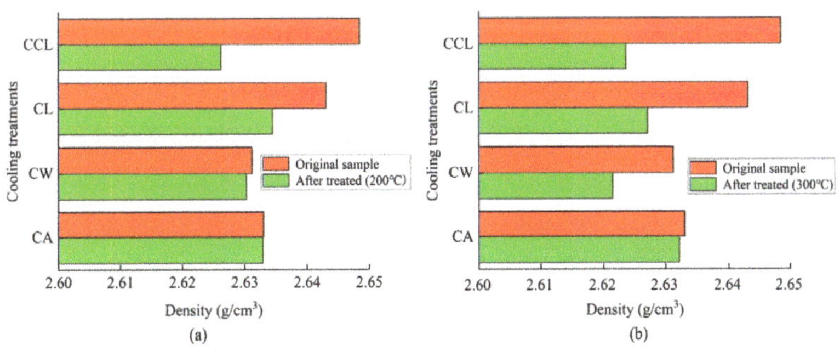

Figure 3. Density changes of granite after different cooling methods. (**a**) Results of the temperature at 200 °C. (**b**) Results of the temperature at 300 °C.

3.1.2. Effect of Different Cooling Treatments on V_p

As a non-destructive testing method, measuring V_p is highly sensitive to internal rock features such as cracks and porosity, thus effectively reflecting the internal damage state of the rock and indirectly estimating its mechanical strength. As shown in Figure 4, under conditions of 200 °C and 300 °C, V_p after all four cooling treatments exhibited significant changes. At 200 °C, compared to pre-treatment speeds, the V_p of samples after CA, CW, CL, and CCL cooling treatments decreased by 20%, 26.3%, 29.9%, and 44.4%, respectively.

Similarly, at 300 °C, V_p samples after CA, CW, CL, and CCL cooling treatments decreased by 25.2%, 26.4%, 38.3%, and 54.5%, respectively, compared to before treatment. This is because as the cooling rate increases, the uneven thermal stress within the rock intensifies, leading to the formation and expansion of microcracks, which, in turn, affects the rock's density and internal structure, resulting in a reduction in wave speed, as described in Section 3.1.1. The density of granite is also significantly affected by different cooling treatments.

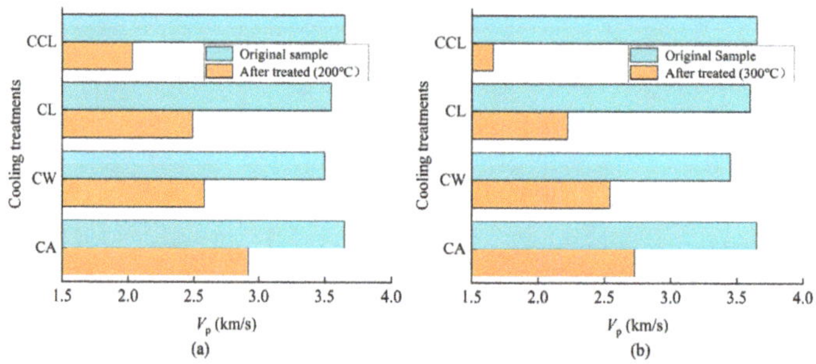

Figure 4. Changes of V_p of granite treated with different cooling methods. (**a**) Results of the temperature at 200 °C. (**b**) Results of the temperature at 300 °C.

3.2. Mechanical Properties

3.2.1. Effect of Different Cooling Treatments on UCS

As shown in Figure 5, the UCS (uniaxial compressive strength) of granite generally decreases with increasing cooling rates. At 200 °C, compared to the UCS of samples after CA (cooling in air) treatment, the UCS of samples after CW (cooling in water), CL (cooling in liquid nitrogen), and CCL (cycle cooling in liquid nitrogen) treatments decreased by 2.6%, 13.4%, and 26.7%, respectively. Similarly, at 300 °C, compared to the UCS of samples after CA treatment, the UCS of samples after CW, CL, and CCL treatments decreased by 7.9%, 17.9%, and 32.2%, respectively. It is evident that different heating temperatures and cooling rates significantly affect the UCS of granite, particularly after CCL treatment, where the reduction in UCS is especially pronounced. This is due to the differences in the coefficients of thermal expansion of the various mineral particles composing the granite. During the cooling process, an increase in the cooling rate exacerbates the mismatched deformation between different mineral particles and also increases the temperature gradient between the inner and outer layers of the granite, thus generating higher thermal stresses, which further intensify the internal damage to the granite, reducing its mechanical strength.

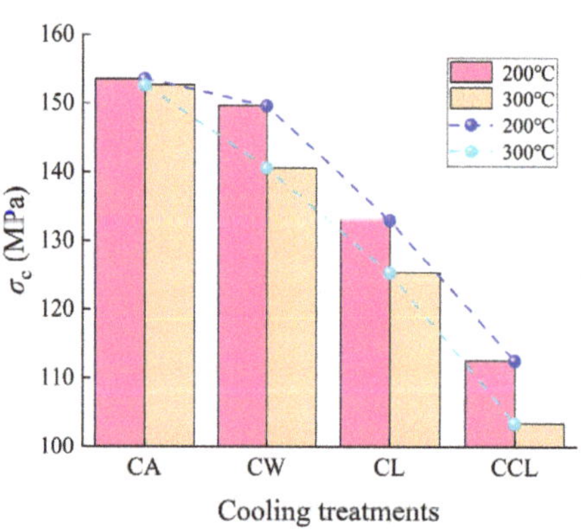

Figure 5. Uniaxial compressive strength of granite after different cooling treatments.

3.2.2. Effect of Different Cooling Treatments on σ_t

As shown in Figure 6, the σ_t (tensile strength) of granite samples tends to decrease with increasing cooling rates. Particularly at 300 °C, the σ_t of granite samples subjected to rapid cooling significantly decreases. Specifically, at 200 °C, compared to the σ_t of samples after CA, the σ_t of samples after CW, CL, and CCL treatments decreased by 9.5%, 20.8%, and 25.5%, respectively. Similarly, at 300 °C, compared to the σ_t of samples after CA, the σ_t of samples after CW, CL, and CCL treatments decreased by 25.9%, 34.4%, and 37%, respectively. The reduction in σ_t is mainly due to various degrees of internal damage caused by temperature differences. During rapid cooling, the sharp decrease in temperature causes rapid contraction of the mineral grains within the rock. Due to the heterogeneity of mineral composition, differences in the coefficients of thermal expansion among different minerals lead to uneven distribution of internal stresses, thereby causing the formation and expansion of microcracks, which, in turn, reduces the σ_t of granite.

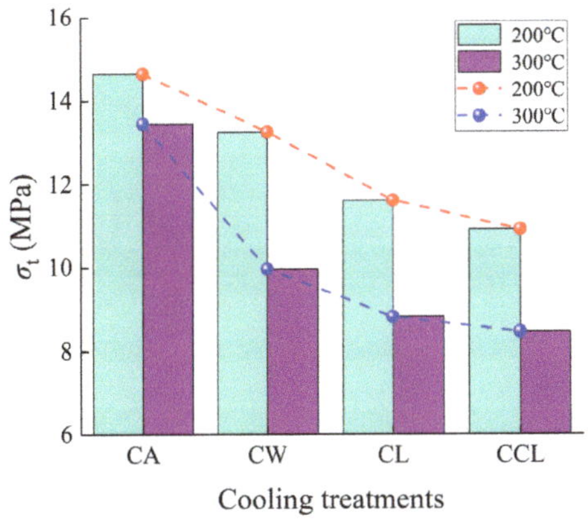

Figure 6. Changes in the tensile strength of granite under different cooling methods.

3.2.3. Effect of Different Cooling Treatments on E

E represents the rock's ability to resist deformation. As shown in Figure 7, except for natural cooling, the E of granite at 200 °C is higher than at 300 °C. With increasing cooling rates, the E of granite shows a decreasing trend. At 200 °C, compared to the E of samples after CA, the E of samples after CW, CL, and CCL treatments decreased by 4.9%, 7.3%, and 11.3%, respectively. At 300 °C, compared to the E of samples after CA, the E of samples after CW, CL, and CCL treatments decreased by 6.2%, 13.5%, and 19%, respectively. At lower cooling rates, the internal temperature distribution of granite is relatively uniform, producing relatively minor thermal stresses; therefore, the development of microcracks inside the rock is slow, and changes in the granite's E are minimal. However, rapid cooling can cause a significant temperature difference between the surface and the interior of the granite, leading to substantial thermal stress, which in turn rapidly generates and expands internal microcracks, thereby reducing the rock's E. Since E is closely related to the integrity of its internal structure, an increase in microcracks and porosity means the internal structure of the rock becomes more loose and fragile. Therefore, we further analyzed the changes in the micro-porous structure of granite before and after different cooling treatments.

3.3. Microstructural Features

3.3.1. Effect of Different Cooling Treatments on Heated Granite Pore Size

We used nuclear magnetic resonance (MacroMR12-150H-I) to test the porosity of granite. Drawing on current research on micro-porous structures, we have categorized the porosity of granite into large pores (>10 μm), mesopores (1–10 μm), small pores (0.1–1.0 μm), and micropores (<0.1 μm), as shown in Figure 8. The porosity of granite is primarily composed of micropores, small pores, and mesopores. With increasing cooling rates, the proportion of micropores and small pores in granite gradually increases, and there is a clear trend of increasing pore size. For instance, at 200 °C, the proportion of small pores in granite subjected to cyclic liquid nitrogen treatment is slightly lower than that observed under the other three treatments, but the volume proportion of these pores is significantly larger, notably higher than the mesopore proportion under other treatments. Similarly, at 300 °C, the mesopore volume proportion in granite samples treated with cyclic liquid nitrogen is the largest. Specifically, at 200 °C, the proportions of micropores in samples after CA, CW, CL, and CCL treatments are 3.95%, 3.24%, and 3.62%, respectively; small pores are 71.11%, 73.39%, and 73.37%, respectively; and mesopores are 24.94%, 23.37%, and 53.42%, respectively. At 300 °C, the proportions of micropores in samples after CA, CW, CL, and CCL treatments are 6.14%, 5.23%, and 4.97%, respectively; small pores are 66.20%, 64.39%, 45.96%; and mesopores are 27.66%, 30.38%, 49.08%. This indicates that an increase in cooling rate facilitates the formation of small pores and promotes the transformation of small pores into mesopores in heated granite, thereby altering the physical and mechanical properties of the rock and causing varying degrees of reduction.

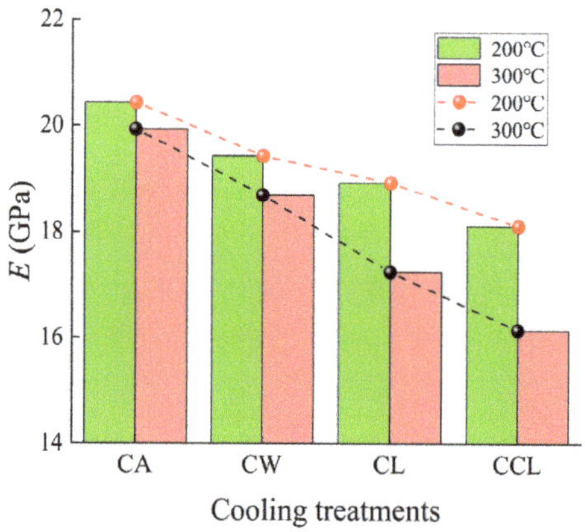

Figure 7. Changes of elastic modulus of granite under different cooling methods.

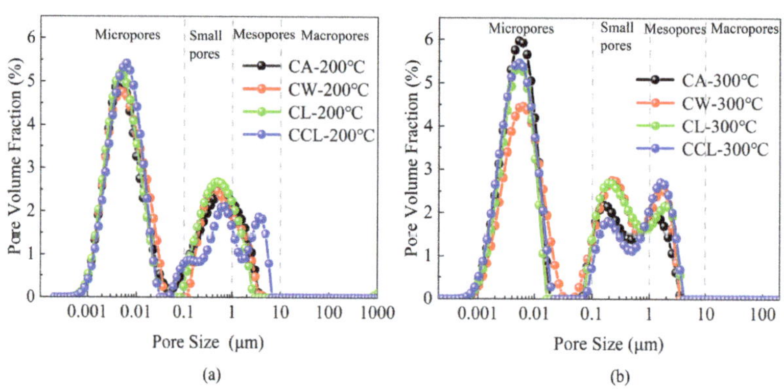

Figure 8. Pore size distribution of heated granite samples under different cooling methods. (**a**) Results of the temperature at 200 °C. (**b**) Results of the temperature at 300 °C.

3.3.2. Effect of Different Cooling Treatments on Porosity

Porosity is an important parameter reflecting the pore structure characteristics of rocks, significantly influencing their physical properties, mechanical strength, and permeability. Porosity can be defined as the ratio of pore volume to the total volume, known as the porosity rate, as shown in Figure 9. With increasing cooling rates, the porosity of granite shows an upward trend, with more significant changes in porosity rate observed at 300 °C. This is because rapid cooling can be considered a form of cold shock to the rock. Since there is a greater temperature difference between the rock heated to 300 °C and LN_2, undergoing LN_2 treatment can generate a greater cold shock. This shock can lead to rapid development and interconnection of microcracks within the rock, forming larger pores. Additionally, the liquid-to-gas phase transition of LN_2 may also generate pressure within microcracks and pores of rock, promoting the connectivity of the pores.

3.3.3. Effect of Different Cooling Treatments on the Micromorphology of Heated Granite

From the results shown in Figures 10 and 11, it can be observed that with increasing cooling rates, the microstructure of granite gradually transitions from dense to loose, and the characteristics of the fracture surfaces reflect a transition from brittle to ductile failure. After CA treatment, the fracture surfaces of granite samples are relatively smooth, with large blocky structures at 500× magnification and a relatively sparse distribution of crack networks. When granite samples undergo CW treatment, the blocky structures are reduced in size at 500× magnification, and the crack network becomes denser. After CL treatment, the granite samples display visible cracks at 500× magnification, and multiple pores and fragments appear at 2000× magnification. Following CCL treatment, distinct through-going cracks can be observed at 500× magnification, with significantly widened crack openings. At 2000× magnification, the fracture surface shows a net-like distribution of cracks after CCL treatment, with microcracks significantly denser than those seen with the other three cooling treatments. These results of SEM provide a visual reference for understanding the impact of different cooling methods on the microstructure of granite. It can be seen that cooling treatments can alter the microstructure of heated granite, especially CCL treatment, which significantly increases the roughness of the fracture surfaces, effectively deteriorating the fine microstructure of the granite.

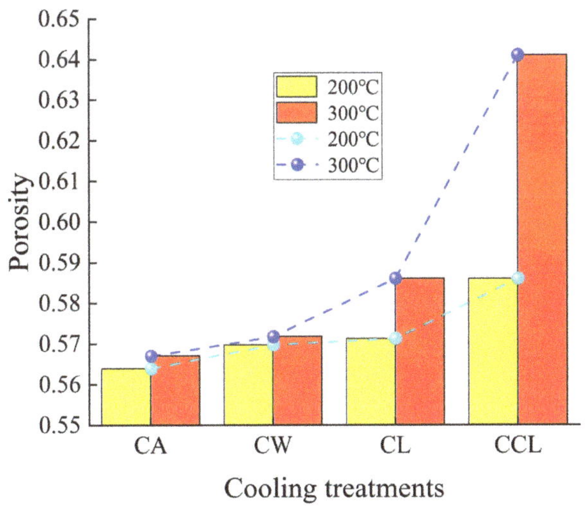

Figure 9. Changes in granite porosity under different cooling methods.

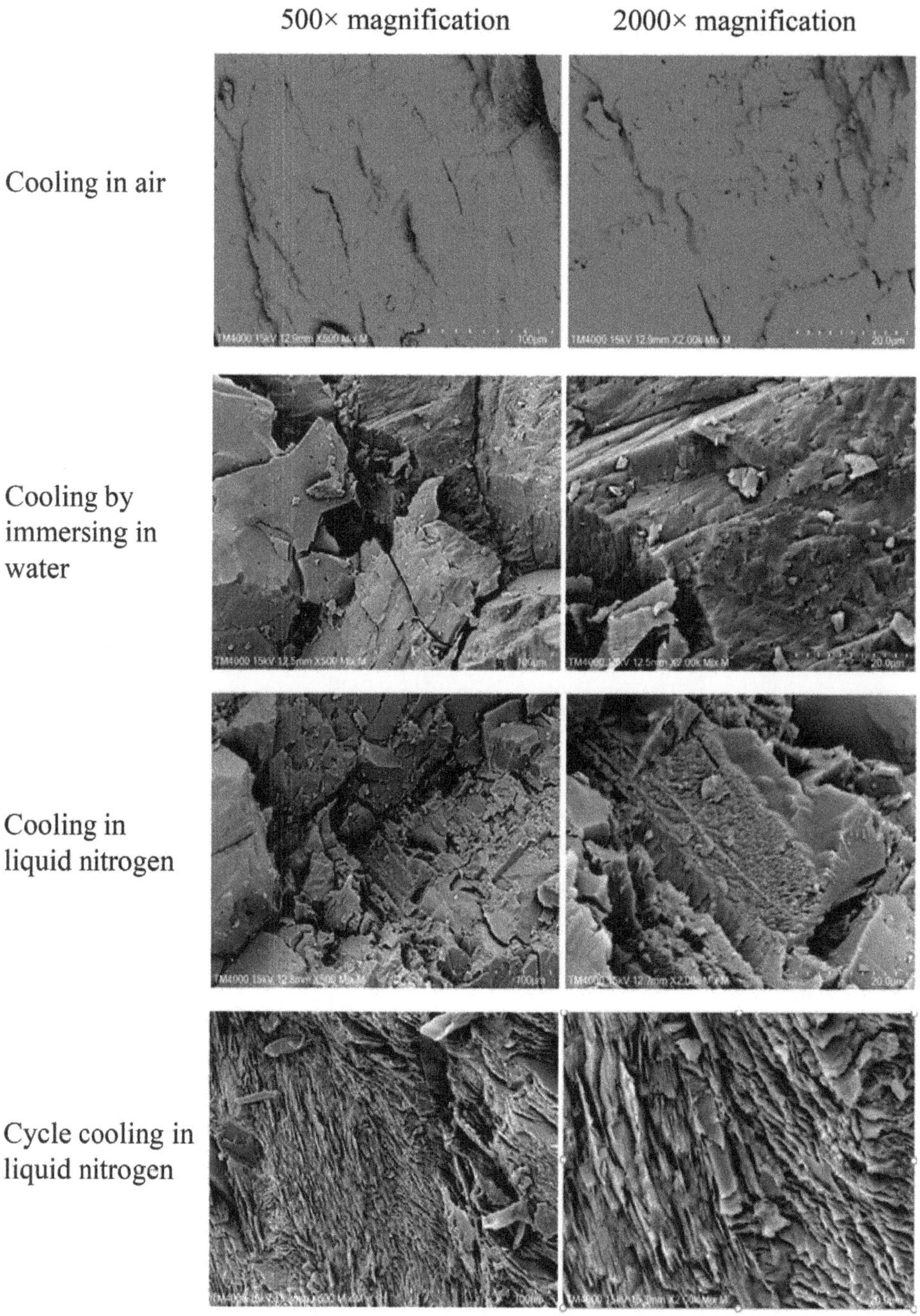

Figure 10. Morphology of fracture surfaces of heated granite under different cooling treatments at 200 °C.

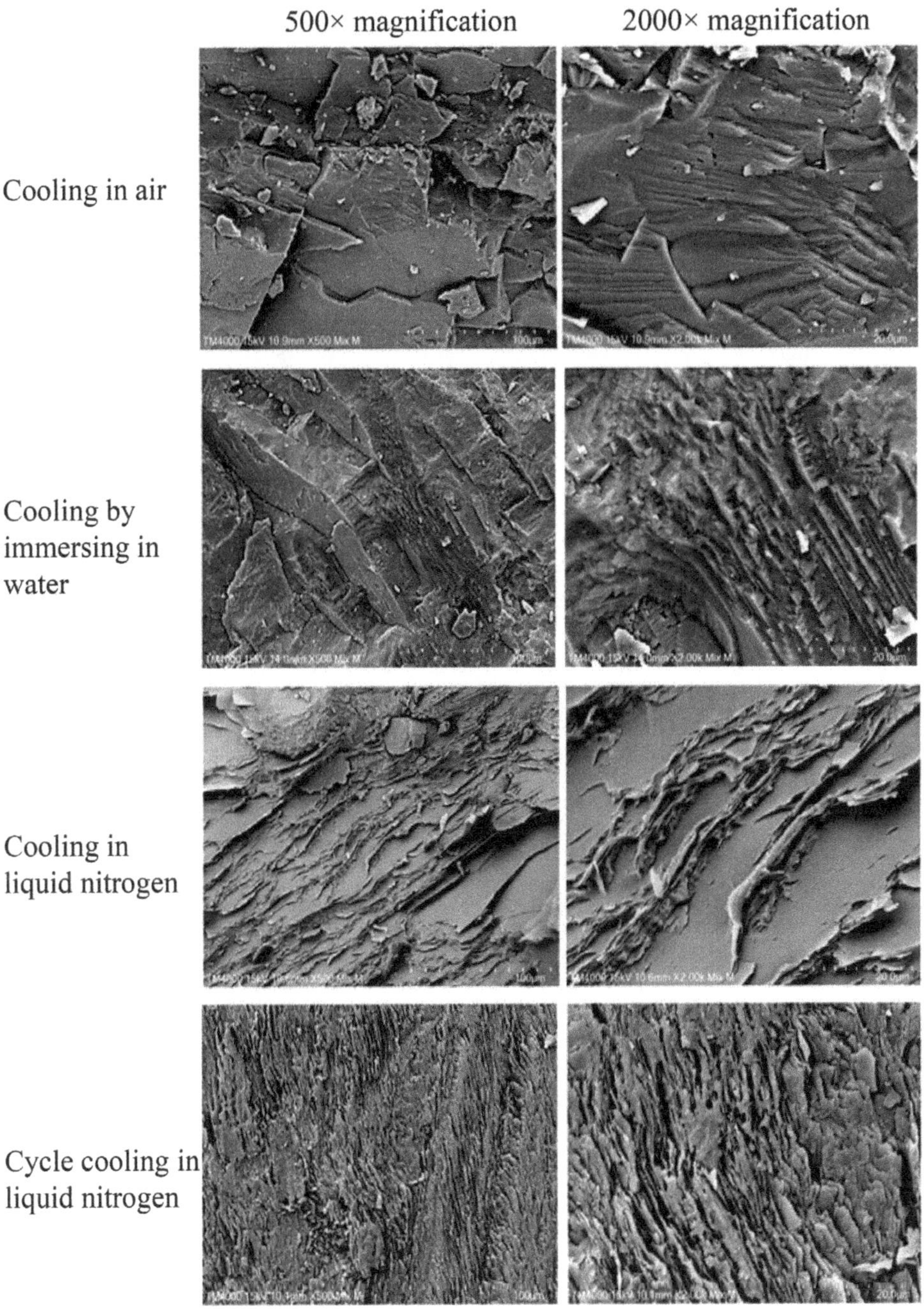

Figure 11. Morphology of fracture surfaces of heated granite under different cooling treatments at 300 °C.

4. Discussion

4.1. Damage Factor

It has been demonstrated that the V_p of rocks can provide a significant indication of their damage situation [43]. As suggested by some researchers, a damage factor is defined in this research, which is related to the V_p, shown as:

$$D = 1 - \left(\frac{V_{post}}{V_{before}}\right)^2 \quad (4)$$

where D is the damage factor, and V_{post} and V_{before} are the P-wave velocities of the heat/cooling-treated specimens and untreated specimens, respectively.

Figure 12 shows the trend of D in granite following different cooling treatments. According to the results shown in the figure, different cooling methods significantly affect the extent of damage in granite. As the cooling rate increases, D at different temperatures shows an upward trend. At 200 °C, the D of granite after CA, CW, CL, and CCL treatments are, respectively, 0.3582, 0.4397, 0.4274, and 0.6544. At 300 °C, the D of granite after CA, CW, CL, and CCL treatments are, respectively, 0.5077, 0.6184, 0.6809, and 0.7924. Comparisons reveal that at heating temperatures of 200 °C and 300 °C, the increase in the damage variable after CCL treatment is significantly greater, being 45.3% and 35.9% higher than after CA treatment, respectively. This may be due to the rapid contraction of any existing microcracks or pores within the granite under the extremely low temperatures of CCL treatment, causing stress concentration, which can rapidly lead to material damage in localized areas. Generally, D can reflect the internal damage of rock specimens with temperature and different cooling treatments, consistent with the experimental results. As described in Sections 3.2.1–3.2.3, the mechanical properties of granite have been significantly weakened.

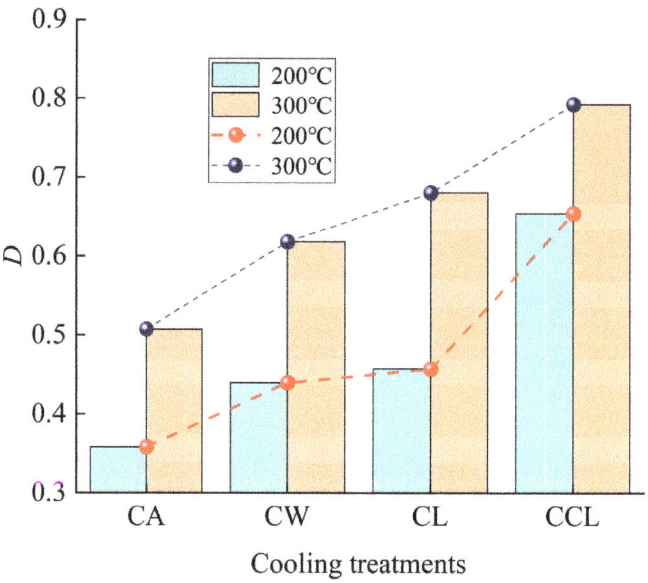

Figure 12. Changes in D of heated granite under different cooling methods.

4.2. Damage Mechanism of Heating and Different Cooling

In this section, we aim to reveal the mechanisms of thermal cracking from both perspectives of heating and different cooling treatments. The damage to granite in this study is primarily caused by two factors: high temperatures during the heating process, causing rock damage, and thermal stress from cooling treatments, causing additional rock damage. Granite is composed of various mineral grains, such as quartz, feldspar, and mica, arranged in a disordered manner within the rock. After the samples are heated, incompatible deformations between these minerals, along with thermal stress caused by the evaporation of free and bound water, lead to the formation and expansion of microcracks. Increasing the temperature gradually increases the thermal stress within the rock, providing energy for the expansion of these microcracks. Therefore, in the temperature changes set in this study, granite samples heated to 300 °C exhibited a more pronounced decrease in wave speed and an increase in porosity. This further leads to a larger damage factor in granite treated at 300 °C.

Moreover, when high-temperature granite undergoes rapid cooling, the intense cold shock generates a considerable temperature gradient, further increasing thermal stress. According to research by Kim et al. [44], when heated rocks come into contact with a coolant, the temperature gradient near the surface of the sample is much higher than near the center of the sample. This creates tensile stress in the external regions of the sample and compressive stress internally. Granite, being a typical brittle material, is more susceptible to damage under tensile stress than under compressive stress. Thus, the tensile stress generated in the external areas of the sample is the primary cause of microcrack expansion and the subsequent deterioration of the rock's properties.

To quantitatively describe the impact of thermal shock on high-temperature granite, Collin et al. proposed a formula for calculating the tensile stress generated in brittle materials due to thermal shock:

$$\sigma_t^T = \frac{E\alpha\Delta T}{1-\mu} f_h \tag{5}$$

$$f_h = \frac{1}{\frac{ak}{hr} + b - ce^{-\frac{dk}{hr}}} \tag{6}$$

where k, E, μ, α, and r represent the thermal conductivity, elastic modulus, Poisson ratio, thermal expansion coefficient, and radius of the sample, respectively; ΔT and h represent the temperature difference and the heat transfer coefficient between the sample and the coolant, respectively; the coefficient a, b, c, and d for the brittle material can be obtained from Collin and Rowcliffe.

For granite heated to a specific temperature, the constants k, E, μ, α, and r remain unchanged. Therefore, according to Equation (5), the tensile stress is determined by ΔT and h. The heat transfer coefficients, h, for water and liquid nitrogen exhibit significant differences. For cylindrical granite, this coefficient ranges from 200 to 1400 W/m^2 K during water cooling and escalates to between 3000 and 39,000 W/m^2 K during liquid nitrogen cooling [45]. Collin and Rowcliffe point out that f_h increases monotonically with the heat transfer coefficient h [45]. Therefore, this results in higher values of f_h for the LN_2 cooling condition compared to the CW treatment condition. Moreover, due to the markedly low temperature of LN_2 (−196 °C), the temperature difference ΔT between the heated granite and the LN_2 far exceeds that between the heated granite and water. Therefore, compared to CW treatment, the tensile stress generated internally by CL treatment is higher, resulting in more pronounced damage to the samples treated with LN_2. This also explains the reason that heated granite shows more significant deterioration in density, V_p, UCS, E, and σ_t, as well as rougher fracture surfaces after CL and CCL treatments compared to samples treated with CA and CW treatments. These results indicate a substantial decline in the physical and mechanical properties of granite treated with CL treatment, particularly after CCL treatments. Therefore, in future HDR extraction processes, LN_2 fracturing or cyclic LN_2 fracturing holds significant potential, likely enhancing the production efficiency of HDR.

5. Conclusions

Investigating the physical and mechanical properties of heated rock under different cooling methods has significant implications for the exploitation of HDR. Ultrasonic testing, uniaxial strength compression experiments, Brazilian splitting tests, nuclear magnetic resonance (NMR), and SEM were conducted on heated granite after different cooling methods, including cooling in air, cooling in water, cooling in liquid nitrogen, and cycle cooling in liquid nitrogen in this study. The special conclusions are as follows:

(1) As the cooling rate increases, the mass loss rate and volumetric expansion rate of granite gradually increase, while the density and V_p of granite tend to decrease.
(2) With an increasing cooling rate, the UCS, σ_t, and E of heated granite show an overall decreasing trend. Samples subjected to CCL treatment not only exhibit a more pronounced decline in mechanical properties but also a higher degree of damage. These results can provide a theoretical reference for the liquid nitrogen cycle fracturing process in geothermal production, thereby optimizing fracturing parameters and improving geothermal production efficiency.
(3) Both the pore size and porosity of granite tend to increase with an increasing cooling rate, and the morphology of the fracture surfaces transitions from smooth to rough, forming more complex surface cracks.

Author Contributions: Writing—original draft, Methodology, Q.L.; Supervision, Validation, G.H.; Validation, Methodology, J.H.; Supervision, Visualization, Funding acquisition, J.Z.; Validation, Y.W. and Q.C. All authors have read and agreed to the published version of the manuscript.

Funding: The authors gratefully acknowledge the financial support of the "Tianchi Talent" project funded by the Xinjiang Uygur Autonomous Region Introduction Plan.

Institutional Review Board Statement: Not applicable.

Informed Consent Statement: Not applicable.

Data Availability Statement: The original contributions presented in the study are included in the article; further inquiries can be directed to the corresponding author.

Conflicts of Interest: The authors declare no conflicts of interest.

References

1. Lee, C.-C.; Wang, C. Does Natural Resources Matter for Sustainable Energy Development in China: The Role of Technological Progress. *Resour. Policy* **2022**, *79*, 103077. [CrossRef]
2. Qin, M.; Su, C.-W.; Zhong, Y.; Song, Y.; Lobonț, O.-R. Sustainable Finance and Renewable Energy: Promoters of Carbon Neutrality in the United States. *J. Environ. Manag.* **2022**, *324*, 116390. [CrossRef]
3. Adebayo, T.S.; Ullah, S. Towards a Sustainable Future: The Role of Energy Efficiency, Renewable Energy, and Urbanization in Limiting CO2 Emissions in Sweden. *Sustain. Dev.* **2024**, *32*, 244–259. [CrossRef]
4. Chitgar, N.; Hemmati, A.; Sadrzadeh, M. A Comparative Performance Analysis, Working Fluid Selection, and Machine Learning Optimization of ORC Systems Driven by Geothermal Energy. *Energy Convers. Manag.* **2023**, *286*, 117072. [CrossRef]
5. Zhong, C.; Xu, T.; Gherardi, F.; Yuan, Y. Comparison of CO_2 and Water as Working Fluids for an Enhanced Geothermal System in the Gonghe Basin, Northwest China. *Gondwana Res.* **2023**, *122*, 199–214. [CrossRef]
6. Li, Z.; Luo, Z.; Wang, Y.; Fan, G.; Zhang, J. Suitability Evaluation System for the Shallow Geothermal Energy Implementation in Region by Entropy Weight Method and TOPSIS Method. *Renew. Energy* **2022**, *184*, 564–576. [CrossRef]
7. Li, S.; Xu, T.; Zhang, S.; Jia, X.; Tian, P.; Yue, D.; Zhang, Q.; Feng, B.; Feng, Z. Hot Dry Rock Geothermal Resource Potential in the Wudalianchi Volcanic Field, NE CHINA: Implications from Geophysical Exploration. *Energy Explor. Exploit.* **2019**, *37*, 663–676. [CrossRef]
8. Kumari, W.G.P.; Ranjith, P.G.; Perera, M.S.A.; Li, X.; Li, L.H.; Chen, B.K.; Isaka, B.L.A.; De Silva, V.R.S. Hydraulic Fracturing under High Temperature and Pressure Conditions with Micro CT Applications: Geothermal Energy from Hot Dry Rocks. *Fuel* **2018**, *230*, 138–154. [CrossRef]
9. Blackwell, D.D.; Negraru, P.T.; Richards, M.C. Assessment of the Enhanced Geothermal System Resource Base of the United States. *Nat. Resour. Res.* **2007**, *15*, 283–308. [CrossRef]
10. Bruel, D. Heat Extraction Modelling from Forced Fluid Flow through Stimulated Fractured Rock Masses: Application to the Rosemanowes Hot Dry Rock Reservoir. *Geothermics* **1995**, *24*, 361–374. [CrossRef]
11. Xu, F.; Shi, Y.; Song, X.; Li, G.; Song, Z.; Li, S. The Characteristics and Laws of Fracture Damage in the Long-Term Production Process of High-Temperature Geothermal Resources. *Rock Mech. Rock Eng.* **2023**, *56*, 275–299. [CrossRef]
12. Feng, G.; Wang, X.; Kang, Y.; Zhang, Z. Effect of Thermal Cycling-Dependent Cracks on Physical and Mechanical Properties of Granite for Enhanced Geothermal System. *Int. J. Rock Mech. Min. Sci.* **2020**, *134*, 104476. [CrossRef]

13. Hao, J.; Qiao, L.; Liu, Z.; Li, Q. Effect of Thermal Treatment on Physical and Mechanical Properties of Sandstone for Thermal Energy Storage: A Comprehensive Experimental Study. *Acta Geotech.* **2022**, *17*, 3887–3908. [CrossRef]
14. Fan, L.F.; Wu, Z.J.; Wan, Z.; Gao, J.W. Experimental Investigation of Thermal Effects on Dynamic Behavior of Granite. *Appl. Therm. Eng.* **2017**, *125*, 94–103. [CrossRef]
15. Wang, D.; Dong, Y.; Wang, Q.; Sun, D.; Yu, B. Experimental Study on the Evolution of Mechanical Properties of Hot Dry Rocks under Alternating Temperature Loads. *Geothermics* **2023**, *107*, 102599. [CrossRef]
16. Zhang, W.; Sun, Q.; Hao, S.; Geng, J.; Lv, C. Experimental Study on the Variation of Physical and Mechanical Properties of Rock after High Temperature Treatment. *Appl. Therm. Eng.* **2016**, *98*, 1297–1304. [CrossRef]
17. Sun, H.; Sun, Q.; Deng, W.; Zhang, W.; Lü, C. Temperature Effect on Microstructure and P-Wave Propagation in Linyi Sandstone. *Appl. Therm. Eng.* **2017**, *115*, 913–922. [CrossRef]
18. Lü, C.; Sun, Q.; Zhang, W.; Geng, J.; Qi, Y.; Lu, L. The Effect of High Temperature on Tensile Strength of Sandstone. *Appl. Therm. Eng.* **2017**, *111*, 573–579. [CrossRef]
19. Guo, Y.; Li, X.; Huang, L. Changes in Thermophysical and Thermomechanical Properties of Thermally Treated Anisotropic Shale after Water Cooling. *Fuel* **2022**, *327*, 125241. [CrossRef]
20. Yang, S.-Q.; Ranjith, P.G.; Jing, H.-W.; Tian, W.-L.; Ju, Y. An Experimental Investigation on Thermal Damage and Failure Mechanical Behavior of Granite after Exposure to Different High Temperature Treatments. *Geothermics* **2017**, *65*, 180–197. [CrossRef]
21. Zhang, Y.; Sun, Q.; He, H.; Cao, L.; Zhang, W.; Wang, B. Pore Characteristics and Mechanical Properties of Sandstone under the Influence of Temperature. *Appl. Therm. Eng.* **2017**, *113*, 537–543. [CrossRef]
22. Yavuz, H.; Demirdag, S.; Caran, S. Thermal Effect on the Physical Properties of Carbonate Rocks. *Int. J. Rock Mech. Min. Sci.* **2010**, *47*, 94–103. [CrossRef]
23. P.G., R.; Viete, D.R.; Chen, B.J.; Perera, M.S.A. Transformation Plasticity and the Effect of Temperature on the Mechanical Behaviour of Hawkesbury Sandstone at Atmospheric Pressure. *Eng. Geol.* **2012**, *151*, 120–127. [CrossRef]
24. Carpenter, M.A.; Salje, E.K.H. Elastic Anomalies in Minerals Due to Structural Phase Transitions. *ejm* **1998**, *10*, 693–812. [CrossRef]
25. McKnight, R.E.A.; Moxon, T.; Buckley, A.; Taylor, P.A.; Darling, T.W.; Carpenter, M.A. Grain Size Dependence of Elastic Anomalies Accompanying the α–β Phase Transition in Polycrystalline Quartz. *J. Phys. Condens. Matter* **2008**, *20*, 075229. [CrossRef]
26. Moarefvand, A.; Gasc, J.; Deldicque, D.; Labrousse, L.; Schubnel, A. P-Wave Velocities Across the $\alpha \rightarrow \beta$ Quartz Transition at Lower Continental Crust Pressure and Temperature Conditions. *JGR Solid Earth* **2024**, *129*, e2023JB027850. [CrossRef]
27. Wu, Y.; Huang, L.; Li, X.; Guo, Y.; Liu, H.; Wang, J. Effect of Cooling Methods on Mechanical Behaviors and Thermal Damage Distributions of Granite: Experiments and Simulations. *Geothermics* **2023**, *114*, 102796. [CrossRef]
28. Xue, Y.; Liu, S.; Chai, J.; Liu, J.; Ranjith, P.G.; Cai, C.; Gao, F.; Bai, X. Effect of Water-Cooling Shock on Fracture Initiation and Morphology of High-Temperature Granite: Application of Hydraulic Fracturing to Enhanced Geothermal Systems. *Appl. Energy* **2023**, *337*, 120858. [CrossRef]
29. Zhu, Z.; Tian, H.; Mei, G.; Jiang, G.; Dou, B. Experimental Investigation on Physical and Mechanical Properties of Thermal Cycling Granite by Water Cooling. *Acta Geotech.* **2020**, *15*, 1881–1893. [CrossRef]
30. Cui, Y.; Xue, L.; Zhai, M.; Xu, C.; Bu, F.; Wan, L. Experimental Investigation on the Influence on Mechanical Properties and Acoustic Emission Characteristics of Granite after Heating and Water-Cooling Cycles. *Geomech. Geophys. Geo-energ. Geo-resour.* **2023**, *9*, 88. [CrossRef]
31. Wang, H.; Zhou, L.; Zhu, Z.; Chen, J.; Ma, L.; Shui, X.; Zhang, H. Investigation of Fracture Characteristics of Cracked Granite Suffered from Different Thermal Treatments and Water-Cooling Time. *J. Mater. Res. Technol.* **2024**, *29*, 286–302. [CrossRef]
32. Fu, H.; Yu, X.; Zeng, L.; Luo, J.; Liu, J. Mechanical Properties and Damage Constitutive Model of Silty Mudstone under Heating and Water-Cooling Cycles. *Acta Geotech.* **2024**, *19*, 5031–5050. [CrossRef]
33. Ding, R.; Sun, Q.; Jia, H.; Tang, L.; Li, D. Experimental Study on the Compressive Strength and AE Characteristics of High-Temperature–Treated and LN_2-Cooled Sandstone. *Int. J. Geomech.* **2024**, *24*, 04024040. [CrossRef]
34. Wang, L.; Zhang, W.; Cao, Z.; Xue, Y.; Liu, J.; Zhou, Y.; Duan, C.; Chen, T. Effect of Weakening Characteristics of Mechanical Properties of Granite under the Action of Liquid Nitrogen. *Front. Ecol. Evol.* **2023**, *11*, 1249617. [CrossRef]
35. Su, S.; Hou, P.; Gao, F.; Liang, X.; Ding, R.; Cai, C. Changes in Mechanical Properties and Fracture Behaviors of Heated Marble Subjected to Liquid Nitrogen Cooling. *Eng. Fract. Mech.* **2022**, *261*, 108256. [CrossRef]
36. Ding, R.; Sun, Q.; Jia, H.; Yang, D. Effects of Liquid-Nitrogen Cooling on Pore Characteristics of High-Temperature Limestone Based on NMR. *Acta Geod. Geophys.* **2022**, *57*, 509–527. [CrossRef]
37. Zhou, C.; Gao, F.; Cai, C.; Du, M.; Wang, Z.; Huo, L.; Bai, Y. Effect of Different Cooling Treatments on the Tensile Properties and Fracture Modes of Granite Heated at Different Temperatures. *Nat. Resour. Res.* **2022**, *31*, 817–833. [CrossRef]
38. Xi, Y.; Wang, H.; Li, J.; Dong, W.; Li, H.; Guo, B. Experimental Comparison of Mechanical Properties and Fractal Characteristics of Geothermal Reservoir Rocks after Different Cooling Treatments. *Energy Rep.* **2022**, *8*, 5158–5176. [CrossRef]
39. Sun, Y.; Zhai, C.; Xu, J.; Yu, X.; Cong, Y.; Zheng, Y.; Tang, W.; Li, Y. Damage and Failure of Hot Dry Rock under Cyclic Liquid Nitrogen Cold Shock Treatment: A Non-Destructive Ultrasonic Test Method. *Nat. Resour. Res.* **2022**, *31*, 261–279. [CrossRef]
40. Fan, L.; Li, H.; Xi, Y. Evaluation of the Effects of Three Different Cooling Methods on the Dynamic Mechanical Properties of Thermal-Treated Sandstone. *Bull. Eng. Geol. Environ.* **2022**, *81*, 154. [CrossRef]

41. Singh, S.B.; Dastgheib, S.A. Characteristics of Graphene Oxide-like Materials Prepared from Different Deashed-Devolatilized Coal Chars and Comparison with Graphite-Based Graphene Oxide, with or without the Ultrasonication Treatment. *Carbon* **2024**, *228*, 119331. [CrossRef]
42. Singh, S.B.; De, M. Thermally Exfoliated Graphene Oxide for Hydrogen Storage. *Mater. Chem. Phys.* **2020**, *239*, 122102. [CrossRef]
43. Lai, Y.; Zhai, C.; Sun, Y.; Xu, H.; Zheng, Y.; Wang, S.; Huang, J. Study on Damage Characteristics of Hot Dry Rock by Liquid Nitrogen Cyclic Cold Shocks Based on Ultrasonic Testing. *Energy Fuels* **2023**, *37*, 16573–16587. [CrossRef]
44. Kim, K.; Kemeny, J.; Nickerson, M. Effect of Rapid Thermal Cooling on Mechanical Rock Properties. *Rock Mech. Rock Eng.* **2014**, *47*, 2005–2019. [CrossRef]
45. Collin, M.; Rowcliffe, D. Analysis and Prediction of Thermal Shock in Brittle Materials. *Acta Mater.* **2000**, *48*, 1655–1665. [CrossRef]

MDPI AG
Grosspeteranlage 5
4052 Basel
Switzerland
Tel.: +41 61 683 77 34

Materials Editorial Office
E-mail: materials@mdpi.com
www.mdpi.com/journal/materials

www.ingramcontent.com/pod-product-compliance
Lightning Source LLC
Chambersburg PA
CBHW061839210926
51850CB00006B/926
9783725877355